PHYSICS RESEARCH AND TECHNOLOGY

NEW DEVELOPMENTS IN THE STANDARD MODEL

Physics Research and Technology

Additional books in this series can be found on Nova's website under the Series tab.

Additional E-books in this series can be found on Nova's website under the E-book tab.

PHYSICS RESEARCH AND TECHNOLOGY

NEW DEVELOPMENTS IN THE STANDARD MODEL

RYAN J. LARSEN
EDITOR

Nova Science Publishers, Inc.
New York

For permission to use material from this book please contact us:
Telephone 631-231-7269; Fax 631-231-8175
Web Site: http://www.novapublishers.com

LIBRARY OF CONGRESS CATALOGING-IN-PUBLICATION DATA

New developments in the standard model / [edited by] Ryan J. Larsen.
p. cm.
Includes index.
ISBN 978-1-61209-989-7 (hardcover)
1. Physics--Mathematical models--Standards. 2. Mathematical physics. I. Larsen, Ryan J.
QC20.7.S73N49 2011
530.1--dc22

2011007139

Published by Nova Science Publishers, Inc. † New York

CONTENTS

Preface

The Standard Model of particle physics is one of the most important successful results of the work of the last century physicists. In this new book, the authors present topical research in the study of new developments in the Standard Model. Topics discussed include non-equilibrium theory, fractional dynamics and the physics of the terascale sector; unexplored regions in QFT and the conceptual foundations of the Standard Model; supersymmetric Standard Model, Branes and Del Pezzo surfaces; fermion condensate as Higgs substitute and Lepton flavor violation shedding light on CP-violation.

Space exploration has been identified by several governments as a priority for their space agencies and commercial industry. A good knowledge and specification of the ionosphere and plasmasphere are the key elements necessary to achieve this goal in the design and operation of space vehicles, remote sensing, reliable communication and navigation. The International Standardization Organization, ISO, recommends the International Reference Ionosphere (IRI) for the specification of ionosphere plasma densities and temperatures and lists several plasmasphere models for extending IRI to plasmaspheric altitudes, as described in the ISO Technical Specification, ISO/TS16457:2009. IRI is an international project sponsored jointly by the Committee on Space Research (COSPAR) and the International Union of Radio Science (URSI). The buildup of IRI electron density profile in the bottomside and topside ionosphere and its extension to the plasmasphere are discussed in the paper. The paper is also an important step towards the promotion of this model to full ISO standard. It includes a section on theoretical models used in data assimilation scenarios. Chapter 1 will introduce recent progress in IRI system developments, comparison of results provided by its different options and prospects of future improvements. Specification of the ionospheric weather index ranging from quiet conditions to severe storm in the ionosphere and plasmasphere is provided. Development of the ISO international standard would harmonize national approaches in this subject area that may serve as barriers to international trade. The ISO information will provide an open source model system to those organizations that are concerned with space vehicle design and operations, plasma environment specification, and communication / navigation services for the common specification of the Earth ionosphere and plasmasphere.

Quantum Field Theory (QFT) lies at the foundation of the Standard Model for particle physics (SM) and is built in compliance with a number of postulates called consistency conditions. The remarkable success of SM can be traced back to a unitary, local, renormalizable, gauge invariant and anomaly-free formulation of QFT. Experimental observations of recent years suggest that developing the theory beyond SM may require a careful revision of conceptual foundations of QFT. As it is known, QFT describes interaction of stable or quasi-stable fields whose evolution is deterministic and time-reversible. By contrast, behavior of strongly coupled fields or dynamics in the Terascale sector is prone to become unstable and chaotic. Non-renormalizable interactions are likely to proliferate and prevent full cancellation of ultraviolet divergences. A specific signature of this transient regime is the onset of long-range dynamic correlations in space-time, the emergence of strange attractors in phase space and transition from smooth to fractal topology. Our focus in Chapter 2 is the impact of fractal topology on physics unfolding above the electroweak scale. Arguments are given for perturbative renormalization of field theory on fractal space-time, breaking of discrete symmetries, hierarchical generation of particle masses and couplings as well as the potential for highly unusual phases of matter that are ultra-weakly coupled to SM. A surprising implication of this approach is that classical gravity emerges as a dual description of field theory on fractal space-time.

As explained in Chapter 3, massive quantum matter of prescribed spin permits infinitely many possibilities of covariantization in terms of spinorial (undotted/dotted) pointlike fields, whereas massless finite helicity representations lead to large gap in this spinorial spectrum which for s=1 includes vectorpotentials. Since the nonexistence of such pointlike generators is the result of a deep structural clash between modular localization and the Hilbert space setting of QT, there are two ways out: gauge theory which sacrifies the Hilbert space and keeps the pointlike formalism and the use of stringlike potentials which allows to preserve the Hilbert space. The latter setting contains also string-localized charge-carrying operators whereas the gauge theoretic formulation is limited to point-like generated observables.

This description also gives a much better insight into the Higgs mechanism which leads to a revival of the more physical "Schwinger-Higgs" screening idea.

The new formalism is not limited to m=0, s=1, it leads to renormalizable interactions in the sense of power-counting for all s in massless representations.

The existence of stringlike vectorpotentials is preempted by the Aharonov-Bohm effect in QFT; it is well-known that the use of pointlike vectorpotentials in Stokes theorem would with lead to wrong results. Their use in Maxwell's equations is known to lead to zero Maxwell charge. The role of string-localization in the problem behind the observed invisibility and confinement of gluons and quarks leads to new questions and problems.

Even though the Standard Model of particles has been confirmed by several experiments, many questions require improvements. Beyond the problem of Grand Unification, the mass gap problem, the question of hierarchies, low boson masses and dynamical soft supersymmetry breaking, there is the really hard difficulty in including gravity in a full quantum paradigm of the Standard Model.

The most famous scheme elaborated in order to solve the last and, possibly, all this points is String Theory. Dualities, mirror symmetry, M-theory and AdS/CFT are some of the powerful tools which permit to perform several progresses in all the mentioned directions, at least in principle. However, interactions of String Theory with phenomenology are really recent results. A way to get a contact between theory and phenomenology is the so called bottom-up approach. The authors will present in Chapter 4 a possible String Theory approach to the (Minimal Supersymmetric) Standard Model based on the geometric engineering construction first proposed in [H. Verlinde and M. Wijnholt, JHEP 0701, 106]. The authors will study the relevant geometry along the lines of [S.L. Cacciatori and M. Compagnoni,

JHEP 1005:078,2010], and the related physics. The authors will study the singular orbifold $\mathbb{C}^3/\Delta_{27}$, with Δ_{27} a suitable non abelian group, its geometry and show how it can be desingularized. To render technical computations as simple as possible the authors will work also with a simplified toric version, studing its main properties at K-theory level, and they will discuss how such calculations should be extended to the non abelian case. The associated relevant physics will be discussed.

In Chapter 5 the authors propose and analyze an alternative model of dynamical electroweak symmetry breaking. In the Standard Model of electroweak interactions the elementary Higgs field and the Higgs sector are replaced by vector-like fermions and their interactions. The new fermions are a weak doublet and a singlet. They have kinetic terms with covariant derivatives and gauge invariant four-fermion interactions. The model is a low energy effective one with a natural cutoff in the TeV regime. Due to the quartic fermion couplings the new fermions form condensates. The new fermions mix in one condensate and the mixing breaks the electroweak symmetry. The condensates contribute to the masses of the new femions, which may or may not have mass terms in the original Lagrangian. Gap equations are derived for the masses of the new fermions and the conditions are presented for mass generations and electroweak symmetry breaking. In the spectrum there are two neutral fermions and a charged one with mass between the neutral ones. The new sector can be described by three parameters, these are the two neutral masses and the mixing angle. These parameters are further constrained by the unitarity of two particle scattering amplitudes, providing an upper bound for the lighter neutral mass depending on the cutoff of the model. The standard chiral fermions get their masses via interactions with the condensing new fermions, but there is no mixing between the standard and the new fermions. There is an effective composite scalar in the model at low energies, producing the weak gauge boson masses in effective interactions. The ρ parameter is one at leading order. The model can be constrained by one-loop oblique corrections. The Peskin-Takeuchi S and T parameters are calculated in the model. The parameters of the model are only slightly constrained, the T parameter requires the new neutral fermion masses not to be very far from each other, allowing higher mass difference for higher masses and smaller mixing. The S parameter gives practically no constraints on the masses. The new fermions can give positive contributions to T allowing for a heavy Higgs in the precision electroweak tests. It is shown that the new fermions will be copiously produced at the next generation of linear colliders and cross sections are presented for the Large Hadron Collider. An additional nice feature of the model is that the lightest new neutral fermion is an ideal and natural dark matter candidate.

Search for Lepton Flavor Violating (LFV) rare decay $\mu \to e\gamma$ has played a key role in forming the standard model. Null result for searches was a hint for the fact thatmore than one type of neutrinos exist and $\nu_e$and $\nu_\mu$are two distinct particles. Search for $\mu \to e\gamma$ is still a powerful tool to look for unknown physics. In fact, the present bound on Br($\mu \to e\gamma$) already probes energy scales that is beyond the reach of the LHC. TheMEG experiment at PSI of Switzerland is currently collecting data to probe values of Br($\mu \to e\gamma$) two orders of magnitude below the present bound. If Br($\mu \to e\gamma$) is close to the present bound, the MEG collaboration will enjoy collecting large amount of data. Considering that decaying muons are almost 100 polarized, it would be possible to study the angular distribution of the final particles and derive information on the parity structure of the underlying theory. Recently, it is shown that by measuring the polarization of the final particles in $\mu \to e\gamma$ as well as in other lepton flavor violating processes such as $\mu \to$ eeeor $\mu -$ econversion on nuclei, one can derive information on the CP-violating parameters of the underlying theory. Remembering that CP-violation is one of the key ingredients in explaining the fundamental question of matter anti-matter asymmetry of the universe, the importance of such a measurement becomes more evident. The authors review this novel method in Chapter 6.

In: New Developments in the Standard Model
Editor: Ryan J. Larsen
ISBN: 978-1-61209-989-7

Chapter 1

TOWARDS ISO STANDARD EARTH IONOSPHERE AND PLASMASPHERE MODEL

Tamara Gulyaeva[1,*] and Dieter Bilitza[2]
[1]IZMIRAN, RAS, Troitsk, Moscow Region, Russia
[2]George Mason University, Fairfax, VA, US and
Goddard Space Flight Center,
Heliospheric Physics Laboratory,
Greenbelt, MD, US

Abstract

Space exploration has been identified by several governments as a priority for their space agencies and commercial industry. A good knowledge and specification of the ionosphere and plasmasphere are the key elements necessary to achieve this goal in the design and operation of space vehicles, remote sensing, reliable communication and navigation. The International Standardization Organization, ISO, recommends the International Reference Ionosphere (IRI) for the specification of ionosphere plasma densities and temperatures and lists several plasmasphere models for extending IRI to plasmaspheric altitudes, as described in the ISO Technical Specification, ISO/TS16457:2009. IRI is an international project sponsored jointly by the Committee on Space Research (COSPAR) and the International Union of Radio Science (URSI). The buildup of IRI electron density profile in the bottomside and topside ionosphere and its extension to the plasmasphere are discussed in the paper. The paper is also an important step towards the promotion of this model to full ISO standard. It includes a section on theoretical models used in data assimilation scenarios. The paper will introduce recent progress in IRI system developments, comparison of results provided by its different options and prospects of future improvements. Specification of the ionospheric weather index ranging from quiet conditions to severe storm in the ionosphere and plasmasphere is provided. Development of the ISO international standard would harmonize national approaches in this subject area that may serve as barriers to international trade. The ISO information will provide an open source model system to those organizations that are concerned with space vehicle design and operations, plasma environment specification, and communication / navigation services for the common specification of the Earth ionosphere and plasmasphere.

* E-mail address: gulyaeva@izmiran.ru

Keywords: International Reference Ionosphere, Plasmasphere, Standard Model, Electron Density, Total Electron Content

1. Introduction

The ionosphere and plasmasphere are conductive, ionized regions of the Earth's atmosphere consisting of free electrons and ions. The ionosphere and plasmasphere are embedded within the Earth's magnetic field and thus are constrained by interactions of the ionized particles with the magnetic field. The ionization levels in this near-Earth space plasma are controlled by solar extreme ultraviolet (EUV) radiation and particle precipitation. The dynamics of the neutral atmosphere plays a significant role in causing movement of the ionized particles by collisions with neutral atoms and molecules from the surrounding thermosphere. The ionosphere extends in altitude from about 65 km to 2000 km and exhibits significant variations with local time, altitude, latitude, longitude, solar cycle, season, and geomagnetic activity. At middle and low latitudes the ionosphere is contained within a region of closed field lines, whereas at high latitudes the geomagnetic field can reconnect with the interplanetary magnetic field and thus open the ionosphere to the driving force of the solar wind.

Plasma flowing upwards from the oxygen-dominated topside ionosphere remains at the lines of force co-rotating with the Earth and comprises the hydrogen-dominated plasmasphere extended up to a few Earth's radii (Carpenter and Park, 1973; Kotova, 2007). These two regions of upper atmosphere are strongly coupled through diffusion and resonant charge exchange reactions between O^+ and H^+. At quiet conditions, H^+ in the plasmasphere typically diffuses down to the topside ionosphere at night and undergoes resonant charge exchange reactions with atomic oxygen to produce O^+ (downward flux). The O^+ produced in this way can make a significant contribution to the maintenance of the nighttime ionosphere, and works in combination with the meridional component of the neutral wind. The depleted nighttime plasmasphere can be refilled during the day through the reverse process; that is, the O^+ ions flow up from the ionosphere, exchange charges with the neutral hydrogen atoms to produce protons, and the protons are then stored in the plasmasphere (upward flux). During geomagnetically disturbed conditions, the flux situation can be changed. Under these conditions, the plasmaspheric plasma can be eroded by the enhanced magnetospheric electric fields, and consequently, the flux becomes upward both during the day and night, due to the reduced plasmaspheric pressure, to refill the empty plasmaspheric flux tubes. While the low-latitude flux tubes refill relatively quickly due to their small volumes, most of the mid-latitude flux tubes are always in a partially depleted state, since the average time between consecutive geomagnetic storms is not long enough for the upflowing ionospheric flux to completely refill the flux tubes.

Terrestrial HF communications rely entirely on reflections from the ionized layers in the upper atmosphere, but the ionosphere also distorts Earth-space and spacecraft-to-spacecraft links. Although empirical models of the ionosphere are now accessible via electronic networking, most of them are far from reliable in predicting the average ionospheric conditions, not to mention their limitations in forecasting the ionospheric "space weather". In particular, a reliable and standard ionosphere-plasmasphere model is required for calibration of trans-ionospheric signals of the high altitude Global Positioning System (GPS) and Global

Navigation Satellite System (GLONASS) satellites at 20,200 km above the Earth which could in turn supply total electron content per m^2, TEC, in the column from the bottom of the ionosphere to the plasmapause.

For many years real-time updates of median ionospheric climatological models for various military and civilian users have been limited by the availability of globally distributed real-time ionospheric measurements. This situation has now improved greatly because of the increasing needs of civilian users for governing the operation of thousands of world-wide receivers of signals from GPS and other satellites navigational systems. Satellite navigation satellites such as GPS, Glonass, Galileo, can detect the delay due to the integrated total electron content (TEC) a radio signal is experiencing during its transit from spacecraft to receiver through the ionosphere. The ionospheric range error correction is required by the geodetic community which in turn is providing continuing flow of total electron content (TEC) data by simple Internet transfer to users anywhere in the world. There are still some limitations in coverage over the oceans. Uncertainties also remain in regards to accurately accounting for the plasmaspheric content and related to the techniques used to convert the slant path measurements into vertical TEC. A small uncertainty also remains due to difficulties to fully account for and determine the instrument bias factors for the ground receiver and satellite transmitter.

Ionospheric modeling using TECgps data has been the focus of numerous studies during the past decade. The range error caused by ionospheric delay in GPS signals is currently the largest component that affects the accuracy of positioning and navigation determination using single frequency GPS measurements. Ionospheric modeling is an effective approach for correcting the ionospheric range error and improving the GPS positioning accuracy. The abundance of GPS measurements from worldwide-distributed GPS reference networks, which provide 24-h uninterrupted operational services to record dual-frequency GPS measurements, provides an ideal data source for ionospheric modeling research.

Ionospheric remote sensing is in a rapid growth phase, driven by an abundance of ground and space-based GPS receivers and the advent of data assimilation techniques for space weather (Bust et al., 2004; Jorgensen et al., 2010). The horizontal resolution can be achieved by a local dense array of ground instruments such as GPS receivers. Vertical resolution can be achieved by GPS occultations for a constellation of satellites such as the current constellation observing system for meteorology ionosphere and climate (COSMIC) array (http://www.cosmic.ucar.edu). COSMIC system, a constellation of six satellites, nominally provides up to 3000 ionospheric occultations per day with an unprecedented global coverage of GPS occultation measurements (between 1400 and 2400 good soundings per day as of January 2010). Calibrated measurements of ionospheric delay, the total electron content or COSMIC-TEC measured from the satellite altitude to GPS orbit, suitable for input into assimilation models (Angling et al., 2004) are currently made available in near real-time from the COSMIC with a latency of 30 to 120 minutes. Similarly, TECgps data are available from worldwide networks of ground GPS receivers. The combined ground and space-based GPS datasets provide a new opportunity to more accurately specify the 3-dimensional ionospheric density with a time lag of only 15 to 120 minutes. It is, however, important to use these data cautiously and with awareness for limitations and uncertainties (Hysell, 2007; Kelley et al., 2009; Vergados and Pagiatakis, 2010; Liu et al, 2010).

TECgps data together with near real-time supply of the ionosonde data provide an outstanding capability for near world-wide monitoring and imaging of the ionosphere and plasmasphere using the relevant model operating in assimilative regime (Bust and Mitchel, 2008; Scherliess et al., 2004). The TEC measurements can be used directly or organized into two-dimensional TEC maps to infer information on the horizontal structuring of the electron density. However, information on how plasma can be lifted to high altitudes and transported to other regions, polar outflow, and other vertical dynamical changes is lost with such simple mapping algorithms. In order to obtain information on the vertical structure of the electron density, its temporal variation, and transport, the three-dimensional time-evolving (hence four-dimensional) spatial field of electron density tomographic reconstruction is necessary. The International Reference Ionosphere extended to the plasmasphere is capable to serve as the initial condition (background ionosphere) in process of 3D tomographic imaging of electron density (Bust et al., 2001; Arikan et al., 2007).

A technique for IRI-2000 implementation for the near real-time US-TEC reconstruction of the three-dimensional distribution of electron density is developed with Gauss-Markov Kalman filtering of ground-based GPS observations (Fuller-Rowell et al., 2006). NOAA's TEC specification methodology (Gauss-Markov Kalman filter) over the continental United States has been expanded to the multi-regional domains and to the entire globe. The ensemble Kalman Filter (EnKF) is a Monte-Carlo approximation of a sequential Bayesian filtering process. Ensemble (Monte-Carlo) samples are used to estimate the covariance of the prior distribution of the model state and of observations. The algorithm consists of recursive application of an analysis (update) step in which the prior ensemble estimate of the state is updated by observations to produce a posterior (analysis), and a forecast step in which the posterior sample is propagated forward in time with a dynamical model to the next observation time. There is no need to compute explicitly the enormous prior covariance matrices that are associated with large dynamical models. The EnKF has been shown to work well with both nonlinear model dynamics and nonlinear relationships between observations and model state variables, and there is no need to linearize a forecast model or a forward (observation) operator. The resulting ease of implementation has led to a number of atmospheric assimilation applications by groups that may not have the resources to develop variational systems like those used for operational numerical weather prediction. To avoid filter divergence (in which ensemble samples diverge gradually from the truth or the observation) due to insufficient variance in the sample posterior/forecast covariance the sample forecast covariance is artificially inflated.

Due to highly temporal and spatial variability of space plasma surrounding the Earth and the requirements of its representation in the design and operation of space vehicles, remote sensing, reliable communication and navigation, modeling of the ionosphere and plasmasphere has been and still is a research focus within the worldwide space science communities. Among these efforts an outstanding part plays the International Reference Ionosphere (IRI) extended to the plasmasphere recognized as a candidate model for an international standard of the specification of ionosphere and plasmasphere plasma densities and temperatures by the International Standardization Organization, ISO (ISO/TS16457:2009).

The International Reference Ionosphere Project was established in 1968 jointly by the Committee on Space Research (COSPAR) and the International Union of Radio Science (URSI). The IRI is an observation-based climatological standard model of the ionosphere that is widely used to predict and mitigate the significant effects the ionosphere has on the performance of communication and global positioning systems. The model is designed to provide vertical profiles of the main ionospheric parameters for suitably chosen locations over the globe, hours, seasons, and levels of solar activity, representing monthly mean conditions based on experimental evidence. The IRI Task Group brought together a distinguished team of experts representing the different ground and space measurement techniques and the different countries interested in ionospheric research. The truly international spirit of the IRI project is demonstrated by the typically more than 20 countries represented at the annual workshops jointly sponsored by COSPAR, URSI and IAGA (International Association on Geomagnetism and Aeronomy). A well balanced team both in terms of regional interest and in terms of science expertise is one of the secrets behind the success of the IRI mission.

One of the most challenging tasks in developing an empirical model is solving the right data question for regions and/or time periods where conflicting results exist for a certain parameter. Early measurements of plasma temperatures, for example, were difficult to interpret since the results of in-situ probes were up to a factor of two higher than observations by incoherent scatter radars. Improvements in probe design and in radar data reduction led to good agreement in the early seventies. In such cases much sensitivity and insight is required to get the experimenters to compare their results and to discuss possible error sources. The main sources of information on which the IRI system was built are the ground based and topside sounding ionosondes, incoherent scatter radars, in-situ rocket and satellite measurements (Bilitza et al., 1993a,b). Plasmaspheric electron densities have been obtained indirectly from trans-plasmaspheric VLF measurements (whistler) and from highly sensitive in-situ instruments. The improvement of the IRI representation of ionospheric parameters, such as electron density, electron temperature, ion composition and ion temperatures, and total electron content (TEC) through the ionosphere and plasmasphere still remains a challenge for the IRI Project.

In the next section we will briefly outline physical models which are important as tools to explore and understand the physical processes that shape the ionospheric environment.

2. Physical Models

Physical models typically use a numerical iterative scheme to solve the Boltzmann equations for the ionospheric gas including the continuity, energy, and momentum equations. They are solved along field-lines of the Earth magnetic field where the field is represented either by a simple tilted dipole or a multiplex model like the International Geomagnetic Reference Field (IGRF). Solving the equations along a full flux tube would take account of the plasmaspheric flux in a self-consistent way for a given geophysical condition, however, to keep control over the flux as a free parameter, the plasmaspheric flux is provided as a top boundary condition. The effects of the geomagnetic field on the transport of the ionospheric plasma are introduced by the magnetic dip (I) and declination (D) angles from the International Geomagnetic Reference Field (IGRF).

The ionosphere is strongly coupled with the neutral atmosphere, chemically as well as dynamically. In addition to the effects of the neutral wind, the neutral atmosphere significantly affects the ionospheric plasma density distribution through neutral composition and temperature. The neutral composition is a crucial factor not only for the production and loss of the plasma, but also for the diffusion of the ionospheric plasma through the neutral atmosphere. The neutral temperature effect on the ionosphere usually comes from the changes of the neutral densities caused by the neutral temperature change. Below the F-region peak, chemical equilibrium prevails and the plasma density profile is largely controlled by the neutral composition through the production and loss. As altitude increases, plasma diffusion becomes important and well above the F-region peak, the plasma density profile is primarily determined by diffusion. However, the diffusion of the plasma through the neutral atmosphere strongly depends on the neutral densities, mainly the O density in the topside ionosphere, via collisions between the plasma and the neutrals.

A physical model requires several input parameters, including the neutral densities and temperature, neutral wind, and plasma temperatures. For these inputs, empirical models are adopted. In assimilative mode of operation, up to six free model parameters should be adjusted to measurements within physically reasonable ranges, and this cannot be reached straightforward under certain conditions.

For the last several years, Utah State University has been developing the Global Assimilative Ionospheric Model, GAIM (Scherliess et al., 2004; Wang et al., 2004a). GAIM uses a physics-based ionosphere-plasmasphere-polar wind model and a Kalman filter as a basis for assimilating a diverse set of real-time (or near real-time) measurements (Schunk and Nagy, 2000). GAIM is a data assimilation model that specifies and forecasts the state of the ionosphere. There are several versions of GAIM under development at USU. One version, a Gauss-Markov Kalman filter uses the Ionosphere Forecast Model as the background model specification. A second version of GAIM being developed by USU is a physics-based, reduced-state Kalman filter assimilation algorithm (Scherliess et al., 2004). This reduced-state approach was tested under simulation, though for the simulation only predictions of electron density obtained from the first principle model (with driver adjustment) were used. The full analyzed electron density and error covariance was not used in the simulation.

The University of Southern California and the Jet Propulsion Laboratory (USC/JPL) physics model (Pi et al., 2003; Hajj et al., 2004; Wang et al., 2004a) is derived from the Sheffield University Plasmasphere Ionosphere Model, SUPIM (Bailey et al., 1997). In physical models, such as SUPIM, the time-dependent equations of continuity, momentum (ignoring the time variation and inertial terms in the momentum equation), and energy balance are solved along eccentric-dipole magnetic field lines for the densities, field-aligned fluxes and temperatures of the ions and the electrons. Its application relies on accurate estimate of the solar EUV, ExB drift, neutral wind, and neutral densities. The ion momentum equation is further broken into a field-parallel and field perpendicular component. The velocity component perpendicular to the magnetic field is considered to be due entirely to ExB and is an input driver. The parallel component of velocity also has input drivers due to neutral winds and electron and ion temperatures. Thus in the USC/JPL system the only state variable solved for is the O^+ density; the rest are input drivers to the system.

The Coupled Thermosphere Ionosphere Model (CTIM) of Fuller-Rowell et al. (1996) was developed from the ionospheric part of the Sheffield model. As with many of the theoretical model, the global atmosphere is divided into a series of elements in geographic latitude, longitude, and pressure (or altitude). Each grid point rotates with Earth to define a non-inertial frame of reference in an Earth-centered coordinate system. The magnetoispheric input is provided with statistical models of auroral precipitation (Fuller-Rowell and Evans, 1987) and electric fields (Foster et al., 1986). Both inputs are keyed to a hemispheric power index (PI), based on the TIROS/NOAA auroral particle measurements. A recent upgrade of this model, including self-consistent plasmasphere and low latitude ionosphere models is in the Coupled Thermosphere-Ionosphere-Plasmasphere model (CTIP; Millward et al. 1996). The effects of E×□B drift at lower latitudes are incorporated by the inclusion of an empirical low-latitude electric field model. The new ionosphere-plasmasphere component of CTIP solves the coupled equations of continuity, momentum and energy to calculate the densities, field-aligned velocities and temperatures of the ions O^+ and H^+ and the electrons, along a total of 800 independent flux-tubes arranged in magnetic longitude and L value (20 longitudes and 40 L values).

The Field Line Interhemispheric Plasma (FLIP) model (Richards, 1996) is a first-principles, one-dimensional, time-dependent, chemical, and physical model of the ionosphere and plasmasphere. It couples the local ionosphere to the overlying plasmasphere and conjugate ionosphere by solving the ion continuity and momentum, ion and electron energy, and photoelectron equations along entire magnetic flux tubes. The interhemispheric solutions yield densities and fluxes of H^+, O^+, He^+, and N^+ as well as the electron and ion temperatures.The neutral densities, temperature, and wind are supplied by the empirical MSIS (Hedin, 1991) and HWM (Hedin et al., 1996) models. During quiet times the error in the inputs for the solar EUV flux, MSIS neutral parameters, reaction rates, and cross sections are typically about 20%. During magnetic storms uncertainties may be much larger. The set of nonlinear, second-order, partial differential equations for continuity, momentum, and energy is transformed into finite difference equations and solved by a Newton-Raphson iterative scheme. The current FLIP model is basically a midlatitude model because it neglects convection electric fields, which are important at equatorial and auroral latitudes.

As described in the previous paragraphs driver inputs must be obtained from empirical models including the following: thermospheric densities from the Mass Spectrometer Incoherent Scatter model (Hedin, 1991), neutral winds from the horizontal wind model (Hedin et al., 1996), solar EUV as described by Tobiska (1991), electric fields (e.g., Fejer, 1991; Heppner and Maynard, 1987; Scherliess and Fejer, 1999), and electron energy precipitation flux (Fuller-Rowell and Evans, 1987). The interested reader can refer to Pi et al. (2003) and references therein. In the 2003 model validation experiment, only vertical drift at the geomagnetic equator was simulated and estimated, while all the other inputs were held at their empirical values. The vertical drift was parameterized by nine coefficients at different local times.

The Open Geospace General Circulation Model (OpenGGCM) is a global model of the magnetosphere-ionosphere system. It solves the magneto-hydrodynamic (MHD) equations in the outer magnetosphere and couples via field aligned current (FAC), electric potential, and electron precipitation to an ionosphere potential solver and the Coupled Thermosphere Ionosphere Model (CTIM) (Raeder et al., 2008). This code coupling enables studies of the global energy budget of the magnetosphere-ionosphere-thermosphere system. The CTIP model (Coupled Thermosphere Ionosphere – Plasmasphere) is a self-consistent first-principles model of the inner magnetosphere and thermosphere-ionosphere-plasmasphere system. It solves the continuity equations and the steady-state momentum equations for densities and velocities of O^+ and H^+ ions along the geomagnetic flux tubes using the finite difference schemes (Wang et al., 2004b).

The NCAR thermospheric general circulation model (TGCM) is extended to include a self-consistent aeronomic scheme of the thermosphere and ionosphere (Roble et al., 1988). The model now calculates total temperature, instead of perturbation temperature about some specified global mean, global distributions of N(2D), N(S) and NO, and a global ionosphere with distributions of O^+, NO^+, $O2^+$, N•', N^+, electron density, and ion temperature as well as the usual fields of winds, temperature and major composition. Mutual couplings between the thermospheric neutral gas and ionospheric plasma occur at each model time step and at each point of the geographic grid. Steady state results for this first Eulerian model of the ionosphere are presented for solar minimum equinox conditions. The calculated thermosphere and ionosphere global structure agrees reasonably well with the structure of these regions obtained from empirical models. This suggests that the major physical and chemical processes that describe the large-scale structure of the thermosphere and ionosphere have been identified and a self-consistent aeronomic scheme, based on first principles, can be used to calculate thermospheric and ionospheric structure considering only external sources. Global empirical atmospheric models, such as the mass spectrometer/ incoherent scatter models (e.g., Hedin, 1991), were used to specify atmospheric properties for ionospheric model. Equations describing the ionosphere and thermosphere are both solved on the TGCM geographic grid. Ion drift for the ionospheric calculation is obtained from the empirical model of Richmond et al. (1980) for low- and mid-latitudes and the empirical model of Heelis et al. (1982) for high latitudes. Consideration of displaced geomagnetic and geographic poles is included. Results for solar minimum equinox conditions are presented that show good agreement with MSIS-86 (Hedin, 1991). The self-consistent model requires only specifications of external sources as solar EUV and UV fluxes, aurora particle precipitation, ionospheric convection pattern, and the amplitudes and phases of semi-diurnal tides from the lower atmosphere.

The models of the ionospheric plasma density distribution and TEC depend on a number of upper atmospheric and ionospheric parameters, such as the neutral density, neutral wind, neutral and plasma temperatures, plasmaspheric flux, and ion-neutral collision frequencies. In the numerical modeling of the ionosphere, these parameters are generally often only roughly known and can cause significant uncertainties in the model results (Jee et al., 2005). The physical models are also tested for implementation in the ionosphere tomography though a numerical model is often derived to give a close approximation to the full theoretical calculations under all conditions. The physical model can be used to determine the qualitative relationship, but we do not have to rely on the physical model to provide the quantitative dependence for operational use (Fuller-Rowell et al., 2001). The physical models can match empirical models in accuracy provided accurate drivers are available, but their true value comes when combined with data in an optimal way (data assimilative scenario).

3. International Reference Ionosphere

3.1. General

A preliminary set of IRI tables was presented at the 1972 URSI General Assembly (Rawer et al., 1972) and COSPAR Scientific Assembly (Ramakrishnan and Rawer, 1972). Composite ionosonde profiles, incoherent scatter data, and total electron content (TEC) Faraday measurements played a dominant role in establishing this first precursor of an IRI model.

From the beginning the computer-accessibility of the evolving model was established, a foresightedness that paid off in the long run and contributed considerably to the popularity of IRI in the user community. One should keep in mind that those were the days of punched cards, paper tape, large mainframes and with still a considerable computer illiteracy in the science community. Making the model software accessible in computer-readable form to a wide user community was still a relatively novel approach; for example, CIRA (COSPAR Reference Atmosphere) continued to be presented primarily in the form of published tables until the release of CIRA-86 in the late eighties. The first widely distributed edition of IRI was released in 1978 (IRI-78) as an URSI Special Report (Rawer et al., 1978) and also as ALGOL and FORTRAN computer codes on punched cards and tape.

The IRI task group was established jointly by COSPAR and URSI in the late sixties to develop and improve a standard model of the ionospheric plasma parameters (electron and ion densities, temperatures, and velocities). The model should be primarily based on experimental evidence using all available ground and space data sources; theoretical considerations can be helpful in bridging data gaps and for internal consistency checks. Where discrepancies exist between different data sources, the IRI team should promote critical discussion to establish the reliability of the different data bases. IRI should be updated as new data become available and as old databases are fully evaluated and exploited.

Using the (CCIR, 1983) world maps for the F2 peak parameters foF2 and M(3000)F2, IRI-78 was a major step forward towards a truly global representation of the ionosphere. By incorporating the maps of the Consultative Committee on International Radiopropagation (CCIR) recommended for international use by the International Telecommunication Union (ITU-R) and URSI, the IRI group underlined its support and concern for radio propagation studies and applications. A special URSI working group was established and succeeded in developing a new set of world maps based on a much better extrapolation scheme for the data-sparse ocean regions (Rush et al., 1989). The improved accuracy over the oceans, however, came at the expense of somewhat less accurate maps for the continents. Different from CCIR who decided to stick with its older maps, the IRI model now provides access to both the CCIR and URSI maps, thus allowing users to utilize the superior accuracy of the URSI maps over the oceans.

For their global distribution both maps rely on the modified dip latitude (modip) that was introduced by Rawer (1963) and is defined as tan (modip) = $\mathrm{dip}/(\cos(\mathrm{lat}))^{1/2}$ where dip is the magnetic inclination and lat is the geodetic latitude. Modip was found to better organize features of the ionospheric F2 layer parameters than other magnetic coordinates. This is due to the fact that the ionosphere is controlled by both the orientation of the Earth's rotation axis and the configuration of the geomagnetic field. Therefore, its variation depends on both the geographical and geomagnetic latitudes, which is embedded in modip depending on both the magnetic inclination and the geographical latitude. In recent years the IRI group has taken a more active role in the mapping of F2 peak parameters. The cumulative data volume from the worldwide network of ionosondes has increased significantly since the CCIR and URSI models were developed. Oyeyemi et al. (2007) [M(3000)F2], Oyeyemi and McKinnell (2008) [foF2], McKinnell and Oyeyemi (2009, 2010) [foF2] have trained Neural Networks (NN) with all available ionosonde data and find increased accuracy compared to the older models. Their models for M(3000)F2 and foF2 are now planned for inclusion in IRI as new options. The propagation factor M(3000)F2 is the MUF divided by foF2, where MUF is defined as the highest frequency at which a radio wave can propagate from a given point over a distance of 3000 km. M(3000)F2 can be deduced from vertical-incidence ionograms by the use of standard methods and global spherical harmonics models similar to the foF2 models were developed for M(3000)F2 under the auspices of CCIR and ITU. M(3000)F2 is closely correlated with the F2 peak height, hmF2, and empirical relationships have been developed that describe this dependencies (Dudeney, 1974; Bilitza et al., 1979). The Bilitza et al (1979) formula is used in IRI because it showed best agreement with ionosonde as well as incoherent scatter measurements.

Guided by the knowledge gained from previous data analysis, and from simulations with a physically-based model (Rodger et al., 1989), observations of the F-region critical frequency foF2 (related with the peak electron density NmF2) from all available sites and from many storms were selected by Fuller-Rowell et al. (2000) sorted as a function of local time, season and magnetic latitude, and by the magnitude of the storm as described by the geomagnetic Ap index. Their empirical ionospheric STORM model is incorporated in IRI.and it provides prediction of the dominant ionospheric storm effects on the peak electron density corrected for effects of the planetary geomagnetic 3-hour Ap index from the preceding 39 hours. Relevant empirical model for the storm-time ionosphere peak height hmF2 associated with changes of peak electron density NmF2 has been deduced from the topside sounding database of ISIS1, ISIS2, Intercosmos-19 and Cosmos-1809 satellites which is included in coupled IRI / Plasmasphere code, IRI-Plas (Gulyaeva et al., 2002a,b, 2011; Gulyaeva, 2011a,b).

The structure of the IRI electron density profile is shown in Figure 1 (Bilitza and Rawer, 1990). The electron density profile is divided into eight height regions from the D layer in the lower ionosphere to the topside and the plasmasphere above the F2 peak.

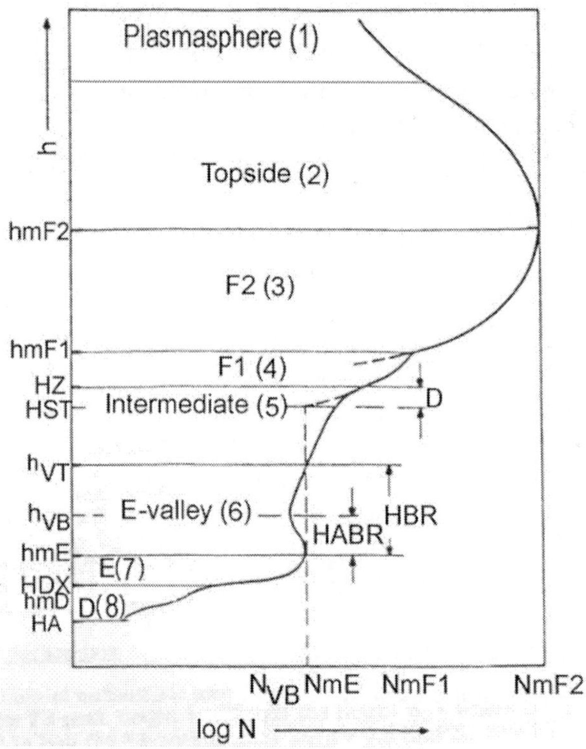

Figure 1. Buildup of IRI electron density profile.

3.2. IRI Topside

By the early seventies the highly successful Alouette 1 and 2 satellites had accumulated a large data base of global topside soundings so that empirical modelling of the topside could be seriously considered. The first major effort was undertaken by Bent and his colleagues (Bent et al., 1972; Llewellyn and Bent, 1973) using more than 50,000 Alouette 1 topside soundings covering the period 1962 to 1966 (low to medium solar activity). For high solar activity they relied on Ariel 3 in-situ measurements for 1967 and 1968 which were combined with F2-peak densities obtained from ground-based ionosondes. Their model is given in graphical form providing plots of the linear variation of their model parameters with daily solar 10.7 cm radio flux (F10.7) for four foF2 classes (2, 5, 8, 11 MHz) and three ranges in geomagnetic latitude (0° to ±30°, ±30° to ±60°, ±60° to ±90°).

In the very first version of IRI (Rawer et al., 1972; Ramakrishnan et al., 1972), topside profiles (range 2 in Figure 1) were based on incoherent scatter data from Malvern (U.K.) and Arecibo (Puerto Rico). The thickness of the upper F-layer was chosen in such a way that the total electron content (TEC) calculated for the IRI profile agreed with TEC measurements. However, the determination of the thickness parameter should really be based on the so-called slab thickness, which is the TEC value normalized with the simultaneously measured F2-peak density. Unfortunately, very little information was available at the time about the global variation of slab thickness. For IRI-1978 Rawer et al. (1978) developed an analytical description of the data base contained in Bent's model using Epstein step- and transition-functions. As in the Bent model the IRI model coefficients are provided as functions of F2 peak plasma frequency (foF2), geomagnetic latitude, and solar activity (F10.7). The analytical representation helped to smooth out some of the unreasonable sharp transitions seen in the original Bent model. An important result of this newer model is a smoothly varying scale height, which is more acceptable than the very irregular scale height behavior obtained with the original Bent model.

As more topside data became available the IRI topside model was evaluated extensively. ISIS 1 and 2, the followon topside sounder satellites to Alouette, were a particular valuable data source (Bilitza et al., 2003). The comparative studies found discrepancies between the data and the IRI model, especially in the upper topside (Iwamoto et al., 2002; Coisson et al., 2002; Bilitza, 2004). Many different profile functions were tested in an effort to improve the IRI model (Bilitza et al., 2006). Simplified aeronomic arguments lead to a Chapman-type profile (Rishbeth and Garriott, 1969)

$$N(h)/NmF2 = \exp\{c[1-z-\exp(-z)]\},\ z = (h-hmF2)/Hm \qquad (1)$$

where NmF2 and hmF2 are the F2 peak density and height, Hm is the Chapman layer topside scale height and c=0.5 or 1 for a Chapman α- or β- function. The α-Chapman scale height Hm is about 3 times less than the topside exponential scale height, Hsc, corresponding to 1/e decay of the peak electron density (Gulyaeva, 2011a).

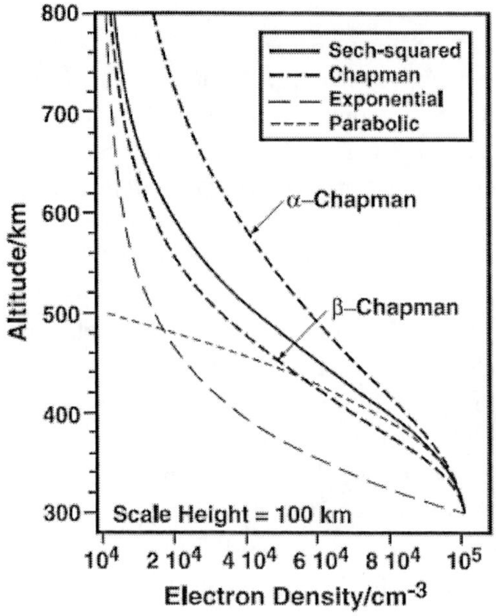

Figure 2. Density profile functions for the topside electron density for a scale height of 100 km (Stankov et al., 2003; reproduced by permission of American Geophysical Union).

Other functions often used in topside modeling include parabolic, exponential, and Epstein-layer (sech-squared) functions (Stankov et al., 2003). Figure 2 shows the typical altitudinal variation behavior of these functions. The Ne-Quick model of Radicella and Leitinger (2001) and Coisson et al. (2006), for example, combined an Epstein F-layer with a height-varying scale height. The merits of different functions in reproducing measured topside profiles and TEC values have been evaluated in a number of studies using different data sources (Stankov et al., 2003; Bilitza et al., 2006; Bilitza, 2009). Stankov et al. (2003), for example, find that average nighttime profiles obtained from AE-C satellite in situ measurements are best represented by the Epstein formulas, whereas the daytime profiles are better approximated by exponential or a Chapman functions. Examples of α-Chapman profile with a constant scale height (dashed curves in Figure 3) demonstrate insufficient plasma density in the topside ionosphere and plasmasphere as compared to full IRI-Plas electron density profile (Gulyaeva et al., 2002a,b).

As a result of these studies two new options were introduced in IRI starting with 2007 version (Bilitza and Reinisch, 2008). The first is a correction factor for the 2001 model based on over 150,000 topside profiles from Alouette 1, 2, and ISIS 1, 2. This term varies with altitude, modified dip latitude, and local time (Bilitza, 2004). The second option is the NeQuick topside model that was developed by Radicella and his collaborators over the last decade (Radicella and Leitinger, 2001; Coisson et al., 2006) using Intercosmos 19 topside sounder data in addition to the ISIS 1 and 2 data. The model fit analytical functions on a set of anchor points, namely the E; F1 and F2 layer peaks, to represent these principal ionospheric layers and compute the electron density profile. NeQuick is adopted by the ITU-R recommendation for TEC modeling (Hochegger et al., 2000).

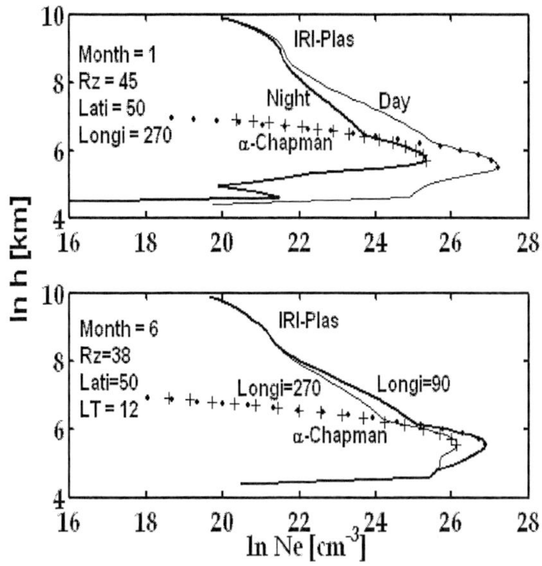

Figure 3. The IRI-Plas model electron density profile in comparison with α-Chapman profile. Noon and midnight examples (upper panel), noon ptofiles at different longitudes (lower panel).

The NeQuick model is divided into two regions: the bottomside, up to the F2-layer peak, consists of a sum of five semi-Epstein layers-1 (Rawer, 1984; 1991) and the topside is described by means of an only sixth semi-Epstein layer with a height-dependent thickness parameter, and in this way produces a smooth transition from an atomic oxygen ionosphere near the F-peak to a light ion ionosphere higher up. The implementation of NeQuick topside into IRI is the only part of IRI model which requires the use of M(3000)F2 in addition to foF2 and hmF2 values. This parameter can be computed either from the CCIR model or inverting the IRI formula that provides hmF2 in terms of M(3000)F2 (Bilitza et al., 1979).

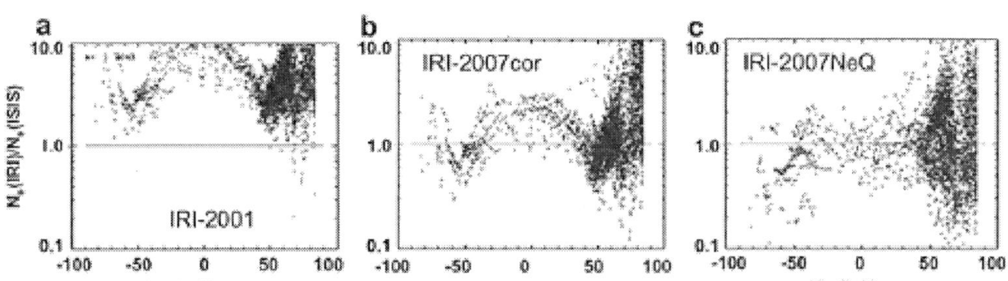

Figure 4. (a–c) Ratio of ISIS-2 topside sounder data versus model predictions at 1000 (±50) km above the F2 peak using IRI-2001 (a), IRI-2007-cor (b), and IRI-2007-NeQ (c). The total number of data points is n = 3043 (Bilitza, 2009; reproduced by permission of Elsevier).

Figure 4 shows the model-data ratio for the three IRI topside options (IRI-2001, IRI-2001 corrected, IRI-NeQuick) at 1000 km altitude using ISIS-2 sounder measurements. A significant improvement (values closer to 1) is achieved with the inclusion of the correction term and even more so with the NeQuick option. The results for all available ISIS 1 and 2 data are summarized in Table 1 indicating a factor of 4 improvement when using the correction term and a factor 7 improvement with the NeQuick model.

Two recent modeling activities promise future improvements for the representation of topside electron density profiles in IRI. Reinisch et al (2007) have developed the Vary-Chap approach. While the standard Chapman profile function (Eq 1) is deduced for a single constituent gas with a constant scale height, Reinisch et al. (2007) have determined the profile function for a more realistic multi-constituent case with height-varying scale height expanding on earlier work by Rishbeth and Garriott (1969). Figure 5 shows how well this new function represents an ISIS-2 topside sounder profile.

Table 1. Mean and standard deviation of the percentage difference between the IRI model and ISIS-2 topside sounder data using 7 separate Alouette/ISIS data sets and all three options available in IRI-2007: IRI-2001 [IRI], corrected IRI-2001 [cor], NeQuick [NeQ] (Bilitza, 2009; reproduced by permission of Elsevier)

		ISIS-2	ISIS-1	Alouette -2	Alouette-1u	Alouette-1t	Alouette-1p	All
IRI	Std	10.39	7.44	9.90	4.50	1.83	2.73	
	Mean	-2.12	-1.79	-2.23	-0.50	-0.58	-1.01	-1.65
Cor	Std	2.97	2.72	2.04	2.44	1.11	1.55	
	Mean	-0.61	-0.43	-0.38	-0.22	-0.25	-0.41	-0.46
NeQ	Std	1.92	1.74	0.98	1.96	1.04	1.48	
	Mean	-0.31	-0.21	-0.04	-0.11	-0.20	-0.36	-0.24
	n	25,214	20,105	5166	19,434	37,240	12,900	120,059

IRI = IRI-2001, cor = IRI-2007-corrected, NeQ = IRI-2007-NeQ.
n – number of profiles, All – all 6 data sets together.

Gulyaeva's (2003) modeling efforts have focused on the topside half-density point, h0.5top. This is the height where the topside electron density has dropped down to half the F2 peak density, a point that can be easily scaled from topside profiles. The ISIS1, ISIS2 and IK19 topside sounder profiles were used to develop a model for h0.5top in terms of sunspot number, local time, and geomagnetic latitude. This model, IRI-Plas, uses a coefficient, 'q-factor', which serves for a correction of the IRI-2001 topside model forcing it to fit the h0.5top parameter (Gulyaeva et al., 2002a,b; Gulyaeva and Titheridge, 2006).

Comparisons of IRI-Plas and NeQuick with Topex-TEC measurements are plotted in Figure 6. The dual-frequency altimeters on board the Topex/Poseidon (T/P) and Jason satellites can provide the ionospheric electron content (TEC) over the oceans for altitudes ranging from 65 to 1336 km (Fu et al., 1994). TEC-model is calculated from the bottom of ionosphere to the TOPEX orbit altitude of 1336 km (Gulyaeva et al., 2009). Figures 6a and b show the diurnal/seasonal variation of TEC as measured by Topex and as given by the models for high solar activity (2002) and low solar activity (2007), respectively. All model results are obtained with CCIR peak maps, and Topex TEC are averaged for each hour ±0.5 h, within longitude/latitude bins in step of 5° and compiled/averaged for each month/season. The models are capable of reproducing the main variations of the Topex TEC but in the worst case depart from the observations up to 20 TECU (TEC unit, TECU=10^{16}el/m^3).

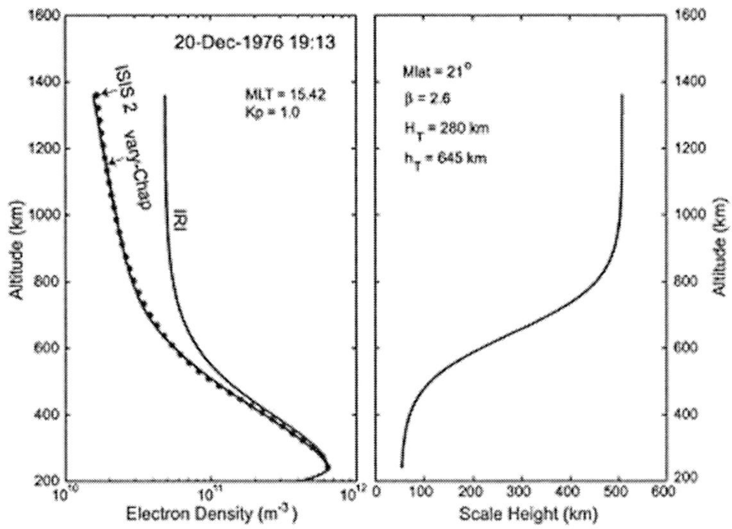

Figure 5. The vary-Chap function (solid line) closely matches the measured ISIS 2 profile (dots); the IRI profile largely overestimates the densities at higher altitudes. The right panel shows the corresponding scale height function H(h) showing a clear transition from an O^+ to an H^+ ion gas higher up (Reinisch et al., 2007; reproduced by permission of Elsevier).

A particular challenge in modeling the topside electron density profile is the accurate representation of the fountain effect at equatorial latitudes. At F-region heights this effect produces the well-known crests at about 18 degrees dip latitude on each side of the magnetic equator (equator anomaly, EA) and a relative minimum at the magnetic equator. With increasing height the two latitudinal peaks move closer towards the magnetic equator and merge into a single peak at the magnetic equator at a height of about 1000 km (Bilitza et al., 2006). Almost all topside models are normalized to the F2 peak density, NmF2. Models for NmF2 are well established and provide the typical camelback signature of the equator anomaly. To reach a single equatorial peak at high altitudes, the topside profile function therefore has to counterbalance the EA signature imprinted by the NmF2 model. An improved representation of the IRI topside electron density profile is also a necessary step towards a successful merging of the ionosphere and plasmasphere models because plasmaspheric models often use the IRI topside density at a certain fixed height as a footpoint for their model.

Figure 7 illustrates the representation of the EA region as it is given by the four options available in IRI for the topside electron density (a – IRI-2001, b – IRI-corr., c – IRI-NeQuick, d – IRI-Plas). The contour plots show the distribution of ionization with geomagnetic latitude and height. At F-region altitudes the models are very similar with the two maxima at about 15 degrees and with the neutral wind pushing ionization up the field line in the northern (summer) hemisphere and down the field line in the southern (winter) hemisphere resulting in the hemispheric asymmetry in F-peak height. However, drastic differences appear between the four options with growing altitudes. The IRI-corr model (Figure 7b) and IRI-Plas model (Figure 7d) produce the expected merging of the EA double peak into a single peak at higher altitudes while the other two models exhibit unrealistic features: IRI-2001 (Figure 7a) shows almost vertical profiles, and IRI-NeQuick (Figure 7c) shows still separate EA peaks at high altitudes.

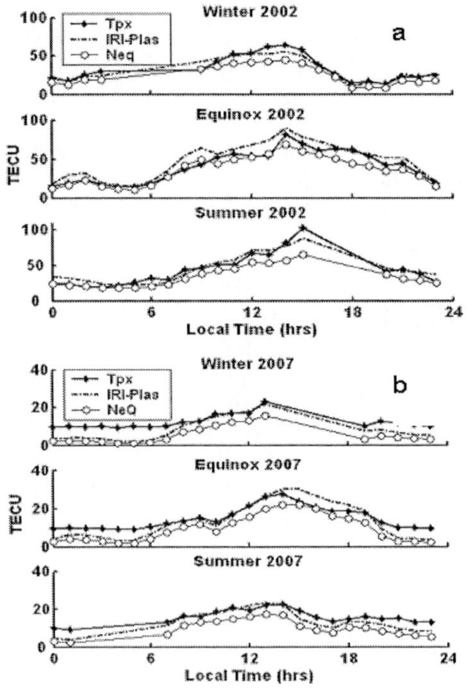

Figure 6. Seasonal-diurnal variation of TEC-model (IRI-Plas and NeQuick) compared with Topex data: (a) high solar activity (2002); (b) low solar activity (2007).

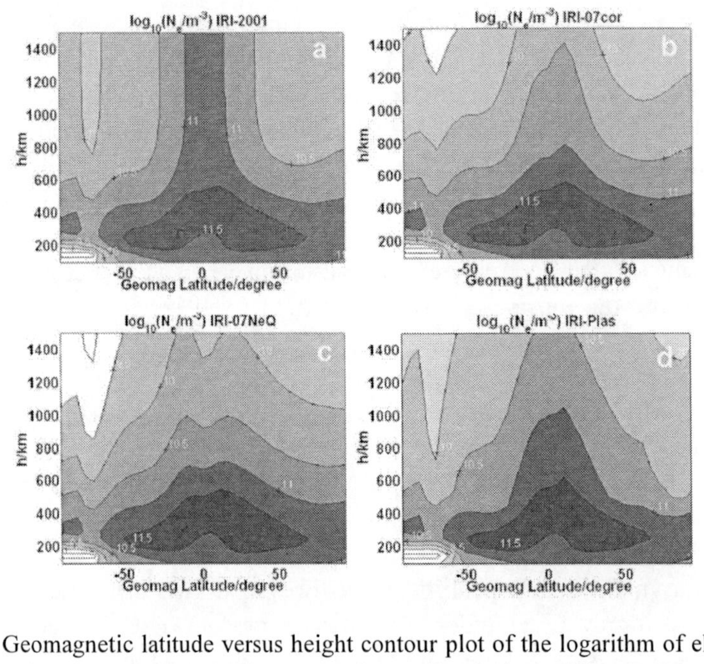

Figure 7. (a–d) Geomagnetic latitude versus height contour plot of the logarithm of electron density at Longitude = 0 and Universal Time (UT) = 16:00 for medium solar activity (R12 = 50) and northern summer for the different IRI model options: IRI-2001 (a), IRI-2007-cor (b), IRI-2007-NeQ (c) and IRI-Plas (d). The increment between contour lines is 0.5 starting from 11.5 (black) down to 9.5 (white).

3.3. IRI Bottomside

The bottomside of the F2 region (range 3 in Figure 1) is represented by the function

$$N(h) = NmF2*\exp(-x^{B1}) /\cosh(x) \qquad (2)$$

where

$$x = (hmF2-h)/Bo$$

and B0 and B1 are profile parameters that describe the thickness and shape of the bottomside layer. This function was proposed by Ramakrishnan and Rawer (1972) who found that with it they could closely reproduce a set of representative profiles obtained from ionosonde measurements. For B0 two model options are offered in IRI. The standard Table option is based on a table of values that was assembled for different latitudes, times of day, levels of solar activity, and seasons (Bilitza et al., 2000).

The second option is based on work by (Gulyaeva (1987; 2007) and therefore is called the "Gulyaeva" option. Gulyaeva (1987, 2007) found a close correlation between the F2 peak height hmF2 and the height $h_{0.5}$ where the electron density has decreased to half the F2 peak density, $N_{0.5}$ = 0.5 NmF2 This relation in a function of the solar zenith angle and season. The shape parameter B_1 is fairly constant and shows only significant differences from day to night. The typical daytime value is 1.9 and the nighttime value is 2.6. Epstein step functions are used to model the day-night transitions at dawn and dusk. Both model options, however, are based on a relatively small data base and have shown shortcomings in comparisons with ionosonde and incoherent scatter data. Altadill et al. (2008, 2009) have applied spherical harmonics analysis to data from 27 globally distributed ionosonde stations obtaining new models for B_0 and B_1 that more accurately describes the observed variations with latitude, local time, month, and sunspot number. Overall the improvements over the IRI-2007 model options are of the order of 15 to 35%. The largest improvements are seen at low latitudes. This new model will be included in IRI-2011 as a third option.

A parabolic F1 layer is included in the bottomside electron density profile when the NmF1 model (Ducharme et al., 1973) predicts the existence of this layer. If no F1 feature is predicted (e.g. at night) the F region is represented b a single F2 layer. The F1 layer (range 4 in Figure 1) starts from the F1 layer peak height, hmF1, where the bottomside profile reaches the F1 peak density, NmF1. The models for NmF1 and for the F1 layer thickness C_1 were developed based on ionosonde data.

While in the F region the IRI electron density profile is normalized to the F peak density and height, it is normalized to the E region peak density and height in the E region. A merging algorithm is used to join the two profile parts. This transition region extends from the top of the E valley to the bottom of the Fl layer (range 5 in Figure 1). In the intermediate region the bottomside profile is merged parabolically with the E-F valley profile. First, the height HST is found where the bottomside profile density is equal to the E peak density NmE. Starting from a point HZ slightly above HST, the profile is then bent downwards so that it meets the vally top. The valley between the E and F regions (range 6 in Figure 1) is derived mostly from the incoherent scatter radar data since it is unvisible with the vertical sounding of the ionosphere. Analytical expressions are used to describe the variation of the valley parameters (depth and width) with solar zenith angle and latitude (Gulyaeva, 1987). The valley marks the region between the E and F layers and reaches great depth at nighttime due to the fast recombination of electrons and ions but it is narrow for the daytime.

The current formalism could lead to discontinuities or artificial valleys under conditions when merging was particularly difficult to accomplish, because of large differences between E and F peak densities and/or small differences between the E and F peak heights. A better functional description for the merging region was developed by Reinisch and Huang (1999). The new algorithm overcomes these problems and provides a much better representation of profile shape observed by ionosondes.

The E region is the region of peak ion/electron production, and the electron distribution is largely controlled by the solar zenith angle. The E-peak height is remarkably constant, and is typically found at about 110 km. The E- peak density, NmE, is about an order of magnitude lower than the F2-peak density. Kouris and Muggleton (1973) developed a model for NmE based on ionosonde data from representative low and middle latitude stations. In addition to the solar zenith angle dependence the model describes the variation with solar activity, season, and geographic latitude. The model was adopted for radiowave propagation applications by CCIR. For IRI the model was improved for large solar zenith angles (nighttime) with the help of incoherent scatter radar results (Bilitza and Rawer, 1990). The off-equator E region electron density enhancements are found to be closely connected with the bottomside of the F region equatorial anomaly crests, where the component of the electron density parallel to the magnetic field line is maximum (Chu et al., 2009). It appears that the off-equator E region electron density enhancements are very likely the footprints of the F region equatorial anomaly crests.

The IRI generally describes the E-region electron density (range **7** in Figure 1) well, but the ionization enhancement at auroral latitudes caused by precipitating particles has not been provided till recently. One of the new features in IRI-2007 (Bilitza and Reinisch, 2008) is a Neural Network (NN) model for this auroral region that was trained with a large volume of EISCAT incoherent scatter data (~700,000 data points) and also with 115 profiles obtained from rocket borne wave propagation experiments (McKinnell et al., 2004; McKinnel and Friedrich, 2006). The model describes the density variations in terms of local magnetic time, riometer absorption, local magnetic index K, solar zenith angle, and atmospheric pressure, the latter accounts for variations with height and season. Newer efforts by Fernandez et al. (2010) and Zhang et al. (2010) using TIMED SABER and GUVI data, respectively, promise major improvements in the representation of the high latitude E region in IRI. As a result of the modeling work with GUVI data by Zhang et al. (2010) the next version of IRI will for the first time include a representation of auroral boundaries and their movement with magnetic activity.

The lower ionosphere is defined below the D region peak density NmD and height hmD. The D region (range 8 in Figure 1) is characterized by large variability and a very small database for modeling studies. The only data sources are rocket experiments, because the region is too low for satellites and the densities are too low for ground ionosondes and radars. IRI includes three options for the description of D region electron densities, thus reflecting the large uncertainties that still exist in this region. Option 1 is the current D region model that was developed by Mechtley and Bilitza (1974) on the basis of a rather limited set of representative rocket data.

Option 2 is the FIRI model by Friedrich and Torkar (2001) derived from a database of the most reliable D region rocket measurements (-200 profiles) collected by these authors

(Friedrich et al., 2001). They experimented with different modeling approaches, including the strong dependence on solar zenith angle and considering to various degrees dependencies on season, latitude, solar activity, and neutral density and even an extension to high latitudes. Their most recent modeling concept combines the rocket data with the results of a theoretical model. The FIRI model is given as a table of values and model profiles are obtained by suitably interpolating using the user-specified zenith angle, latitude, season, and solar activity.

Option 3 is the model of Danilov et al. (1995) based primarily on Russian rocket data. In addition to the dependence on solar zenith angle, latitude, and solar activity it also provides users with an estimate of the changes observed in the D region during disturbed conditions for winter daytime. Although their rocket database is quite limited in volume, they find that the data can be grouped into five distinct classes using the following criteria: (1) undisturbed conditions, (2) weak winter anomaly (WA) defined by an increase of the absorption in the 2-2.8 MHz range at short A3 paths by 15 dB, (3) strong WA defined by an increase of 30 dB, (4) weak stratospheric warming conditions defined by a temperature increase at the 30 hPa level by 10 degmes, and (5) strong stratospheric warming conditions defined by an increase of 20 degrees.

Thus, the IRI electron density profile below the F2 peak is determined from the key parameters:

Peak densities: NmD, NmE, NmF1, NmF2
Peak heights: hmD, hmE, hmF1, hmF2
Layer thickness: Bo, B_1, C_1
Valley parameters: h_{VT}, h_{VB}, N_{VT}

3.4. Additional IRI Parameters

For many applications of ionospheric models the single most important parameter is the electron density and in particular the electron content along a radio wave signal path. But with the increased utilization of space other parameters have become of importance as well. The IRI model includes specifications for many of these parameters including the percentages of the major ions (also called ion composition), the temperatures of electrons and ions, the ion drift at the magnetic equator that is important for the development of the Equatorial Ionization Anomaly, and the occurrence probability for spread-F a phenomena that can cause havoc to communications and navigation primarily at low latitudes.

The ion composition models in IRI are based mostly on rocket measurements in the bottomside ionosphere and on satellite measurements in the topside ionosphere. Incoherent scatter radar measurements have been used to evaluate and improve these models. The dominant ion in the F region is O^+, at higher altitudes light ions, mostly H^+ with some He^+ and N^+, become more important and than dominant. Towards lower altitudes the percentage of molecular ions, $O2^+$ and NO^+ increases and they become the dominant constituent in the E region (100-150 km). Even lower down, in the relatively dense D-region (~80-90 km) Cluster ions can form and make up most of the ion population. For the topside IRI relies on the modeling work of Triskova et al. (2003) and Truhlik et al. (2004) who have compiled a large data base of satellite ion data (AE-C, -D, -E, Interkosmos 24, ISIS-2, ISS-b) and analyzed these data to establish the dominant global and temporal variation patterns.

The ion composition in the bottomside is the result of modeling work by Danilov and Smirnova (1995) who used their compilation of Russian rocket data together with data from the AE-C, S3-1, AEROS-B, Sputnik-3 and Cosmos-274 satellites. As a second option IRI also offers the earlier models by Danilov and Semenov (1978) for the bottomside and Danilov and Yaichnikov (1985) for the topside. All of these models describe variations with altitude, latitude, solar zenith angle, season, and solar activity. Discrepancies were found when using IRI molecular densities to compute airglow and EUV response (Nicolls et al., 2006; Vlasov et al., 2010). To overcome these problems Richards et al. (2010) proposed a new bottomside model for IRI that is based on the well established photochemistry in this region using the FLIP (Richards, 2002) and NRLMSIS00 (Picone et al., 2002) models and normalizing the outcome to the IRI electron density profile. The Richards et al. (2010) model will be introduced in IRI as a new option for the bottomside ion composition.

Electron temperature (T_e) and ion temperature (T_i) have been measured by incoherent scatter radars (ISRs) from the ground and insitu by satellite experiments. Satellite data provide the global morphology of temperature variations and the ISRs are the ideally resource for studying diurnal, seasonal, and solar cycle variations. A first global representation of T_e and T_i for IRI was developed by Bilitza (1981) based on a combination of satellite and ISR data. More recently Truhlik et al. (2000) have develop a model based on their large satellite data base. Both models describe variations with altitude, dip latitude, and magnetic local time and both models are given as separate options in IRI. A recent extension of the Truhlik et al (2000) model now also includes the dependence on the F10.7 solar flux based on work by Bilitza et al. (2007) and Truhlik et al. (2009) using satellite data and comparisons with the theoretical FLIP model and the empirical ISR model of Zhang et al. (2005). Like for T_e the global representation of T_i relies heavily on satellite data (Bilitza, 1981). Below 120 km thermal equilibrium is assumed and both plasma temperatures coincide with the neutral temperature as given by the NRLMSIS00 model (Picone et al., 2002). Higher up the heating of the electron gas through photo-electrons keeps T_e above T_i and the transfer of energy from the electrons to the ions through Coulomb collisions keeps T_i above T_n. For these reasons IRI enforces $T_e \geq T_i \geq T_n$ throughout the ionosphere.

3.5. IRI Plasmasphere Extension

An increasing number of users of ionospheric models also require information about the plasma conditions above the ionosphere in the plasmasphere (region 1 in Figure 1). The plasmasphere is the region from the top of the ionosphere up to a boundary, the plasmapause, where a sharp drop in plasma density occurs. The plasmasphere is a torus of relatively cold (~1-50 eV) and relatively dense (> 10 cm^{-3}) plasma that consists mostly of H^+ ions trapped along Earth's magnetic field lines and thus co-rotating with Earth. The plasmasphere structure and dynamics are driven by ionospheric sources and the plasmasphere feeds back the ionosphere by night and during the post-storm recovery. The Global Positioning System (GPS) receivers measure the total electron content through the ionosphere and plasmasphere so any use of GPS data needs to account for the plasmaspheric contribution to the total electron content between ground station and satellite (Yizengaw et al., 2007)..Low-earth orbit measurements include the upper reaches of the ionosphere and the thermosphere, which must be correctly modeled in order that the measurements properly constrain the outer plasmasphere (e.g. Heise et al., 2002).

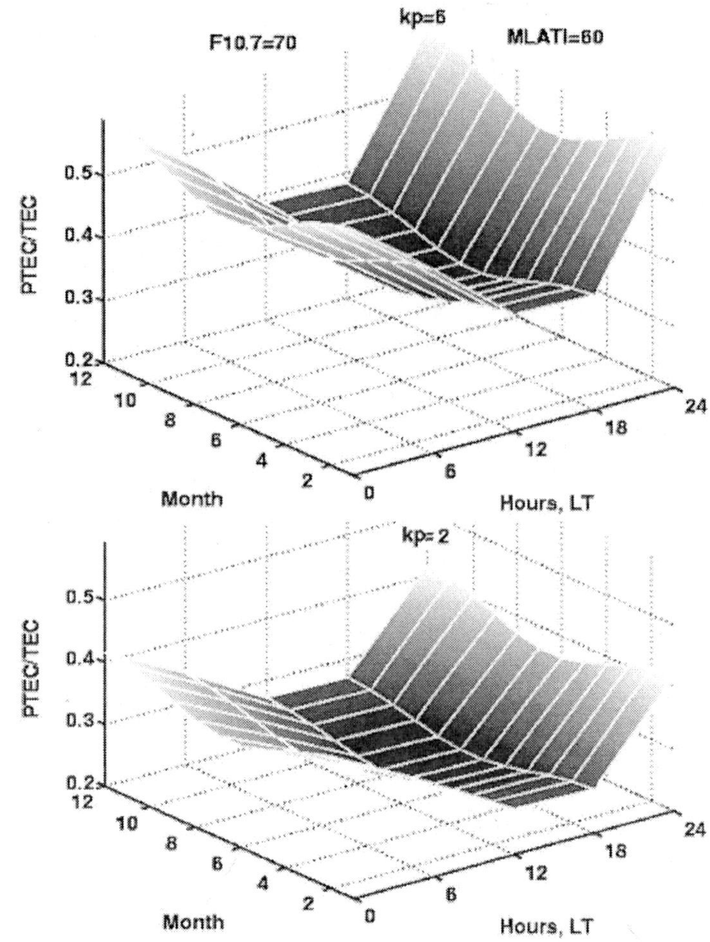

Figure 8. Seasonal/local time variation of plasmaspheric contribution to TECgps simulated with IRI-Plas model under quiet magnetic conditions (lower panel) and disturbed conditions (upper panel).

These TEC measurements can be used directly or organized into two-dimensional 2D TEC maps to infer information on the horizontal structuring of the electron density. However, information on how plasma can be lifted to high altitudes and transported to other regions, polar outflow, and other vertical dynamical changes is lost with such simple mapping algorithms. In order to obtain information on the vertical structure of the electron density, its temporal variation, and transport, 4D model is necessary. For the point profiles, measured by ionosondes or incoherent scatter radars (ISR), there is no information about large-scale horizontal gradients and convection of plasma that causes structuring at the profile position. The time-evolving nature of 4D modeling is crucial due to temporal and spatial changes of electron density distribution within a given region.

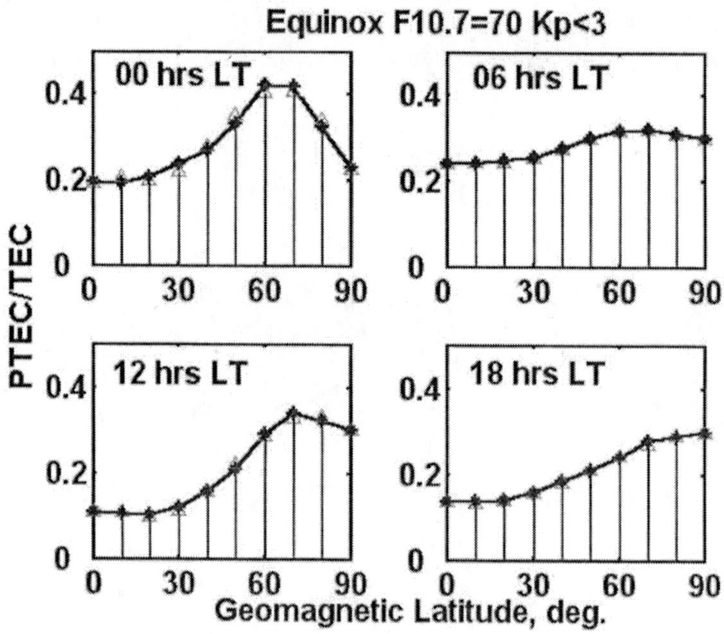

Figure 9. Plasmasphere contribution to TECgps varying with magnetic latitude simulated with IRI-Plas model during equinox under quiet magnetic conditions at low solar activity.

For IRI serving as a background model for tomographic reconstruction of 4D plasma structure from GPS-derived TEC, the IRI topside profile should be extended towards few Earth's radii (Bust and Mitchell, 2007). Figure 8 provides hints on plasmasphere contribution in GPS-derived TEC. Ratio of plasmaspheric electron content, PTEC, to the TEC (from the ground surface to 20,200 km at the GPS orbit) at sub-auroral latutude can comprise 40% towards midnight under quiet magnetosphere (bottom section, kp=2) and can exceed 50% during magnetic storm (upper section, kp=5). The diurnal variation of PTEC/TEC is illustrated for four selected local times in Figure 9 where this ratio is plotted versus modified dip latitude (modip).

A number of approaches have been proposed for extending IRI to plasamspheric altitudes. We will discuss briefly four of these models.

The **Global Core Plasma Model (GCPM-2000)** of Gallagher et al. (2000) is an empirical description of thermal plasma densities in the plasmasphere, plasmapause, magnetospheric trough, and polar cap. It has been developed from retarding ion mass spectrometer data collected by the Dynamic Explorer satellite, includes several previously published regional models, and represents the low energy plasma distribution along the field lines from 0 to 24 hours magnetic local time world-wide. GCPM-2000 is smoothly coupled to IRI in the transition region of 400-600 km altitude. It was applied also for the plasmasphere extension of NeQuick model (Cueto et al., 2006).

The **Global Plasmasphere Ionosphere Density (GPID)** model is a semi-empirical representation that was developed by Webb and Essex (2000, 2004). GPID includes IRI below about 500 km to 600 km and extends with a theoretical plasmasphere electron density description along the magnetic field lines. Authors report on drawbacks of merging of the IRI with the plasmasphere part of GPID.

The **IMAGE/RPI plasmasphere model** (Huang et al., 2004) is based on radio plasma imager (RPI) (Reinisch et al., 2000) measurements of the electron density distribution along magnetic field lines. A plasmaspheric model is evolving for up to about four earth radii. The depletion and refilling of the plasmasphere during and after magnetic storms is described in Reinisch et al (2004). A power profile model as function of magnetic activity was developed from RPI observations for the polar cap region (Nsumei et al., 2003).

Table 2. List of IRI-Plas model output parameters

Symbol	Designation
Year	Year
MN	Month
DY	Day of month
UT	Universal time (h)
LT	Local time (h)
MLT	Magnetic local time (h)
XHI	Solar zenith angle (°)
Rz	Sunspot number
Cov	Solar radio flux F10.7 (Covington index)
Kp	Planetary geomagnetic index, Kp
GLAT	Geodetic latitude (°)
GLON	Geodetic longitude (°)
MLAT	Geomagnetic latitude (°)
MLON	Geomagnetic longitude (°)
MODIP	Modified Dip latitude (°)
foF2	The F2 layer critical frequency (MHz)
hmF2	The F2 layer peak height (km)
NmF2	Peak electron density (m^{-3})
h05bot	Bottomside half peak density height (km) at Ne=0.5 NmF2
h05top	Topside half peak density height (km) at Ne = 0.5 NmF2
Nes	Electron density at O^+/N^+ transition height (m^{-3})
Nepl	Electron density at 20,200 km (m^{-3})
ECbot	Electron content from 65 km to hmF2 (TECU)
ECtop	Electron content from hmF2 to 1364 km, Topex orbit (TECU)
ECpl	Electron content from 1364 to 20,200 km (TECU)
TEC	Total electron content from 65 to 20,200 km, GPS orbit (TECU)
TAU	Slab-thickness, km (TEC/NmF2)
Hsc	Topside exponential scale height (km)
h	Altitude over the Earth (km)
Ne	Electron density at altitude h (m^{-3})
Te	Electron temperature at altitude h (°K)
Ti	Ion temperature at altitude h (°K)
Tn	Neutral temperature at altitude h (°K)

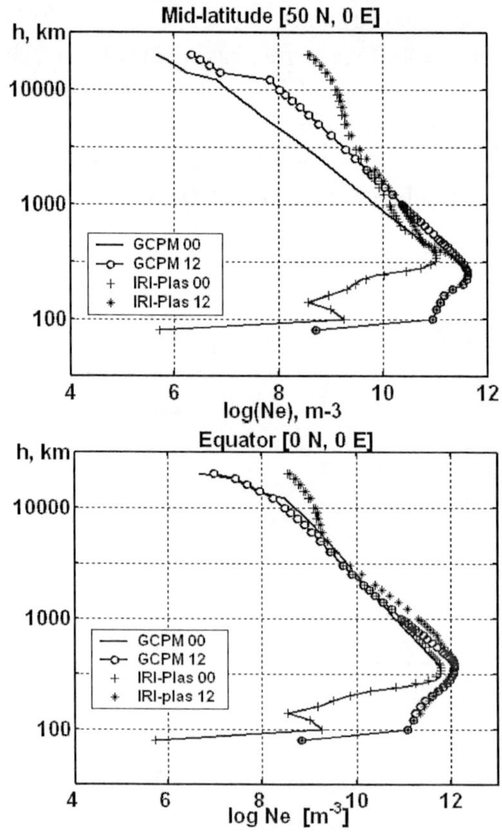

Figure 10. Comparison of GCPM and IRI-Plas plasmasphere electron density profile for noon and midnight: (a) mid-latitude; (b) megnetic equator.

The **IZMIRAN plasmasphere model** (Chasovitin et al., 1998; Gulyaeva et al., 2002a,b) is an empirical model based on whistler and satellite observations. IZMIRAN is the Institute of Terrestrial Magnetism, Ionosphere and Radiowaves Propagation. The IZMIRAN plasmasphere model presents global vertical analytical profiles of electron density (Ne) smoothly linked with the IRI electron density profile at altitude of one basis scale height above the F2 peak (400 km for electron temperature) and extended towards the plasmapause up to 36,000 km (IRI-Plas). For the smooth fitting of the two models, the shape of the IRI topside electron density profile was changed based on ISIS 1, ISIS 2 and IK19 satellite inputs (Gulyaeva 2003). ISIS is the International Satellites for Ionospheric Studies satellite program and IK-19 is the abbreviation for the Russian Intercosmos-19 satellite. The plasmasphere model depends on solar activity and magnetic activity (kp-index). List of IRI-Plas output parameters is given in Table 2. Figures 8 and 9 presents results of calculations with IRI-Plas model (Gulyaeva et al., 2002a,b) which are consistent with observations (Gulyaeva and Gallagher, 2007).

In order to evaluate a measure of the plasmasphere contribution to the transionospheric-transplasmaspheric TEC, two plasmaspheric extensions of the International Reference Ionosphere model, IRI-Plas (IRI-IZMIRAN) and IRI-GCPM, are compared (Gulyaeva and Gallagher, 2007). The both models differ from the original IRI electron density height distribution in the topside ionosphere. The GCPM algorithm includes a search for the best fitting transition height between the ionosphere and plasmasphere, which is usually found at altitudes of 400 to 600 km depending on location and conditions. The model of the IRI-Plas topside profile semi-thickness, equal to the altitude range from the ionosphere peak height to the topside half peak density height, in terms of geomagnetic latitude, diurnal, seasonal, and solar cycle variations has been included in the IRI-Plas code (Gulyaeva, 2003; Gulyaeva and Titheridge, 2006). Further improvement to the analytical adjustment of the IRI topside profile to the half peak density point has been made by incorporation of TECgps input in data assimilative mode of operation updating three model parameters: (1) an instantaneous peak electron density (Gulyaeva et al., 2011); (2) an instantaneous F2 layer peak height (Gulyaeva, 2011b); (3) the electron density scale height at the lower topside (Gulyaeva, 2011a).

Comparison of IRI-Plas and GCPM electron density profiles near magnetic equator and middle latitude at low solar activity and quiet magnetic conditions is shown in Figure 10a,b for noon and midnight. The IRI part in the bottomside and the lower topside coincide in the both models fitted to CCIR (ITU-R) F2 peak predictions. The plasma density at 20,000 km is coincident for day and night with IRI-Plas and GCPM near the magnetic equator, they are slightly different with GCPM at mid-latitude but they differ up to the order of magnitude of the absolute values near the GPS orbit in two models. With more plasmasphere modeling these differences could be resolved (Jorgensen, 2010).

Both GCPM and IRI-Plas are statistical models derived from many years of measurements designed to represent typical conditions as a function of geomagnetic and solar activity. Much like a weather forecast, these are not capable of always representing dynamically driven conditions at any given time. To the extend that TEC and new satellite measurements are accumulated into a statistical quantification of plasma densities at differing solar and geomagnetic conditions, a more meaningful comparison can be made to a statistical model.

Data assimilation procedures are extensively being developed for near real time forecasting of the ionospheric weather. They are organized by merging, by any means, a model which is a physical description or an empirical (analytical) description of a system with measurements which constrain the state or evolution of the system in some relevant way. The free model parameters are then adjusted to maximize the agreement between the model and the measurements. Final product of the ionosphere/plasmasphere standard model would be incomplete if it could not provide guidance on quantitative measure of the ionospheric weather ranging from quiet conditions to severe storm in the ionosphere and plasmasphere. Such specification is proposed in the next Section.

4. Ionospheric Weather Indices

The short-term perturbations of the ionospheric parameters vary from a few seconds to a few hours, induced by the solar flares, the solar wind, the coronal mass ejection, affecting the Earth's magnetosphere, plasmasphere and ionosphere. Apart from the monthly average variations provided by IRI-Plas model, the daily assessment and forecast of the ionosphere variability are required for many applications (Jakowski et al., 2006). The negative or positive percentage deviation of the current value of foF2 from the quiet background value can serve as an "ionospheric activity", AI, index, characterizing a measure of the ionosphere disturbance (Kutiev and Muchtarov, 2001; Bremer et al., 2006). So defined index, however, does not provide a uniform measure of negative and positive phases of the ionospheric storm, because positive deviations are deeper than negative ones (Kouris et al., 1999). In particular, the depletion of the peak electron density cannot reach 100%, because a decrease in foF2 to zero would imply an annihilation of the ionosphere. On the other hand, there is no limit imposed on an increase in foF2 during the positive phase of the ionospheric storm when the critical frequency could exceed the median value by a few hundreds of percent.

To avoid these disproportions the decimal logarithm of the hourly value of NmF2 (or TEC), normalized by the quiet reference (median), NqF2 (or TECq), is taken as a measure of the NmF2 (or TEC) variability (Gulyaeva, 1996, 2002c; Field and Rishbeth, 1997; Fuller-Rowell et al., 2001; Gulyaeva et al., 2008). For the quiet reference, 27 days (solar rotation) daily-hourly median of foF2 or TECgps can be taken from the observations, or IRI-CCIR (ITU-R) predictions can serve as a reference.

The logarithmic scale of the deviations is presented as a decimal logarithm of the ratio of the current hourly value of NmF2 (TEC) to the quiet background values NqF2 (TECq):

$$DNmF2 = \log(NmF2/NqF2) \tag{3a}$$

$$DTEC = \log(TEC/TECq) \tag{3b}$$

The sign of DNmF2 (DTEC) specifies the positive or negative phase of the ionospheric perturbation. We assume that the period of 27 days corresponding to the solar rotation yields median values that might also be valid for day 28. This appears to be a reasonable solution for the forecasting purposes, since one has a reference value one day in advance as distinct from the monthly median available only after the month has passed.

For indexing the ionosphere variability similar to geomagnetic k-indices (Menveielle and Berthelier, 1991) we introduce the ionospheric weather index W with thresholds specified in Table 3. We use a non-uniform logarithmic scale similar to Gulyaeva (1996). The intervals for the positive and negative deviation of (Eq. 3a,b) are equal to each other but the relevant threshold of the changes in NmF2 or TEC would be different for the negative and positive deviations. Index W=±1 is used for the quiet state, W=±2 for the moderate disturbance, W=±3 for the moderate ionospheric storm or sub-storm, and W=±4 for the intense ionosphere storm. Criteria for selection of such intervals are based on the conventional evaluation of the negative ionospheric NmF2 deviation within [0, −10%] from the quiet reference for the quiet state, [−10, −30%] for the moderate disturbance, [−30, −50%] for the moderate ionospheric storm, and a depletion of NmF2 greater than −50% for the intense storm conditions. It is found that the moderate disturbance (W = ±2) is a prevailing state of the ionospheric weather. The stormy conditions comprise 1 to 20% of times which occur more frequently at high latitudes, by night, during equinox and winter.

Example of W index inferred from GPS-TEC global ionospheric map, GIM, at solar maximum on 15-17 July 2000 (Figure 11) demonstrates electron content depletion (upper panel) at grid point in the Northern hemisphere [50°N, 0°E] but positive storm effect (TEC enhancement) at the magnetic conjugate point [42°S, 19°E] (middle panel) in the Southern hemisphere. The W index reached storm peak at the main phase of magnetosphere storm, registered with Disturbance Storm Time, Dst, index (lower panel).

Table 3. Ionospheric weather index W, relevant thresholds of logarithmic deviations (X=NmF2 or X=TEC, Xq - quiet reference) and corresponding state of the ionosphere

W	DN=log(X/Xq)	State
4	DN > 0.301	Intense positive W^+ storm
3	0.155<DN≤0.301	Moderate W^+ storm or W^+ sub-storm
2	0.046<DN≤0.155	Weak W^+ disturbance
1	0.0<DN≤0.046	Quiet W^+ state
-1	-0.046≤DN<0.0	Quiet W^- state
-2	-0.155≤DN<-0.046	Weak W^- disturbance
-3	-0.301≤DN<-0.155	Moderate W^- storm or W^- sub-storm
-4	DN<-0.301	Intense negative W^- storm

Taking advantage of availability of worldwide TECgps data in the Global Ionospheric Maps, GIM, the planetary ionosphere-plasmasphere storms have been identified from the GIM-TEC maps, and they are provided online from 1999 up to present (Gulyaeva and Stanislawska, 2008; 2010). The planetary ionospheric storm Wp index is obtained from the W-index map as a latitudinal average of the distance between maximum positive and minimum negative W-index weighted by the latitude/longitude extent of the extreme values on the map.

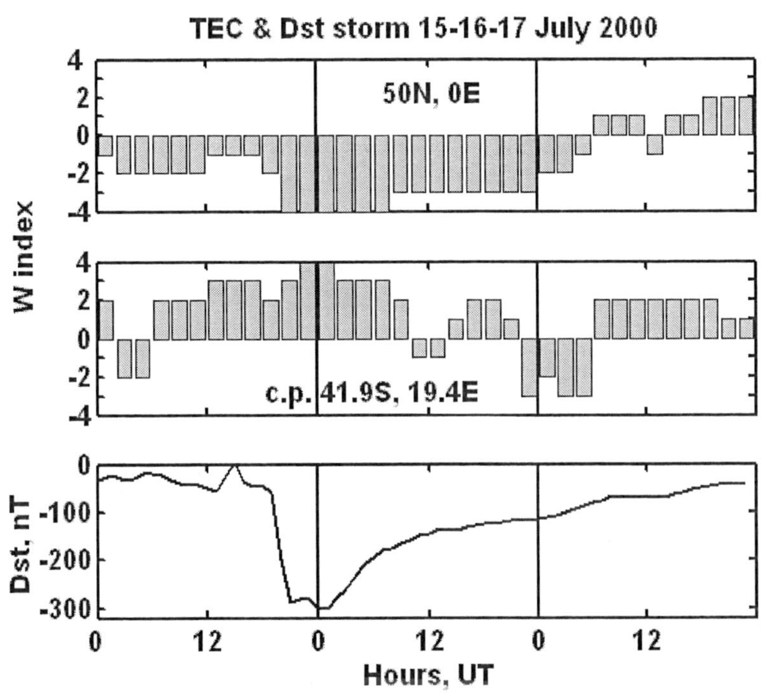

Figure 11. W index of ionospheric weather obtained from TECgps map at two magnetic conjugate locations (upper and middle sections), and magnetic Dst index during the storm on 15-17 July 2000.

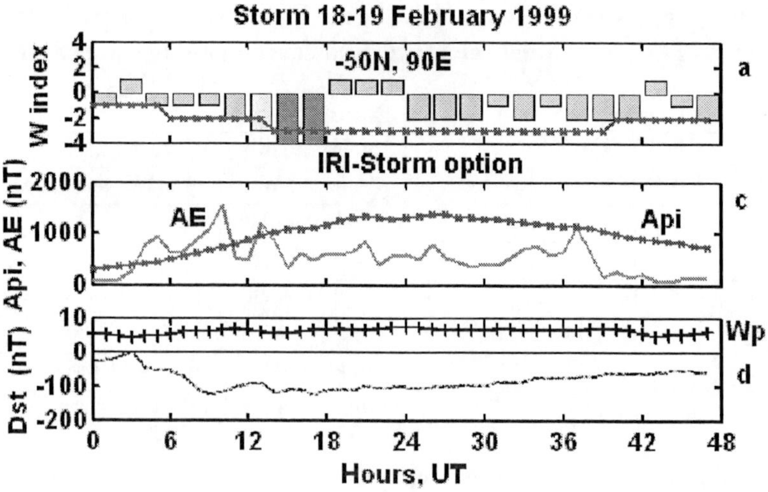

Figure 12. W index of ionospheric weather obtained from TECgps map at [50°S, 90°E] and IRI-Storm prediction (upper section), auroral electrjet AE index and integrated Api magnetic index (middle section), planetary ionospheric Wp index and Dst index (lower section) during the space weather storm on 18-19 February 1999.

The threshold Wp exceeding 4.0 index units and the peak value Wpmax≥6.0 specify the duration and the power of the planetary ionosphere-plasmasphere storm. W index variation during the space weather storm on 18-19 February 1999 is plotted in Figure 12 (upper panel) at [50°S, 90°E] accompanied by IRI-Storm prediction (solid curve with asterisks). Geomagnetic Auroral Electrojet AE index and integrated Api index (Fuller-Rowell et al., 2000; Araujo-Pradere et al., 2002) are shown in middle panel. The planetary ionospheric Wp index mirrors variation of Dst index in the lower panel.

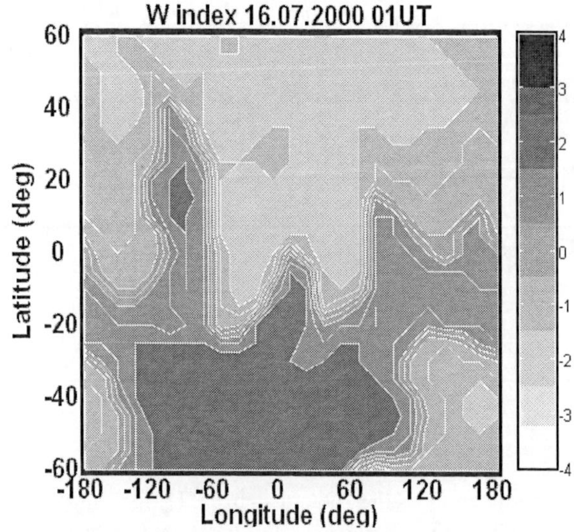

Figure 13. Global map of TEC-based W index during the peak of the space weather storm on 16 July 2000 at 1h UT.

The TEC-based global W index map is calculated at each grid point of TECgps map at latitudes from 60°S to 60°N in step of 5°, longitudes from -180° to 180°E in step of 15° (corresponding to 1h step in local time worldwide). Example of global W index map is given in Figure 13 on 16th July 2000 at the peak of Dst index at 1h UT during the storm (see Figure 11). Prevailing negative storm effects are seen at nighttime in the Northern hemisphere while the positive storm dominated at the Southern hemisphere. Product of W index derived from the F2 layer peak electron density (critical frequency foF2) is provided online in near real time at the Ionospheric Weather page (http://www.izmiran.ru/services/iweather).

It can be calculated and provided along with other IRI standard output parameters (see Table 2) when the IRI-Storm option is used or when IRI is used in data assimilative mode (Fuller-Rowell et al., 2000; 2006; Bilitza, 2001; Gulyaeva et al., 2011).

Conclusion

An international standard for the representation of the Earth's plasma parameters in the ionosphere and plasmasphere is important for a wide spectrum of applications. Examples are Altimetry, Radioastronomy, Satellite Navigation, Communication and Orbit determination. It's usage for HF Communications implies that the electromagnetic waves traveling through the ionosphere and plasmasphere experience a retardation and refractive effect. Implementation of such standard in Space Exploration Industry is needed because a remote sensing technique relying on signals traversing the ionosphere and plasmasphere therefore needs to account for the plasma influence on satellite operation.

The International Reference Ionosphere (IRI), a joint project of URSI and COSPAR, is the de facto international standard for the climatological specification of ionospheric parameters and as such it is currently undergoing registration as Technical Specification (TS16457) of the International Standardization Organization (ISO). Since its first edition in 1969 the IRI model has been steadily improved with newer data and with better mathematical descriptions of global and temporal variation patterns. A large number of independent studies have validated the IRI model in comparisons with direct and indirect ionospheric measurements not used in the model development.

The IRI model is widely used as a tool both in science and eingineering and also for educational purposes for university course materials. Table 4 demonstrates the distriubution of IRI citations in the science literature year-by-year and by institutions in different countries (total number of 699 for the period of 1993 to 2010). A review of acknowledgements of Radio Science papers in 2009 and 2010 showed that during both years 10% of the science papers in this publication acknowledged the use of the IRI model.

While IRI is the internationally accepted standard for the empirical representation of ionospheric parameters, a similar concensus does not yet exist for the plasmasphere. Several modeling approaches have been proposed for extending IRI to plasmaspheric altitudes and the most well established ones are included in this paper. We present comparisons of two of these plasmaspheric models with measurements.

IRI provides one of the better empirical model specifications freely available online for the users. IRI extension above 2000 km is needed, in particular, because the GPS satellites widely used for monitoring the ionosphere nowadays are located at 20,200 km. While the plasma density above 2000 km is at least two (frequently four or more) orders of magnitude less than the F-region peak density, the altitude range in the plasmasphere above the IRI model is about an order of magnitude greater than the thickness of the ionosphere.

Table 4. Publications in Radio Science related with IRI model

Year	Record count	% of 699		Country / Territory	Record count	% of 699
1993	12	1.7		USA	234	33.5
1994	40	5.7		RUSSIA	77	11.0
1995	45	6.4		INDIA	66	9.4
1996	10	1.4		ITALY	54	7.7
1997	31	4.4		ARGENTINA	51	7.3
1998	28	4.0		JAPAN	51	7.3
1999	33	4.7		PEOPLES R CHINA	46	6.6
2000	10	1.4		ENGLAND	36	5.2
2001	34	4.9		BRAZIL	30	4.3
2002	36	5.2		GERMANY	28	4.0
2003	51	7.3		NIGERIA	28	4.0
2004	52	7.4		CZECH REPUBLIK	26	3.7
2005	27	3.9		POLAND	25	3.6
2006	46	6.6		SOUTH AFRICA	25	3.6
2007	61	8.7		TAIWAN	25	3.6
2008	44	6.3		AUSTRALIA	19	2.7
2009	49	7.0		FRANCE	18	2.6
2010	50	7.2		SPAN	18	2.6
				AUSTRIA	16	2.3
				CANADA	13	1.9
				TURKEY	11	1.6
				BULGARIA	9	1.3

The temporal and spatial variation in ionospheric structures have often frustrated the efforts of communications and radar system operators who base their frequency management decisions on monthly mean predictions of radio propagation in the high frequency (short-wave) band. With the growth of trans-ionospheric and trans-plasmaspheric radio navigation systems, TEC measurements from these systems have become the most numerous ionospheric data set. Treating GPS TEC measurements as the worldwide source of information on the ionosphere and plasmasphere, data assimilative algorithms can more fully determine the spatial structure and dynamics of the space plasma environment. The particular model parameters such as F2 layer peak electron density and total electron content adjusted to observation allows to express the ionospheric weather from quiet state to severe storm by ionospheric W-index characterizing ionosphere and plasmasphere state retrieved from deviation of instant value from the reference median value.

The ionospheric plasma density distribution and TEC depend on a number of upper atmospheric and ionospheric parameters, such as the neutral density, neutral wind, neutral and plasma temperatures, plasmaspheric flux, and ion-neutral collision frequencies. Considerable efforts are currently expended on the development and promotion of assimilative algorithms that allow to combine a physical or empirical model with measured data, and thus allow adapting the model to real-time conditions. In the numerical physical modeling of the ionosphere these assimilative inputs are often only roughly known and can cause significant uncertainties in the physical model results (Jee et al., 2005). The promotion of the empirical IRI model to a full international standard is well justified by the reliability and accuracy of IRI output parameters and the public availability of IRI code for the worldwide user community. Added benefits are the user-friendly interface for online calculation of IRI standard parameters, and the availability of guidance materials for newcomers helping with implementation and adapting of IRI system to the particular user's needs. The high degree of acceptance of IRI in the heliospheric and geospheric community is also the fact that IRI is used as the reference model for determining the skill score of evolving physical models, e.g., for the CEDAR challenge and for the CCMC (Sojka et al., 2007).

It is expected that the significant contributions to international collaboration and coordinated input relative to space vehicle design and operations, plasma environment specification, and communication and navigation services will result from the promotion of this proposed ISO Standard.

APPENDIX: Model Availability

The IRI homepage is at http://IRI.gsfc.nasa.gov. From this page the user has access to information about the model, to the Fortran code (http://nssdcftp.gsfc.nasa.gov/models /ionospheric/iri/) and to a web interface for computation, listing, and plotting of model parameters (http://omniweb.gsfc.nasa.gov/vitmo/iri_vitmo. html). The Fortran code and IRI interface are also accessible from the Community Coordinated Modeling Center (CCMC; http://ccmc.gsfc.nasa.gov/requests/requests.php).

A FORTRAN code implementation of GCPM that includes all except the polar cap is available from dennis.gallagher@msfc.nasa.gov.

The GPID model source code was written using commercial MATLAB software, but is not currently available for release.

Source code for the IZMIRAN IRI ionosphere-plasmasphere version is available from the IZMIRAN web site ftp://ftp.izmiran.ru/pub/izmiran/SPIM/.

Acknowledgments

The IRI model is the result of modeling efforts by a group of ionospheric experts that form the IRI Working Group of the Committee of Space Research (COSPAR) and the International Union of Radio Science (URSI). The efforts by this group are greatly appreciated and we are thankful to COSPAR and URSI for sponsoring this activity. TLG acknowleges the support of the joint grant from RFBR (Project 11-02-91370-CT_a) and TUBITAK (Project EEEAG 110E296) for this work and DB acknowledges support through NASA grants NNX09AJ74G and NNX07A065G and NSF grant AGS-0819440. The ISO/TS16457:2009 Document is provided at (http://www.iso.org /iso/iso_catalogue/catalogue_tc/catalogue_detail.htm? csnumber=51248:). Global Ionospheric Maps, TECgps, and TOPEX data are provided online at (ftp://cddis.gsfc.nasa.gov/pub/gps/products/ionex/).

References

Altadill, D., Arrazola, D., Blanch, E., Buresova, D. Solar activity variations of ionosonde measurements and modeling results, *Adv. Space Res.* 42, 610–616, 2008.

Altadill, D., Torta, J.M., Blanch, E. Proposal of new models of the bottom-side B0 and B1 parameters for IRI, *Adv. Space Res.* 43, 1825 -1834, doi:10.1016/j.asr.2008.08.0144, 2009.

Angling, M.J., Cannon, P.S. Assimilation of radio occultation measurements into background ionospheric models. *Radio Sci.* 39, doi:10.1029/2002RS002819, 2004.

Araujo-Pradere, E.A., Fuller-Rowell, T.J., Codrescu, M.V. STORM: An empirical storm-time ionospheric correction model. 1. Model description. *Radio Sci.*, 37(5), 1070, doi:10.1029/2001RS002467, 2002.

Arikan, O., Arikan, F., Erol, C.B. Computerized ionospheric tomography with the IRI model. *Adv. Space Res.*, 39, 859–866, doi:10.1016/j.asr.2007.02.078, 2007.

Bailey, G.J., Balan, N. and Su, Y.Z. The Sheffield University plasmasphere ionosphere model—A review, *J. Atmos. Sol. Terr. Phys.*, 59, 1541–1552, 1997.

Bent, R.B., Llewellyn, S.K., Walloch, M.K. *SAMSO TR-72-239*, Space and Missile Systems Organization, Los Angeles, California 90045, 1972.

Bilitza, D., Electron density in the D-region as given by the International Reference Ionosphere, World Data Center A for Solar-Terrestrial Physics, *Report UAG-82*, 7-11, 1981.

Bilitza, D., Sheikh, N.M., Eyfrig, R. A global model for the height of the F2 peak using M3000 values from the CCIR. *Telecomm. J.*, 46, 549-553, 1979.

Bilitza, D., Rawer, K. New options for IRI electron density in the middle ionosphere. *Ady. Space Res.* 10(11), (11)7-(11)16, 1990.

Bilitza, D., Rawer, K., Bossy, L., Gulyaeva, T.. International Reference Ionosphere - Past, Present, and Future: I. Electron density. *Ady. Space Res.* 13(3), (3)3-(3)13, 1993a.

Bilitza, D., Rawer, K., Bossy, L., Gulyaeva, T.. International Reference Ionosphere - Past, Present, and Future: II. Plasma temperatures, ion composition and ion drift. *Adv. Space Res.* 13, N 3, 15–23, 1993b.

Bilitza, D., S. Radicella, B. Reinandh, J. Adeniyi, M. Mosert de Gonzalez, S. Zhang, and 0. Obrou, New B_0 and B_1 models for IRI, *Adv. Space Res.,* 25(l), 89-96, 2000.

Bilitza D. International Reference Ionosphere 2000, *Radio Science.* 36, 261–275, 2001.

Bilitza, D., Reinisch, B., Benson, R., Grebowsky, J., Papitashvili, N., Huang, X., Schar, W., Hills, K. Online data base of satellite sounder and in situ measurements covering two solar cycles. *Adv. Space Res.* 31(3), 769-774, 2003.

Bilitza, D. A correction for the IRI topside electron density model based on Alouette/ISIS topside sounder data. *Adv. Space. Res.* 33(6), 838–843, 2004.

Bilitza, D., Reinisch, B.W., Radicella, S.M., Pulinets, S., Gulyaeva, T., Triskova, L. Improvements of the IRI model for the topside electron density profile. *Radio Sci.,* 41(5), RS5S15, doi:10.1029/2005RS003370, 2006.

Bilitza, D., Truhlik, V., Richards, P., Abe, T., Triskova, L. Solar cycle variation of mid-latitude electron density and temperature: Satellite measurements and model calculations, *Adv. Space Res.*, 39(5), 779-789, doi: 10.1016/j.asr.2006.11.022, 2007.

Bilitza, D., Reinisch, B.W. International Reference Ionosphere 2007: Improvements and new parameters. *Adv. Space Res.* 42, 599-609, doi:10.1016/j.asr.2007.07.048, 2008.

Bilitza, D. Evaluation of IRI-2007 model options for the topside electron density. *Adv. Space Res.*, 44 (8), 701-706, doi:10.1016/j.asr.2009.04.036, 2009.

Bremer, J., Cander, Lj. R., Mielich, J., Stamper, R. Derivation and test of ionospheric activity indices from real-time ionosonde observations in the European region. *J. Atmos. Sol.-Terr. Phys.*, 68, 2075–2090, 2006.

Bust, G.S., Coker, C., Coco, D.S., Gaussiran II, T.L., Lauderdale, T. IRI data ingestion and ionospheric tomography. *Adv. Space Res.* 27, 157–165, 2001.

Bust, G.S, Garner, T.W., Gaussiran II, T.L. Ionospheric Data Assimilation Three Dimensional (IDA3D): A Global, Multi-Sensor, Electron Density Specification Algorithm, *J. Geophys. Research*, 109, A11312, doi:10.1029/2003JA010234, 2004.

Bust, G.S., Mitchell, C.N. History, current state, and future directions of ionospheric imaging. *Rev. Geophys.* 46, RG1003, doi:10.1029/2006RG000212, 2008.

Carpenter, D.L., Park, C.G. On what ionospheric workers should know about the plasmapause – plasmasphere. *Rev. Geophys. Space Phys.*, 11, 133-154, 1973.

CCIR, *Atlas of ionospheric characteristics*, Comite Consultatif International des Radio Communications, Rep. 340, International Telecommunication Union, Geneve, 1983.

Chasovitin, Yu.K., Gulyaeva, T.L., Deminov, M.G., Ivanova, S.E. Russian standard model of ionosphere (SMI). In *COST251TD(98)005*, RAL, UK, 161-172, 1998.

Chu, Y.-H., Wu, K.-H., Su, Ch.-L. A new aspect of ionospheric E region electron density morphology. *J. Geophys. Res.*, 114, A12, A12314, DOI:10.10292008JA014022, 2009.

Coisson, P., Radicella, S.M., Nava, B. Comparisons of experimental topside electron concentration profiles with IRI and NeQuick models. *Ann. Geophys.* 45 (1), 125–130, 2002.

Coisson, P., Radicella, S.M., Leitinger, R., Nava, B. Topside electron density in IRI and NeQuick: features and limitations. *Adv. Space Res.* 37 (5), 934–937, 2006.

Cueto, M., Coisson, P., Radicella, S.M., Herraiz, M., Ciraolo, L., Brunini, C. Topside ionosphere and plasmasphere: Use of NeQuick in connection with Gallagher plasmasphere model. *Adv. Space. Res.*, 39, 739–743, doi:10.1016/j.asr.2007.01.073, 2007.

Danilov, A.D., Semenov, V.K. Relative ion composition model at midlatitudes, *J. Atmos. Terr. Phys.*, 40, 1093, 1978.

Danilov, A.D., Yaichnikov, A.P. A New Model of the Ion Composition at 75 to 1000 km for IRI, *Adv. Space Res.* 5(7), 75-79, 107-108, 1985.

Danilov, A.D., Smirnova, N.V. Improving the 75 to 300 km ion composition model of the IRI, *Adv. Space Res.* 15(2), 171-177, 1995.

Danilov, A., Rodevich, A. Smirnova, N. Problems with incorporating a new D-region model into the IRI, *Adv. Space Rex,* 15(2), 165-169, 1995.

Ducharme, E.D., Petrie, L.E., Eyfrig, R. A method for predicting the F1-layer critical frequency. *Radio Sci.*, 6, 369-378, 1971; *Radio Sci.*, 8, 837-839, 1973.

Dudeney, J.R. A Simple Empirical Method for Estimating the Height of the F2-Layer at the Argentine Islands Graham Land, *Science Report No.* 88, London, UK, British Antarctic Survey, 1974.

Fejer, B.G. Low latitude electrodynamic plasma drifts: A review, *J. Atmos. Terr. Phys.*, 53, 677–693, 1991.

Fernandez J.R., Mertens, C.J., Bilitza, D., Xu, X., Russell III, J.M., Mlynczak, M.G. Feasibility of developing an ionospheric E-region electron density storm model using TIMED/SABER measurements, *Adv. Space Res.* 46(8), 1070-1077, doi:10.1016/j.asr.2010.06.008, 2010.

Field, P.R., Rishbeth, H. The response of ionospheric F2-layer to geomagnetic activity: an analysis of worldwide data, *J. Atmos. Sol.-Terr. Phys.*, 59, 2, 163–180, 1997.

Foster, J.C., Holt, J.M., Musgrove, R.G., Evans, D.S. Ionospheric convection associated with discrete levels of particle precipitation. *Geophys. Res. Letters* 13, 656–659, doi:10.1029/GL013i007p00656, 1986.

Friedrich, M., Torkar, K. FIRI: A semiempirical model of the lower ionosphere, *J. Geophys. Res.,* 106, Al0, 21409-21418, 200l.

Friedrich, M., Pilgram, R., Torkar, K. A novel concept for empirical D-region modelling. *Adv. Space Res.*, 27(1), 5-12, 2001.

Fu, L.L., Christensen, E.J., Yamarone Jr, C.A. TOPEX/Poseidon mission overview, *J. Geophys. Res.*, 99, 24369–24381, 1994.

Fuller-Rowell, T.J., Evans, D.S. Height-integrated Pedersen and Hall conductivity patterns inferred from the TIROS-NOAA satellite data, *J. Geophys. Res.*, 92, 7606–7618, 1987.

Fuller-Rowell, T.J., D. Rees, S. Quegan, R.J. Moffett, M.V. Codrescu, G.H. Millward, *STEP Handbook of Ionospheric Models*, edited by R.W. Schunk, Utah State Univ.,1996.

Fuller-Rowell T.J., Araujo-Pradere E.A., Codrescu M.V. An Empirical Ionospheric Storm-Time Correction Model. *Adv. Space Res.*, 25(1), 139-146, doi: 10.1016/S0273-1177(99)00911-4, 2000.

Fuller-Rowell, T.J., Codrescu, M.V., Araujo-Pradere, E.A. Capturing the storm-time ionospheric response in an empirical model. *AGU Geophys. Monograph,* 125, 393-401, 2001.

Fuller-Rowell, T., Araujo-Pradere, E.A., Minter, C., Codrescu, M. , Spencer, P., Robertson, D., Jacobson, A. R. US-TEC: A new data assimilation product from the Space Environment Center characterizing the ionospheric total electron content using real-time GPS data, *Radio Sci.*, 41, RS6003, doi:10.1029/2005RS003393, 2006.

Gallagher, D.L., Craven, P.D., Comfort, R.H.. Global Core Plasma Model. *J. Geophys. Res.*, 105, A8, 18819-18833, 2000.

Gulyaeva T.L. Progress in ionospheric informatics based on electron density profile analysis of ionograms. *Adv. Space Res.* 7, No.6, 39-48, 1987.

Gulyaeva ,T. L.: Logarithmic scale of the ionosphere disturbances, *Geomagn. Aeronomy*, 36, 1, 115–118, 1996.

Gulyaeva T.L., Huang, X., Reinisch, B.W. Plasmaspheric extension of topside electron density profiles. *Adv. Space Res.,* 29, No.6, 825-831, 2002a.

Gulyaeva T.L., Huang, X., Reinisch, B.W. Ionosphere-Plasmasphere Model Software for ISO. *Acta Geod. Geophys. Hungarica,* 37, Nos. 2-3, 143-152, 2002b.

Gulyaeva, T. L.: Daily assessment of the ionosphere variability, *Acta Geod. Geophys. Hungarica*, 37(2–3), 303–308, 2002c.

Gulyaeva, T.L. Variations in the half-width of the topside ionosphere according to the observations of space ionosondes ISIS 1, ISIS 2, and IK19, *Int. J. Geomagn. Aeron.* 4(3), 201-207, 2003.

Gulyaeva, T.L., Titheridge, J.E. Advanced specification of electron density and temperature in the IRI ionosphere-plasmasphere model. *Adv. Space Res.*, 38(11), 2587-2595, doi:10.1016/j.asr.2005.08.045, 2006.

Gulyaeva T.L. Variable coupling between the bottomside and topside thickness of the ionosphere. *J. Atmos. Solar-Terr. Phys.*, 69(4-5), 528-536, doi:10.1016/j.jastp.2006.10.015, 2007.

Gulyaeva, T.L., Gallagher, D.L. Comparison of two IRI electron-density plasmasphere extensions with GPS-TEC observations. *Adv. Space Res.*, 39, 744-749, doi:10.1016/j.asr.2007.01.064, 2007.

Gulyaeva, T.L., Stanislawska, I. Derivation of a planetary ionospheric storm index. *Annales Geophysicae*, 26, N.9, 2645-2648, 2008.

Gulyaeva, T.L., Arikan, F., Delay, S. Scale factor mitigating non-compliance of double frequency altimeter measurements of the ionospheric electron content over the oceans with GPS-TEC maps. *Earth, Planets and Space*, 61, 1103-1109, 2009.

Gulyaeva, T.L., Stanislawska, I. Magnetosphere associated storms and autonomous storms in the ionosphere-plasmasphere environment. *J. Atmos. Solar-Terr. Phys.*, 72, 90-96, doi:10.1016/j.jastp.2009.10.012 , 2010.

Gulyaeva, T.L., Arikan, F., Stanislawska, I. Inter-hemispheric Imaging of the Ionosphere with the upgraded IRI-Plas model during the space weather storms. *Earth, Planets and Space*, 2011.

Gulyaeva T.L. Storm time behavior of topside scale height inferred from the ionosphere-plasmasphere model driven by the F2 layer peak and GPS-TEC observations. *Adv. Space Res.*, 47, 913-920, doi:10.1016/j.asr.2010.10.025, 2011a.

Gulyaeva, T.L. Storm-time model of the ionosphere peak height associated with changes of peak electron density. *Adv. Space Res.*, 2011b (submitted).

Hajj, G. A., Wilson, B.D., Wang, C., Pi, X., and Rosen, I.G. Data assimilation of ground GPS total electron content into a physics-based ionospheric model by use of the Kalman filter, *Radio Sci.*, 39, RS1S05, doi:10.1029/2002RS002859, 2004,.

Hedin, A.E. Extension of the MSIS thermosphere model into the middle and lower atmosphere, *J. Geophys. Res.*, 96, 1159–1172, 1991.

Hedin, A.E., Fleming, E.L., Manson, A.H., Schmidlin, F.J., Avery, S.K., et al. Empirical wind model for the upper, middle and lower atmosphere, *J. Atmos. Terr. Phys.*, 58, 1421–1447, 1996.

Heelis, R.A., Lowell, J.K., and Spiro, R.W. A Model of the High-Latitude Ionospheric Convection Pattern. *J. Geophys. Res.* 87, A8, 6339-6345, doi:10.1029/JA087iA08p06339, 1982.

Heise, S., Jakowski, N., Wehrenpfennig, A., Reigber, Ch., Luehr, H. Sounding of the topside ionosphere/plasmasphere based on GPS measurements from CHAMP: initial results. *Geophys. Res. Lett.* 29 (14), doi:10.1029/2002GL014738, 2002.

Heppner, J.P., Maynard, N.C. Empirical high-latitude electric field models, *J. Geophys. Res.*, 92, 4467-4489, 1987.

Hochegger, G., Nava, B., Radicella, S.M., Leitinger, R. A family of ionospheric models for different uses. *Phys. Chem. Earth (C)*, 25 (4), 307–310, doi:10.1016/S1464-1917(00)00022-2, 2000.

Huang, X., Reinisch, B.W., Song, P., Nsumei, P., Green, J.L., Gallagher, D.L. Developing an empirical density model of the plasmasphere using IMAGE/RPI observations, *Adv. Space Res.*, 33, 6, 829-832, 2004.

Hysell, D.L. Inverting ionospheric radio occultation measurements using maximum entropy, *Radio Sci.*, 42, RS4022, doi:10.1029/2007RS003635, 2007.

Iwamoto, L., Katoh, H., Maruyama, T., Minakoshi, H., Watari, S., Igarashi, I. Latitudinal variation of solar flux dependence in the topside plasma density: comparison between IRI model and observations. *Adv. Space Res*. 29 (6), 877–882, 2002.

Jakowski, N., Stankov, S.M., Schlueter, S., Klaehn, D. On developing a new ionospheric perturbation index for space weather operations. *Adv. Space Res.*, 38(11), 2596-2600, doi:10.1016/j.asr.2005.07.043, 2006.

Jee, G., Schunk, R.W., Scherliess, L. On the sensitivity of total electron content (TEC) to upper atmospheric / ionospheric parameters. *J. Atmos. Terr. Phys.*, 67(11), 1040-1052, 2005.

Jorgensen, A.M., Ober, D., Koller, J., Friedel, R.H.W. Specification of the Earth's plasmasphere with data assimilation. *Adv. Space Res.*, doi:10.1016/j.asr.2010.06.013, 2010.

Kelley, M.C., Wong, V.K., Aponte, N., Coker, C., Mannucci, A.J., and Komjathy, A. Comparison of COSMIC occultation-based electron density profiles and TIP observations with Arecibo incoherent scatter radar data, *Radio Sci.*, 44, RS4011, doi:10.1029/2008RS004087, 2009.

Kotova, G.A. The Earth's Plasmosphere: current state of research. *Geomagnetism and Aeronomy*, 47(4), 1-16, 2007.

Kouris, S.S., Muggleton, L.M. Diurnal variation in the E-layer ionization. *J. Atmos. Terr. Phys.*, 35, 133-137, 1973.

Kouris, S. S., Fotiadis, D. N., Zolesi, B. Specifications of the F region variations for quiet and disturbed conditions, *Phys. Chem. Earth, Part C*, 24(4), 321–327, 1999.

Kutiev, I. and Muchtarov, P. Modeling of midlatitude F-region response to geomagnetic activity, *J. Geophys. Res.*, 106, 15 501–15 509, 2001.

Liu, J.Y., Lin, C.Y., Lin, C.H., Tsai, H.F., Solomon, S.C., Sun, Y.Y., Lee, I.T., Schreiner, W.S., Kuo, Y.H. Artificial plasma cave in the low-latitude ionosphere results from the radio occultation inversion of the FORMOSAT-3/COSMIC, *J. Geophys. Res.*, 115, A07319, doi:10.1029/2009JA015079, 2010.

Llewellyn, S.K., Bent, R.B. Documentation and description of the Bent ionospheric model, *Rep. AFCRL-TR-73-0657*, Air Force Cambridge Res. Lab., Hanscom AFB, Bedford, Mass., 1973.

McKinnell, L.A., Friedrich,M., Steiner, R.J. A new approach to modeling the daytime lower ionosphere at auroral latitudes. *Adv. Space Res.* 34(9), 1943–1948, 2004.

McKinnel, L.A., Friedrich, M. Results from a new lower ionosphere model. *Adv. Space Res.* 37 (5), 1045–1050, 2006.

McKinnell, L.A., Oyeyemi, E.O. Progress towards a new global foF2 model for the International Reference Ionosphere (IRI). *Adv. Space Res.*, 43(11), 1770-1775, doi:10.1016/j.asr.2008.09.035, 2009.

McKinnell, L.A., Oyeyemi, E.O. Equatorial predictions from a new neural network based global foF2 model. *Adv. Space Res.*, 46 (8), 1016-1023, doi:10.1016/j.asr.2010.06.003, 2010.

Mechtley, E., Bilitza, D. Models of D-region electron concentration, *Rep. IPW-WB 1*, Institut fur physikalische Weltraumforschung, Freiburg, Germany, 1974.

Menveielle, M., Berthelier, A. The K-derived planetary indices: description and availability, *Rev. Geophys*., 29(3), 415–432, 1991.

Millward, G.H., Moffett, R.J., Quegan, S., Fuller-Rowell, T.J. "A Coupled Thermosphere Ionosphere-Plasmasphere Model, CTIP," *STEP Handbook of Ionospheric Models*, edited by R.W. Schunk, Utah State Univ, 1996.

Nicolls, M.J., Vlasov, M.N., Kelley, M.C., Shepherd, G.D. Discrepancy between the nighttime molecular ion composition given by the international reference ionosphere model and airglow measurements at low latitudes, *J. Geophys. Res.*, 111, A03304, doi:10.1029/ 2005JA011216, 2006.

Nsumei, P.A., Huang, X., Reinisch, B.W., Song, P., Vasyliunas, V.M., Green, J.L., Fung, S.F., Benson, R.F., and Gallagher, D.L. Electron Density Distribution Over the Northern Polar Region Deduced from IMAGE/RPI Sounding, *J. Geophys. Res.*, 108, A2, 2003

Oyeyemi, E.O., McKinnell, L.A., Poole, A.W.V. Neural network-based prediction techniques for global modeling of M(3000)F2 ionospheric parameter. Adv. Space Res., 39(5), 643-650, doi:10.1016/j.asr.2006.09.038, 2007.

Oyeyemi, E.O., McKinnell, L.A. A new global F2 peak electron density model for the International Reference Ionosphere (IRI). *Adv. Space Res.*, 42(4), 645-658, doi:10.1016/j.asr.2007.10.031, 2008.

Pi, X., Wang, C., Hajj, G.A. , Rosen, G. , Wilson, B.D., Bailey, G. J. Estimation of ExB drift using a global assimilative ionospheric model: An observation system simulation experiment, *J. Geophys. Res.*, 108(A2), 1075, doi:10.1029/2001JA009235, 2003.

Picone, J.M., Hedin, A.E., Drob, D.P., Aikin, A.C. NRLMSIS00 empirical model of the atmosphere: Statistical comparisons and scientific issues, *J. Geophys. Res.*, 107(A12), 1468, doi:10.1029/2002JA009430, 2002.

Radicella, S.M., Leitinger, R. The evolution of the DGR approach to model electron density profiles. *Adv. Space Res.* 27 (1), 35–40, 2001.

Raeder J., Larson D., Li W., Kepko E.L., and Fuller-Rowell T. Open GGCM Simulations for the THEMIS Mission. *Space Science Review*, 141(1-4), 535-555, 2008.

Ramakrishnan, S., Rawer, K. Model electron density profiles obtained by empirical procedures, *Space Research* XII, 1253-1259, Akademie Verlag, Berlin, German Democratic Republic, 1972.

Rawer, K. *Meteorological and Astronomical Influences on Radio Wave Propagation*, ed. B. Landmark, Pergamon Press., Oxford, pp.221-250, 1963.

Rawer, K., Ramakrishnan, S., Bilitza, D.. Preliminary Reference profiles of Electron and Ion Densities and Temperatures, Institute for Space Research, *Report IPW-WB2,* Freiburg, Germany, 1972.

Rawer, K., Ramakrishnan, S., Bilitza, D. International Reference Ionosphere 1978, International Union of Radio Science, *Special Report*, Brussels, Belgium, 1978.

Rawer, K. New description of the electron density profile. *Adv. Space Res.*, 4(1), 11-15, 1984.

Rawer, K. Data needed for an analytical description of the electron density profile. *Adv. Space Res.*, 11(10), (10)57-(10)65, 1991.

Reinisch, B., Huang, X. Redefining the IRI F1 Layer profile, *Adv. Space Res,* 25(l), 81-88, 1999.

Reinisch, B.W., Haines, D.M., Bibl, K., Cheney, G., Galkin, I.A., Huang, X., Myers, S.H., Sales, G.S., Benson, R.F., Fung, S.F., Green, J.L., Boardsen, S., Taylor, W.W.L., Bougeret, J.-L., Manning, R., Meyer-Vernet, N., Moncuquet, M., Carpenter, D.L., Gallagher, D.L., and Reiff, P. The Radio Plasma Imager investigation on the IMAGE spacecraft, *Space Science Reviews*, 91, pp. 319-359, 2000.

Reinisch, B.W., Huang, X., Song, P., Green, J.L., Fung, S.F., Vasyliunas, V.M., Gallagher, D.L., Sandel, B.R. Plasmaspheric mass loss and refilling as a result of a magnetic storm, *J. Geophys. Res.*, 109, A1, A01202, 1-11, doi:10.1029/2003JA009948, 2004

Reinisch, B.W., Nsumei, P., Huang, X., Bilitza, D.K. Modeling the F2 topside and plasmasphere for IRI using IMAGE/RPI and ISIS data. *Adv. Space Res.* 39 (5), 731–738, 2007.

Richards, P.G. The Field Line Interhemispheric Plasma Model. In: *Solar-Terrestrial Energy Program: Handbook of Ionospheric Models*, edited by R.W. Schunk, p. 207, 1996.

Richards, P.G. Ion and neutral density variations during ionospheric storms in September 1974: comparison of measurement and models. *J. Geophys. Res.* 107 (A11), 1361, doi:10.1029/2002JA009278, 2002.

Richards, P.G., Bilitza, D., and Voglozin, D. Ion density calculator (IDC): A new efficient model of ionospheric ion densities, *Radio Sci.*, 45, RS5007, doi:10.1029/2009RS004332, 2010.

Richmond, A.D., Blanc, M., Emery, B.A., Wand, LH., Fejer, B.G., Woodman, R.F., Ganguly, S., Amayenc, P., Behnke, R.A., Calderon, C., and Evans, J.V. An Empirical Model of Quiet-Day Ionospheric Electric Fields at Middle and Low Latitudes. *J. Geophys. Res.* 85, A9, 4658-4664, doi:10.1029/JA085iA09p04658, 1980.

Rishbeth, H., Garriott, O.K. Introduction to ionospheric physics, Academic Press, New York, 331pp., 1969.

Roble, R.G., Ridley, E.C., Richmond, A.D. A Coupled Thermosphere/Ionosphere general circulation model. *Geophys. Res. Lett.*, 15(12), 1325-1328, 1988.

Rodger, A.S., Wrenn, G.L., Rishbeth, H. Geomagnetic storms in the Antarctic F region, II, physical interpretation, *J. Atmos. Terr. Phys.*, 51, 851-866 1989.

Rush, C., Fox, M., Bilitza, D., Davies, K., McNamara, L., Stewart, F., PoKempner, M. Ionospheric mapping – An update of foF2 coefficients. *Telecomm. J.*, 56, 179-182, 1989.

Scherliess, L., Fejer, B.G. Radar and satellite global equatorial F region vertical drift model, *J. Geophys. Res.*, 104, 6829–6842, 1999.

Scherliess, L., Schunk, R.W., Sojka, J.J., Thompson, D.C. Development of a physics-based reduced state Kalman filter for the ionosphere, *Radio Sci.*, 39, RS1S04, doi:10.1029/2002RS002797, 2004.

Schunk, R.W., Nagy, A.F. *Ionosphere: Physics, Plasma Physics, and Chemistry*, Cambridge, New York, pp. 333-335, 2000.

Sojka, J.J., Thompson, D.C., Scherliess, L., Schunk, R.W., Harris, T.J. Assessing models for ionospheric weather specifications over Australia during the 2004 Climate and Weather of the Sun-Earth-System (CAWSES) campaign, *J. Geophys. Res.*, 112, A09306, doi:10.1029/2006JA012048, 2007.

Stankov, S.M., Jakowski, N., Heise, S., Muhtarov, P., Kutiev, I., Warnant, R. A new method for reconstruction of the vertical electron density distribution in the upper ionosphere and plasmasphere. *J. Geophys. Res.* 108, doi:10.1029/2002JA009570, 2003.

Tobiska, W.K. Revised solar extreme ultraviolet flux model, *J. Atmos. Terr. Phys.*, 53, 1005–1018, 1991.

Triskova, L., Truhlik, V., Smilauer, J. An empirical model of ion composition in the outer ionosphere. *Adv. Space Res.* 31 (3), 653–663, 2003.

Truhlik, V., Triskova, L., Smilauer, J., Afonin, V. Global empirical model of electron temperatures in the outer ionosphere for period of high solar activity based on data of three Intercosmaos satellites, *Adv. Space Res.*, 25(1), 163-172, 2000.

Truhlik, V., Triskova, L., Smilauer, J. New advances in empirical modeling of ion composition in the outer ionosphere. *Adv. Space Res.* 33, 844–849, 2004.

Truhlik V., Triskova, L., Bilitza, D., Podolska, K. Variations of daytime and nighttime electron temperature and heat flux in the upper ionosphere, topside ionosphere and lower plasmasphere for low and high solar activity, *J. Atmos. Sol.-Terr. Phys.* 71, 2055–2063, doi:10.1016/j.jastp.2009.09.013, 2009.

Vergados, P., Pagiatakis, S.D. First estimates of the second-order ionospheric effect on radio occultationobservations, *J. Geophys. Res.*, 115, A07317, doi:10.1029/2009JA015161, 2010.

Vlasov, M.N., Kelley, M.C. Crucial discrepancy in the balance between extreme ultraviolet solar radiation and ion densities given by the international reference ionosphere model, *J. Geophys. Res.*, 115, A08317, doi:10.1029/2009JA015103, 2010.

Wang, C., Hajj, G., Pi, X., Rosen, I.G., Wilson, B. Development of the Global Assimilative Ionospheric Model. *Radio Sci.* 39, doi:10.1029/2002RS002854, 2004a.

Wang, W., Wiltberger, M., Burns, A.G., Solomon, S.C., Killeen, T.L., Maruyama, N., Lyon, J.G. Initial results from the coupled magnetosphere-ionosphere-thermosphere model: thermosphere-ionosphere responses. *J. Atmos. Solar-Terr. Phys.*, 66(15-16), 1425-1441, doi:10.1016/j.jastp.2004.04.008, 2004b.

Webb, P.A., Essex, E.A. An ionosphere-plasmasphere global electron density model. *Phys. Chem. Earth (C)*, 25, No. 4, pp. 301-306, 2000.

Webb, P.A., Essex, E.A. A dynamic global model of the plasmasphere. *J. Atmos. Solar-Terr. Phys.*, 66(12), 1057-1073, doi:10.1016/j.jastp.2004.04.001, 2004.

Yizengaw, E., Moldwin, M.B., Galvan, D., Iijima, B.A., Komjathy, A., Mannucci, A.J. Global plasmaspheric TEC and its relative contribution to GPS TEC. *J. Atmos. Sol. Terr. Phys.* 70, 1541–1548, 2008.

Zhang, S.-R., Holt, J.M., van Eyken, A.P., McCready, M., Amory-Mazaudier, C., Fukao, S., Sulzer, M. Ionospheric local model and climatology from long-term databases of multiple incoherent scatter radars. *Geophys. Res. Lett.* 32, L20102, doi:10.1029/2005GL023603, 2005.

Zhang Y-L., Paxton, L.J., Bilitza, D. Near real-time assimilation of auroral peak E-region density and equatorward boundary in IRI. *Adv. Space Res.* 46(8), 1055-1063, doi:10.1016/j.asr.2010.06.029, 2010.

Referred by Lee-Anne McKinnell, lmckinnell@sansa.org.za (a) South African National Space Adency, Hermanus, South Africa (b) Department of Physics and Electronics, Rhodes University, Grahamstown, South Africa.

In: New Developments in the Standard Model
Editor: Ryan J. Larsen
ISBN: 978-1-61209-989-7

Chapter 2

NON-EQUILIBRIUM THEORY, FRACTIONAL DYNAMICS AND PHYSICS OF THE TERASCALE SECTOR

Ervin Goldfain
Photonics CoE, Welch Allyn Inc., Skaneateles Falls, NY, US

Abstract

Quantum Field Theory (QFT) lies at the foundation of the Standard Model for particle physics (SM) and is built in compliance with a number of postulates called consistency conditions. The remarkable success of SM can be traced back to a unitary, local, renormalizable, gauge invariant and anomaly-free formulation of QFT. Experimental observations of recent years suggest that developing the theory beyond SM may require a careful revision of conceptual foundations of QFT. As it is known, QFT describes interaction of stable or quasi-stable fields whose evolution is deterministic and time-reversible. By contrast, behavior of strongly coupled fields or dynamics in the Terascale sector is prone to become unstable and chaotic. Non-renormalizable interactions are likely to proliferate and prevent full cancellation of ultraviolet divergences. A specific signature of this transient regime is the onset of long-range dynamic correlations in space-time, the emergence of strange attractors in phase space and transition from smooth to fractal topology. Our focus here is the impact of fractal topology on physics unfolding above the electroweak scale. Arguments are given for perturbative renormalization of field theory on fractal space-time, breaking of discrete symmetries, hierarchical generation of particle masses and couplings as well as the potential for highly unusual phases of matter that are ultra-weakly coupled to SM. A surprising implication of this approach is that classical gravity emerges as a dual description of field theory on fractal space-time.

1. Introduction

Time and again, experimental observations have confirmed that the Standard Model (SM) is a robust theoretical framework for the description of elementary particle physics up to the scale of electroweak (EW) interaction. Experiments have covered a wide range of direct searches at particle accelerators, as well as precision tests of EW parameters.

It is known that relativistic Quantum Field Theory (QFT) represents the backbone of SM and is built in compliance with a number of postulates called *consistency conditions*. They define the range of applicability of SM. The remarkable success of SM can be attributed to a unitary, local, renormalizable, gauge invariant and anomaly-free formulation of QFT [1-3]. Since SM is based on a renormalizable gauge field theory, the prevailing opinion among theorists is that it can be extrapolated to energies above the EW scale. The underlying assumption is that QFT *stays compliant* to consistency conditions throughout all energy scales.

Despite being confirmed in many independent tests, SM remains an incomplete framework. The root cause of EW symmetry breaking (EWSB) is still unknown. We lack compelling evidence for the Higgs boson that is alleged to break the electroweak $SU(2)_L \times U(1)_Y$ symmetry to its smaller electromagnetic $U(1)_{EM}$ subgroup. The search for the source of EWSB has been one of the main drivers in both experimental and theoretical high-energy physics for the past 25 years.

Beyond our ignorance on the mechanism of EWSB, there are expectations that new phenomena will surface at the Large Hadron Collider (LHC) and other detector sites in the not-so-distant future [4-6]:

- A fundamental scalar Higgs boson is not the only way to induce EWSB. What is certain is that a *light Higgs boson* is consistent with precision EW data, but this does not generally preclude other EWSB scenarios.
- The mass parameter of the Higgs boson — which is closely tied to the scale of EWSB — is extremely sensitive to quantum corrections. As a result, attempts to extrapolate SM to energies much higher than the EW scale lead to the *gauge hierarchy problem*, where an extreme fine-tuning is required to maintain the EW scale at its observed value. Although this is not inconsistent with the underlying principles of QFT, it is at least un-natural.
- SM is unable to account for the presence of *dark matter*. In many theories beyond SM, dark matter consists of stable and weakly coupling states whose existence protects the EW scale.
- SM is unable to account for the asymmetry of visible matter over antimatter. New physics near or above the EW scale can potentially explain the fundamental baryon asymmetry of the universe.

Among other challenges facing SM, we list the origin of fermion replication, a quantum description of gravity, an explanation for the cosmological constant, the source of broken discrete symmetries, the sources of flavor mixing and neutrino masses [4-8]. It is believed that these open questions are likely to be solved by new physics above the EW scale. Irrespective of the particular nature of new physics, it is also generally believed that the outcome at the LHC would contain an excess of observed leptons, photons, jets and missing transverse energy in some combination. Searches for new physics and SM-related phenomenology at the LHC and other detector sites include, but are not limited to, the following items:

- Supersymmetry (SUSY), leptoquarks, hidden valley states, unparticles, extra-dimensions and strings,

- CP violation in the B-meson sector,
- Top quark physics,
- Z' physics and other "would-be" heavy bosons,
- Probing the origin of neutrino mass,
- Physics of heavy quarks, quarkonium and gluon-gluon fusion,
- Understanding the phase diagram of deconfined high-temperature quantum chromodynamics (QCD). The goal is explaining the behavior and properties of quark-gluon plasma (QGP) and color condensates (GLASMA) resulting from hadron collisions,
- Probing for the fourth family quark and the existence of sterile neutrino,
- Probing for exotic phases of matter including dark matter.

Inspired by the ubiquity of nonlinear dynamics and complex behavior in physical phenomena [9-11, 58], we follow here a less explored path to physics beyond SM. To this end, we recall that there is a wealth of alleged *anomalies* and *broken symmetries* which directly or indirectly relate to physics near or above the EW scale: mass generation via gauge symmetry breaking, violation of CP and chiral symmetries [2-3], absence of flavor transitions between charged leptons and their anomalous magnetic moments [12], non-unitarity of lepton mixing matrix due to neutrino masses [13], symmetry violation between neutrinos and anti-neutrinos in Mini-BooNE data [14], dijet asymmetry at the ATLAS detector [15], anisotropic flow of QGP in ultra-relativistic collisions of heavy nuclei [16], top anti-top asymmetry at CDF and D0 [17], CDF muon anomaly [18], PAMELA positron anomaly [19] and so on.

Tying these isolated clues together hints that *time-asymmetric, space-asymmetric* and *non-local* field theories are among the most likely candidates for physics beyond SM. In particular, developing the theory beyond SM may require a careful revision of conceptual foundations of QFT and itsconsistency conditions.

It is known that QFT describes interaction of stable or quasi-stable fields whose evolution is deterministic and time-reversible. Divergence cancellation in UV is tantamount for a successful description of physics beyond SM. By contrast, behavior of strongly coupled fields or dynamics in the Terascale sector is prone to become unstable and chaotic. Non-renormalizable interactions are likely to proliferate and prevent full cancellation of ultraviolet divergences. As a result of incessant fluctuations and the upsurge in entropy arisen from the loss of predictability, any system of fields in nonlinear interaction much above the EW scale is bound to

- Become inherently statistical and dissipative,
- Migrate from stationary to out-of-equilibrium conditions.

A transient regime in nonlinear dynamics opens the door for the emergence of strange attractors in phase space and transition from smooth to fractal topology [20-21]. Drawing from these premises, the goal of this report is to evaluate the likely impact of *fractional dynamics* and *fractal topology* on physics unfolding above the EW scale.

Ideas introduced below are gradually built in self-contained steps. For the sake of concision and clarity, the presentation is often times formatted in a "bulleted" style. The next section develops the motivation for model building using fractional dynamics. A brief review of what fractional dynamics stands for and its array of current applications is outlined in section 3. Section 4 focuses on a series of hints for fractional dynamics stemming from the theoretical structure of SM. The remainder of the report discusses the connection between physics beyond SM and fractional dynamics. Conclusions are presented in the last section.

We caution from the outset that ideas presented here are in their infancy. Since, by construction, SM is an "effective" theoretic framework, any proposed extensions beyond its realm must be approached with a healthy dose of skepticism. At this stage, many controversial issues remain unsettled and successful theoretical developments are yet to come. We believe that the intricate nature of topics and incomplete knowledge from the experimental side preclude a comprehensive and definitive analysis. Model building efforts as well as concurrent testing data are needed to falsify, confirm or expand these tentative findings.

2. Foundational Questions

In our view, there are three foundational questions that need to be answered prior to developing the theory beyond SM:

- Are Terascale phenomena in dynamic equilibrium?

By dynamic equilibrium we mean a condition in which all processes act simultaneously to maintain the system of interacting fields in an overall steady state. Consider a few-body system of interacting classical fields. Its steady state follows from minimization of the interaction energy and is described as stable if sufficiently small perturbations away from it damp out in time. Perturbations may be *internal* to the system or *external*, the latter case describing *open* systems coupled to their environment. The replica of equilibrium states in nonlinear dynamics is the fixed-point solutions of evolution equations [20-21].

- Are Terascale phenomena quantum or classical?

Take an isolated system of interacting quantum fields whose Hamiltonian factors out into three independent contributions,

$$H = H_0 + H_P + H_I \tag{1}$$

H_0 is the term associated with the fields, H_P describes internal perturbations and H_I the coupling between fields and perturbations. Decoherence represents the inherent *loss of phase information* induced by H_I and is responsible for suppressing the quantum nature of fields [22]. The time it takes a generic system of quantum oscillators to decohere is on the order of

$$t_d \sim \frac{1}{\gamma \langle E \rangle T \langle \Delta E_P \rangle^2} \tag{2}$$

Here, γ encodes the dissipative effects produced by perturbations, $\langle E \rangle$ is the average overall energy of the system, T its temperature and $\langle \Delta E_P \rangle$ the average energy spacing in the perturbation spectrum. Since Terascale physics is characterized by large values of parameters appearing in the denominator, transition to classical behavior is bound to occur extremely fast. A similar scenario applies to interacting quantum fields whose dynamics exhibits spontaneous symmetry breaking [22]. It is instructive to note that:

- Decoherence enables non-abelian gauge fields to undergo transition to chaos as classical fields [23].
- Erasing phase information encoded in the quantum description of phenomena is an inherent source of entropy increase [61].

It follows from these premises that at very large temperatures, commensurate with probing the near and deep Terascale sector, many quantum phenomena are likely to decohere almost instantaneously and become unstable[1]. On account of previous points, we adopt the foundational view of [24, 65, 66] that unstable few-body quantum processes are *intrinsically time-asymmetric* and favor the onset of *non-equilibrium dynamics*. This conjecture has been reinforced in recent years by the observation that *complex behavior* in the form of bifurcations and chaos, fractal geometry and random-looking evolution in time and space can occur in low-dimensional as well as in few-body systems [11]. Because chaos is ubiquitous at the level of microscopic dynamics of single particles it should also determine to a large extent the macroscopic behavior of interacting fields.

- What constraints need to be applied to phenomenological models of the Terascale sector?

A successful model of the Terascale sector must be able to recover the physics of SM in its low-energy limit. [8]. In particular:

- Has to be compatible with EW precision data,
- Has to convincingly resolve the unitarity problem at the SM scale,
- Has to maintain gauge invariance and renormalizability at the SM scale.

Next sections indicate how fractional dynamics has the potential of meeting all these constraints as the departure from equilibrium dynamics goes to zero. In a nutshell, transition to equilibrium at low-energies decouples fractional dynamics from the physics of SM.

3. What Is Fractional Dynamics?

Fractional dynamics studies the behavior of nonlinear physical systems that are [25]

---

[1] It is unclear if this assumption remains true regardless of the energy scale. Fast thermalization of QGP may provide a valid counter-argument, if it stands at transition temperatures well above $\tau_t = 175$ GeV [55, 61].

- Out-of-equilibrium *and*
- Described by differential and integral operators of non-integer orders (fractal operators).

Equations containing such operators are used to analyze the behavior of systems characterized by

- Power-law nonlinearity,
- Power-law long-range spatial correlations or long-term memory,
- Fractal or multi-fractal properties.

In the last decade, the number of applications of fractional dynamics in science and engineering has been steadily growing. They include models of fractional-relaxation effects, anomalous transport in fluids and plasma, wave propagation in complex media, viscoelastic materials, universal response in dielectric media, non-Markovian evolution of quantum fields, networks of fractional oscillators, dynamics of non-extensive statistical systems and so on. The reader is referred to [26] for a comprehensive update of fractional calculus and fractional dynamics.

For the sake of convenience, we introduce the next few definitions and properties of fractal operators that are relevant to our context. Let $f(x,\lambda) \in L_p(E^1)$ an arbitrary function of x defined on a one-dimensional Euclidean space E^1 where λ is a parameter and $1 < p < 1/\alpha$. Fractional integration of order α on $(-\infty, y)$ and $(y, +\infty)$ is described by [27]

$$(I_+^\alpha f)(y,\lambda) = \frac{1}{\Gamma(\alpha)} \int_{-\infty}^{y} \frac{f(x,\lambda)\,dx}{(y-x)^{1-\alpha}} \tag{3a}$$

$$(I_-^\alpha f)(y,\lambda) = \frac{1}{\Gamma(\alpha)} \int_{y}^{+\infty} \frac{f(x,\lambda)\,dx}{(x-y)^{1-\alpha}}$$

An alternate formulation is given by the left (L) and right (R) Riemann-Liouville operators,

$${}_0D_L^\alpha f(y,\lambda) = \frac{1}{\Gamma(1-\alpha)} \frac{d}{dy} \int_0^y (y-x)^{-\alpha} f(x,\lambda)\,dx \tag{3b}$$

$${}_0D_R^\alpha f(y,\lambda) = \frac{1}{\Gamma(1-\alpha)} \left(-\frac{d}{dy}\right) \int_y^0 (x-y)^{-\alpha} f(x,\lambda)\,dx$$

There is a close connection between *fractals* and *fractional dynamics* [25-27]. Fractals are metric sets with non-integer dimensionality. Integration over an axially-symmetric fractal space W with *Hausdorff dimension* D is defined as

$$\int_W f(x)d\mu_H(x) = \frac{2\pi^{D/2}}{\Gamma(D/2)}\int_0^\infty f(r)r^{D-1}dr \tag{4}$$

in which $d\mu_H(x)$ stands for the differential *Hausdorff measure* of W [27]. It satisfies the scale-invariance property

$$d\mu_H(x/s) = s^{-D}d\mu_H(x) \tag{5}$$

The same property applies to (4) on account of (5)

$$\int_W f(sx)d\mu_H(x) = s^{-D}\int_W f(x)d\mu_H(x) \tag{6}$$

The Hausdorff dimension for a subset $E \subset W$ is given by

$$D = \dim_H(E) \tag{7a}$$

such that, for any non-negative number α,

$$\mu_H(E) = \infty \text{ if } 0 \le \alpha < D \tag{7b}$$

$$\mu_H(E) = 0 \text{ if } D < \alpha < \infty$$

In section 9 we introduce quantum charges associated with non-abelian gauge theory. In anticipation of that discussion, consider an arbitrary charge distribution on W defined by dimension D. Let $\rho(\mathbf{r},t)$ describe the charge density function. The total charge enclosed within the fractal volume V_D is described by [28]

$$q_D(W) = \int_W \rho(\mathbf{r},t)dV_D,\ dV_D = c_3(D,\mathbf{r})dV_3 \tag{8a}$$

where V_3 represents the ordinary volume of space and

$$c_3(D,\mathbf{r}) = \frac{2^{3-D}\Gamma(3/2)}{\Gamma(D/2)}|\mathbf{r}|^{D-3} \tag{8b}$$

A close connection can also be established between *fractional dynamics* and *nonextensive statistical mechanics*. The latter has led to many successful applications dealing with the study of dynamical systems outside equilibrium [65].

4. Hints for Fractional Dynamics in High-EnergyPhysics

Unusual quantum regimes such as ultra-relativistic nucleus-nucleus collisions, quark-gluon plasma, decay of heavy resonances, strong-coupling in infrared QCD, behavior of non-Fermi liquids, fractional quantum Hall effect, non-extensive behavior of high-temperature or large-density QCD, spin glasses, high-momentum scattering of longitudinally polarized vector bosons are a few representative examples of *out-of-equilibrium processes*[2].

Recent years have consistently shown that fractional dynamics is an indispensable tool for modeling such processes [29]. A natural question to ask is: *What leads are there that suggest using fractional dynamics for model building beyond SM?* Answering this question is our next objective.

- Hints from dimensional regularization

Theoretical challenges associated with divergences of perturbative QFT were first recognized by Heisenberg and Pauli in 1929 and 1930. A viable solution had to wait until 1949 when Dyson realized that divergences could be reabsorbed in a countable number of parameters defining the theory. Models that accommodate this procedure were called "renormalizable". It was later determined that typical non-renormalizable theories contain coupling coefficients having dimensions of inverse powers of mass [30].

Standard renormalization in QFT is conceived as a two-step program: regularization and subtraction. One first controls the divergence present in momentum integrals by inserting a suitable "regulator", and then brings in a set of "counter-terms" to cancel out the divergence. Momentum integrals in perturbative QFT have the generic form

$$I = \int_0^\infty d^4 q F(q) \tag{9}$$

Two regularization techniques are frequently employed to manage (9), namely "momentum cutoff" and "dimensional regularization". In the momentum cutoff scheme, the upper limit of (9) is replaced by a finite mass scale M ,

$$I \to I_M = \int_0^M d^4 q F(q) \tag{10}$$

Explicit calculation of the convergent integral (10) amounts to a sum of three polynomial terms

---

[2] We mention here the pioneering work of Prigogine who conjectured that non-equilibrium microscopic processes cannot be properly described by S-matrix theory and require moving beyond the conventional Hilbert space of quantum theory [24].

$$I_M = A(M) + B + C(1/M) \tag{11}$$

Dimensional regularization proceeds instead by shifting the momentum integral (9) from a four-dimensional space to a continuous D - dimensional space

$$I \to I_D = \int_0^\infty d^D q F(q) \tag{12}$$

Introducing the parameter $\varepsilon = 4 - D$ leads to

$$I_D \to I_\varepsilon = A'(\varepsilon) + B' + C'(1/\varepsilon) \tag{13}$$

It is known that M and ε are not independent regulators and relate to each other via the approximate connection [31]

$$\varepsilon = 4 - D \approx \frac{1}{\log(M/M_0)} \tag{14}$$

where M_0 stands for an arbitrary and finite reference scale. (11) and (13) may be interpreted in two different ways:

a) In the asymptotic limit $M \to \infty$ and $\varepsilon \to 0$, C and A' vanish whilst A and C' become singular.
b) Let E denote the energy scale of phenomena described by a given field theory. If the regulator is chosen to stay finite or non-zero (that is, either $M < \infty$ or $\varepsilon \neq 0$), the theory is no longer meaningful for any $E \geq M$ or for any $\varepsilon' \leq \varepsilon$.

Renormalizability goes along with a) and boils down to the requirement that all momentum integrals (1) are convergent and independent of the regulator as $M \to \infty$ or $\varepsilon \to 0$. For a number of years, this criterion was regarded as a necessary consistency condition that any trustworthy QFT must satisfy [32]. The modern point of view has now shifted to b). According to this interpretation, a field theory that is non-renormalizable represents a *valid low-energy approximation* to a more comprehensive theoretical framework. To understand why this is the case, consider a non-renormalizable theory with a single generic coupling g whose mass dimension is M^{-2}. The renormalized perturbative expansion of an N - point amplitude up to the order $(g^2)^n$ reads [32]

$$A_N(E) = A_N^0(E) \sum_{i=0}^{n} c_i \left(\frac{E}{M}\right)^{2i} \tag{15}$$

Here $c_0 = 1$ and all coefficients c_i, $i = 2,3,....,n-1$ are fixed once renormalization has been carried out for amplitudes with less than N points. Since new divergences may develop at order n, the last coefficient in the series (c_n) cannot be derived from theory. This lack of predictivity on c_n becomes however irrelevant if $E \ll M$ due to the small contribution arisen from the corresponding term in (15). Higher-order divergences can be safely ignored as long as $E \ll M$ or $\varepsilon' \geq \varepsilon$ and the chosen built-in scale M or *continuous dimension* ε sets the limit of validity of the underlying theory.

We conclude this discussion by noting that non-renormalizable interactions may be linked to the fundamental baryon asymmetry of the Universe [33].

- Hints from effective Lagrangian analysis

The goal of the effective Lagrangian method is to represent in a simple way the dynamical content of a field theory in its low energy limit. A generic effective Lagrangian can be presented as [34]

$$L_{EFF} = \sum_i g_i O_i \tag{16}$$

where O_i are *local* operators built with the light fields, and the information on any heavy fields is contained in the couplings g_i. The operators O_i are usually organized according to their dimension (d_i), which fixes the dimension of their coefficients:

$$[O_i] = d_i \rightarrow g_i \sim \frac{1}{\Lambda^{d_i - 4}} \tag{17}$$

with Λ some characteristic heavy scale of the system. At energies below this scale ($E < \Lambda$), the behavior of the different operators is determined by their dimension. There are three types of operators: relevant ($d_i < 4$), marginal ($d_i = 4$) and irrelevant ($d_i > 4$). The effect of irrelevant operators is weak at low energies because it is suppressed by powers of E/Λ. Irrelevant operators usually contain interesting information about the underlying dynamics at higher scales. For example, the SM Lagrangian without the Higgs and Yukawa sectors assumes the generic form

$$L_{SM} = \sum_\alpha \overline{\Psi}^\alpha [i\gamma^\mu (D_\mu \Psi)^\alpha] - \frac{1}{4} F^a_{\mu\nu} F^{a\mu\nu} = L_M[\Psi^\alpha, (D_\mu \Psi)^\alpha] - \frac{1}{4} F^a_{\mu\nu} F^{a\mu\nu} \tag{18}$$

Using (16) and (17) we may write (18) as

$$L_{EFF} = L_{SM(d_i<4)} + \sum_{d_i>4} \frac{g_i O_i}{\Lambda^{d_i-4}} \tag{19}$$

where corrections induced by non-renormalizable interactions $d_i > 4$ are highly suppressed by powers of E/Λ at energies $E < \Lambda$. For example, the dependence of matter Lagrangian L_M on $F^{\alpha\mu\nu}$ as well as higher covariant derivatives $D_\nu D_\mu \Psi$ creates non-renormalizable terms that are absent below the scale of EW interaction.

A basic premise of effective field theory is that *non-local* heavy–particle exchanges can be replaced by a tower of *local* and non-renormalizable interactions among light particles [34]. There are two ways in which this assumption can be violated at large energies:

- Heavy fields that yield relevant interactions near Λ cannot be integrated out and remain coupled to light fields,
- The onset of out-of-equilibrium dynamics prevents non-local heavy particles to be replaced by local interactions among light particles.

The general view is that effective Lagrangian analysis places meaningful bounds on new dynamical structures that may occur in the Terascale sector and beyond [3]. In light of our discussion, it is apparent that this kind of analysis may no longer be a reliable metric for what happens at energies far beyond the EW scale.

- Hints from the requirement of scale invariance

The Lagrangian density for classical massless electrodynamics reads

$$L = -\frac{1}{4} F_{\mu\nu} F^{\mu\nu} + \bar{\psi} i \gamma^\mu D_\mu \psi \tag{20}$$

An arbitrary change in coordinate scale $x \to x' = \lambda x$ along with the corresponding field transformations

$$\psi(x) \to \psi'(x) = \lambda^{3/2} \psi(x),\ A_\mu(x) \to A'_\mu(x) = \lambda A_\mu(x) \tag{21}$$

can be shown to leave the action unchanged [35]. The Noether current associated with the change of scale is given by

$$J^\mu_{scale} = x_\nu \theta^{\mu\nu} \tag{22}$$

in which $\theta^{\mu\nu}$ represents the conserved energy-momentum tensor of the theory, $\partial_\mu \theta^{\mu\nu} = 0$. The conservation of scale current (22) amounts to the vanishing of the trace of the energy-momentum tensor, that is,

$$\partial_\mu J^\mu_{scale} = \theta^\mu_\mu = 0 \tag{23}$$

In D space-time dimensions the trace of massive theory can be cast in the form

$$\theta^\mu_\mu = \frac{\varepsilon}{4} F^{\eta\sigma} F_{\eta\sigma} + m\overline{\psi}\psi + R(\varepsilon, m, \psi, \overline{\psi}, F^{\eta\sigma} F_{\eta\sigma}) \tag{24}$$

where the first two terms explicitly highlight the contribution of electron mass and the deviation from four-dimensionality of underlying space-time. All terms vanish in the limiting case $m = 0$ and $\varepsilon = 0$. The residual term in (24) embodies correction effects not included in the first two terms.

It is known that *scale invariance* of the theory can be interpreted as the independence of the action functional from the choice of measurement units. Scale invariance represents a fundamental symmetry of covariant field theories and is broken in SM by the presence of fermion masses or the mass scale of QCD [3]. Enforcing scale invariance defined by a vanishing trace in (24) implies that electrons gain mass *on account of* deviations from $D = 4$. Since ε is related to the mass scale of the theory M and $\varepsilon \to 0$ is equivalent to $M \to \infty$, the relationship between m and ε amounts to a non-perturbative Renormalization Group(RG) flow. The flow equation can be presented as

$$\frac{dm}{d\varepsilon} = \beta_m(m) \tag{25a}$$

Unlike the electromagnetic field tensor, the field tensor of Yang-Mills theory ($F_{\mu\nu}$) depends explicitly on the coupling charge g_{YM}. The "pure" Yang-Mills term in (24) vanishes in four-dimensional space-time ($\varepsilon \to 0$). This means that, when considering *free* Yang-Mills theories in four-dimensional space-time, there are no grounds to invoke a flow equation similar to (25a). This is no longer the case when $\varepsilon \neq 0$ and gauge fields interact with fermions.In this situation, g_{YM} plays a dynamic role similar to m in (24). One is led to a flow equation for g_{YM} having the form

$$\frac{dg_{YM}}{d\varepsilon} = \beta_g(g_{YM}) \tag{25b}$$

5. From Quantum Field Theory to Fractional Dynamics

It is instructive, for the sake of clarity, to consolidate all arguments developed so far in a mnemonic flowchart. Its purpose is to enable a "bird's eye view" of how description of the Terascale sector of particle physics may evolve from QFT to a framework based on fractional dynamics [36-37].

NONLINEAR QUANTUM FIELDS

⇓

DECOHERENCE

⇓

NONLINEAR CLASSICAL FIELDS

⇓

TRANSITION TO CHAOS AND FRACTALS

⇓

SELF-SIMILARITY

⇓

NON-LOCALITY AND NON-EQUILIBRIUM CONDITIONS

⇓

FRACTIONAL DYNAMICS

This transition may uncover a new layer of reality with its own set of concepts and rules and it may very well emerge in a variety of unexpected ways.

The meaning of the flowchart is as follows: SM describes quantum interaction of non-linear gauge fields with matter fields. Decoherence turns quantum fields into their classical counterparts and triggers the irreversible transition to chaos and fractal topology of underlying space-time and phase-space. Self-similarity associated with fractal structures blurs the traditional distinction between "locality" and "non-locality": fractals are structures living on infinitely many observation scales. Physical processes on fractals are no longer in stationary conditions but in an ever-evolving and random state of change. Adequate modeling of such processes requires use of fractional dynamics and fractal operators.

6. Fractional Dynamics and Consistency Conditions

It is known that *unitarity* and *locality* are two fundamental principles that ensure internal consistency of both QFT and SM [1-3]. Perturbative QFT relies on a unitary S-matrix formulation; regularization of quantum corrections is required to preserve consistency by suppressing infrared or ultraviolet divergences; introduction of unphysical "ghost" states is mandatory for internal consistency of local gauge field theories. Likewise, since QFT is a manifestly relativistic field theory, locality is mandatory to ensure compliance with Lorentz invariance. In a nutshell,

- *Unitarity* enforces conservation of probability. It excludes transitions that fail to be norm-preserving as well as negative-norm solutions of field theory.
- *Locality* precludes the possibility of action-at-a-distance. Lagrangian is forbidden to contain terms depending on two spatially separated points, for example

$$L_{NL} = \int \varphi(x)\varphi(y)\, d^3x d^3y \text{ or } L_{NL} = \int \varphi(x \pm y)\, d^3x d^3y \tag{26}$$

The object of this section is to elaborate upon the relationship between fractional dynamics and these two principles of QFT. To fix ideas, consider the inelastic scattering of longitudinally polarized W bosons. The tree-level scattering amplitude computed in SM without the Higgs boson grows with the square of scattering energy and it threatens to violate unitarity around 1 TeV [38]. The contribution from the Higgs exchange cancels the dangerously growing terms and the full amplitude is well behaving for arbitrary high energies. The unitarity issue in WW scattering at large energies can be, however, approached from a standpoint that goes beyond S-matrix theory. To this end we proceed in two steps:

- We first follow [39] and indicate the difference between "*transient*" and "*persistent*" scattering. The latter leads to violation of unitarity condition.
- Next, we show how fractional dynamics can be used to restore unitarity of persistent scattering upon a suitable re-definition of probability distribution function.

The probability distribution function $\rho(\mathbf{x},\mathbf{p},t)$ in S-matrix theory is localized in phase space and can be normalized to unity

$$\int d\mathbf{p} \int d\mathbf{x}\, \rho(\mathbf{x},\mathbf{p},t) = (const) \int d\mathbf{p} \int d\mathbf{k}\, \rho_{\mathbf{k}}(\mathbf{p},t)\, \delta(\mathbf{k}) = 1 \tag{27}$$

where $\rho_{\mathbf{k}}(\mathbf{p},t)$ represents the Fourier transform of $\rho(\mathbf{x},\mathbf{p},t)$

$$\rho_{\mathbf{k}}(\mathbf{p},t) = \int \rho_{\mathbf{k}}(\mathbf{x},\mathbf{p},t) e^{-i\mathbf{p}\mathbf{x}} d\mathbf{x} \tag{28}$$

Unitarity can be alternatively expressed as

$$\int d\mathbf{x} \int d\mathbf{p}\, \rho(\mathbf{x},\mathbf{p},t) = (const) \int d\mathbf{x} \int d\boldsymbol{\lambda}\, \rho_{\lambda}(\mathbf{x},t)\, \delta(\boldsymbol{\lambda}) = 1 \tag{29}$$

with

$$\rho_{\lambda}(\mathbf{x},t) = \int \rho_{\lambda}(\mathbf{x},\mathbf{p},t) e^{i\mathbf{p}\mathbf{x}} d\mathbf{p} \tag{30}$$

Relations (27) to (30) describe "transient" scattering. Consider now the situation where $\rho(\mathbf{x},\mathbf{p},t)$ is a function which is delocalized in phase space. For example, it fails to vanish either in the infrared limit $|\mathbf{x}| \to \infty$ or in the ultraviolet limit $|\mathbf{p}| \to \infty$. In these asymptotic cases, the Fourier component of $\rho(\mathbf{x},\mathbf{p},t)$ becomes singular at $\mathbf{k}=0$ and $\boldsymbol{\lambda}=0$, respectively, with a delta function singularity. Consider the first case, that is,

$$\lim_{|x|\to\infty} \rho(\mathbf{x},\mathbf{p},t) > 0 \tag{31}$$

The scattering is now "persistent". The Fourier component of the distribution function is singular at $\mathbf{k} = 0$ with a delta-function singularity

$$\rho_{\mathbf{k}}(\mathbf{p},t) = \rho_0(\mathbf{p},t)\delta(\mathbf{k}) + \rho_{\mathbf{k}}^{NS}(\mathbf{p},t) \tag{32}$$

in which $\rho_{\mathbf{k}}^{NS}$ is the non-singular part of the distribution function at $\mathbf{k} = 0$. This distribution function cannot be normalized to unity [39].

One can employ the tools of fractional calculus to restore unitarity [27]. Consider a generic probability distribution function $\rho(x,\lambda)$ depending on parameter λ and defined on one-dimensional Euclidean space E^1, $\rho(x,\lambda) \in L_1(E^1)$. The standard normalization condition corresponding to (27) is given by

$$\int_{-\infty}^{+\infty} \rho(x,\lambda)\, dx = 1 \tag{33}$$

Using (3) we can generalize (33) as follows

$$(I_+^{\alpha}\rho)(y,\lambda) + (I_-^{\alpha}\rho)(y,\lambda) = 1 \tag{34}$$

Fractional equivalent of the normalization condition reads

$$\int_{-\infty}^{+\infty} \overline{\rho}(x,\lambda)\, d\mu_{\alpha}(x) = 1 \tag{35}$$

where

$$\overline{\rho}(x,\lambda) = \frac{1}{2}[\rho(y-x,\lambda) + \rho(y+x,\lambda)] \tag{36}$$

and the Hausdorff measure introduced in section 3 is

$$d\mu_{\alpha}(x) = \frac{|x|^{\alpha-1}}{\Gamma(\alpha)} dx \tag{37}$$

Comparing of (36) with (26) shows that the price paid for restoring unitarity in (35) is a manifest loss of *locality*. To restore locality, we note that *self-similarity* of fractals blurs the distinction between observation scales. Taking advantage of this property, one can simply rescale the distance $|x-y|$ below the spatial measurement resolution Δ with no consequence on results. By definition, coordinates x, y are indistinguishable from each other if and only if

$$|(y-x)-(y+x)| = 2|x| \leq \Delta \tag{38}$$

Divide each term in (36) by an arbitrary large scale $s \gg 1$ such that

$$\frac{2|x|}{s} \ll \Delta \tag{39}$$

Using (5) and (6) leads to a local normalization condition, that is

$$\bar{\rho}(\frac{x}{s},\lambda) = \frac{1}{2}[\rho(\frac{y-x}{s},\lambda) + \rho(\frac{y+x}{s},\lambda)] \tag{40}$$

whose outcome is

$$\int_{-\infty}^{+\infty} \bar{\rho}(x/s,\lambda)\, d\mu_\alpha(x/s) = \int_{-\infty}^{+\infty} \bar{\rho}(x,\lambda)\, d\mu_\alpha(x) = 1 \tag{41}$$

7. PerturbativeRenormalization on Fractal Space-Time

The concept of space-time endowed with non-integer metric can be used for perturbative renormalization of QFT. Here we follow [40] and reproduce a method for renormalization of low-order radiative corrections in quantum electrodynamics (QED) defined on fractal space-time. Consider the full momentum-space propagator S of electron

$$S = \frac{1}{(\gamma p - m_0 - \Sigma + i\varepsilon)} \tag{42}$$

where m_0 stands for the bare electron mass and Σ the proper self-energy. Replacing for Σ its lowest-order contribution yields

$$S(p) = \frac{Z_2}{(\gamma p - m + i\varepsilon)}[1 + Z_2(\gamma p - m)\sigma(p)]^{-1} \tag{43}$$

in which the physical electron mass m and its renormalization constant are given by

$$m = Z_2 m_0 \tag{44}$$

$$Z_2 = 1 + \frac{3}{2\pi}\frac{\alpha_{EM}}{(4-D)}$$

In (43) $\sigma(p)$ is a function defined by (A1.5b) in [40], α_{EM} is the fine-structure constant and it is assumed that the departure from four space-time dimensionality is small ($4-D \ll 1$). Expanding the vacuum polarization $\Pi(q^2)$ around the mass-shell $q^2 = 0$, we obtain

$$(Z_3)^{-2} = 1 - \Pi(q^2 = 0) = 1 + \frac{2}{\pi}\frac{\alpha_{EM}}{(4-D)} \tag{45}$$

$$(\alpha_{EM})_0 = \frac{e_0^2}{4\pi} = \alpha_{EM} Z_3^{-2} = \alpha_{EM}[1 + \frac{2}{\pi}\frac{\alpha_{EM}}{(4-D)}] \tag{46}$$

Here, $(\alpha_{EM})_0$ represents the bare fine-structure constant and Z_3 the charge renormalization factor. It can be also shown that, based on the degree of divergence of QED diagrams, singular behavior of some radiative corrections tends to attenuate or vanish for $0 < D \leq 4$.

We close this section with the general observation that, embedding perturbative field theory on fractal space-time, helps reducing or eliminating divergence of momentum integrals. Place (9) on fractal space-time support characterized by the Hausdorff measure in momentum space $\mu_H(E)$ and Hausdorffdimension D. According to definition (7b), singular behavior of the integrand $f(q)$ can be dampened by choosing $D < \alpha < \infty$ which automatically leads to a vanishing Hausdorff measure, that is, $d\mu_H(E) = 0$.

8. Gauge Bosons and Fermions on Fractals

This section explores the consequences of placing classical SM fields on fractal space-time. This setting may be well suited to describe conditions developing near or above the EW scale.

8.1. Gauge Fields on Fractals

One of the most counter-intuitive properties of fractal space-time is that it carries a topological form of internal *energy*. This contribution stems from the ability of fractal topology to *polarize space-time* and can be quantified in terms of continuous parameter (14). The net result is that fractal space-time can be modeled as an *effective medium* departing from the passive properties of classical vacuum. For instance, classical electrodynamics action on fractals can be built from effective field quantities and reads [28].

$$S_{eff} = -\int d^4x(\frac{1}{4}F_{eff,\mu\nu}F_{eff}^{\mu\nu} + J_{eff}^{\mu}A_{eff,\mu}) \tag{47}$$

Action (47) is invariant to local gauge transformations $A_{eff,\mu} \rightarrow A_{eff,\mu} - \partial_{\mu}\theta$ if and only if the fractional continuity equation holds true, that is, if

$$\partial_{\mu}J_{eff}^{\mu} = \partial_{\mu}(c_{\mu}J^{\mu}) = 0 \tag{48}$$

Here, c_{μ} are coefficients depending on fractal dimension, as listed in [28]. Using the language of effective quantities, Lagrangian of the free Maxwell fields on fractals can be presented as

$$L_{eff} = -\frac{1}{4}F_{eff,\mu\nu}F_{eff}^{\mu\nu} = \frac{1}{2}(\mathbf{E}_{eff}^2 - \mathbf{B}_{eff}^2) = \frac{1}{2}(\mathbf{E}^2 - \mathbf{B}^2) + \Delta L \tag{49}$$

where

$$\mathbf{E}^{eff}(\mathbf{r},t) = c_1(\gamma,\mathbf{r})\mathbf{E}(\mathbf{r},t) \tag{50}$$

$$\mathbf{B}^{eff}(\mathbf{r},t) = c_2(d,\mathbf{r})\mathbf{B}(\mathbf{r},t) \tag{51}$$

$$c_1(\gamma,\mathbf{r}) = \frac{2^{1-\gamma}\Gamma(1/2)}{\Gamma(\gamma/2)}|\mathbf{r}|^{\gamma-1} \tag{52}$$

$$c_2(d,\mathbf{r}) = \frac{2^{2-d}}{\Gamma(d/2)}|\mathbf{r}|^{d-2} \tag{53}$$

Hence the differential contribution of fractality to the Lagrangian is given by

$$\Delta L = L - L_{eff} = \frac{1}{2}[(c_1^2 - 1)\mathbf{E}^2 + (c_3^2 - 1)\mathbf{B}^2] \tag{54}$$

(54) vanishes on smooth space-time $\varepsilon = 0$ however, near $\varepsilon \ll 1$, it may emerge in various forms: it can produce an excess of charges or currents, change the magnitude of the fine-structure constant, generate new particles or make photons massive. The ability of fractal space-time to impart mass to Maxwell fields follows from identification of (54) with the mass term of the Proca Lagrangian

$$\Delta L = \frac{1}{2} M^2 A_\mu A^\mu = \frac{1}{2} M^2 (A_0 A^0 + A_i A^i) \tag{55}$$

Since Proca Lagrangian describes dynamics of a spin-1 massive field, (54) and (55) lead to a novel mechanism of mass generation in the EW sector arising from polarization attributes of fractal space-time. Unlike the Proca model which fails gauge invariance, (47)-(49) lead to a gauge invariant theory containing massive gauge bosons.

Using effective quantities, Maxwell's equations on fractals can be cast in their traditional condensed form,

$$\partial_\mu (F_{eff})^{\mu\nu} = (J_{eff})^\nu \,, \partial_\mu (F_{eff})^{\mu\nu} = 0 \tag{56}$$

where [41]

$$(F_{eff})^{\mu\nu} = \frac{1}{2} \varepsilon^{\mu\nu\rho\sigma} (F_{eff})_{\rho\sigma} \tag{57}$$

$$(F_{eff})^{0i} = -(E_{eff})^i \,, (F_{eff})^{ij} = -\varepsilon^{ijk} (B_{eff})^k$$

Let us next generalize these findings and consider coupling of two-component massless Weyl fermions to Yang-Mills fields on fractals. We posit that, near $\varepsilon \approx 0$, fermion field picks up infinitesimal corrections from fractal topology, which convert massless states into *nearly* massless states, i.e.

$$\Psi_L = \begin{pmatrix} \psi_L (1+\varepsilon) \\ \varepsilon \psi_L \end{pmatrix}, \Psi_R = \begin{pmatrix} \varepsilon \psi_R \\ \psi_R (1+\varepsilon) \end{pmatrix} \tag{58}$$

The interaction of Yang-Mills fields with a system of massless fermions in four-dimensional space-time is represented by [42]

$$L = \sum_\alpha \overline{\Psi}^\alpha [i\gamma^\mu (D_\mu \Psi)^\alpha] - \frac{1}{4} F^a_{\mu\nu} F^{a\mu\nu} \tag{59}$$

Using again the language of effective quantities on fractals, we define effective chromo-electric and chromo-magnetic fields as [43]

$$(E_{eff})^i_a = (F_{eff})^{i0}_a \,, (B_{eff})^i_a = -\frac{1}{2} \varepsilon_{ijk} (F_{eff})^{jk}_a \,, i, j, k = 1, 2, 3 \tag{60}$$

The effective Yang-Mills Lagrangian assumes the form

$$(L_{eff})_{YM} = \frac{1}{2}\sum_a [(\mathbf{E}_{eff})_a \cdot (\mathbf{E}_{eff})_a - (\mathbf{B}_{eff})_a \cdot (\mathbf{B}_{eff})_a] \tag{61}$$

It is seen that (61) contains an extra term due to fractal corrections that can be formally attributed to the emergence of massive gauge fields

$$\boxed{\frac{1}{2}\sum_a M_a^2 (A^{a\mu} A_\mu^a) \to \Delta L_{YM} = (L_{eff})_{YM} - L_{YM}} \tag{62}$$

The coupling term between effective gauge field and Weyl fermions becomes [44]

$$(L_{eff})_{\text{int}} = g_{eff}\,(A_{eff})_\mu^a [\overline{\Psi}^\alpha \gamma^\mu (T^a)_{\alpha\beta} \Psi^\beta] \tag{63}$$

The difference in interaction terms may be attributed to the emergence of massive fermions, that is

$$\boxed{-m\Psi\overline{\Psi} \to \Delta L_{\text{int}} = (L_{eff})_{\text{int}} - L_{\text{int}}} \tag{64}$$

in which the scalar $\overline{\Psi}\Psi$ is built from the nearly massless Weyl fields introduced in (58).

8.2. Fermions on Fractals

We now turn to a model building strategy that highlights how conserved quantities arise on fractals. To this end, consider one of the many fractional generalizations of the free Dirac equation, namely [45]

$$(A\partial_t^\alpha + B\partial_x)\Psi(t,x) = 0\,,\, \Psi(t,x) = \begin{pmatrix} \psi_L(t,x) \\ \psi_R(t,x) \end{pmatrix} \tag{65}$$

Here, $0 < \alpha < 1$, I stands for the identity operator and A and B are 2×2 matrices obeying Pauli's algebra

$$A^2 = I\,, B^2 = -I\,, \{A, B\} = AB + BA = 0 \tag{66}$$

Both components of the Dirac field satisfy the fractional evolution equation

$$\partial_t^{2\alpha} \psi_{L,R}(t,x) - \partial_{xx} \psi_{L,R}(t,x) = 0 \tag{67}$$

Lagrangian of the free Dirac field corresponds to $\alpha = 1$ and is given by

$$L_D = \overline{\Psi} A \partial_t \Psi + \overline{\Psi} B \partial_x \Psi \tag{68}$$

or by its equivalent conventional form

$$L_D = \overline{\Psi}(i\gamma^{\mu}\partial_{\mu})\Psi \tag{69}$$

Dirac Lagrangian is invariant under parity transformation since the parity operator turns ψ_L into ψ_R, ∂_x into $-\partial_x$ and $\overline{\sigma}^{\mu}\partial_{\mu}$ into $\sigma^{\mu}\partial_{\mu}$, in which

$$\sigma^{\mu} = (1, \sigma^{i}), \overline{\sigma}^{\mu} = (1, -\sigma^{i}) \tag{70}$$

Here, σ^{i} represent Pauli matrices. Despite the non-local nature of the fractional time operator ∂_t^{α}, it can be shown that (68) leads to a conserved analogue of the Dirac Hamiltonian defined as

$$H_{\alpha}(t,x) = -\int_{-\infty}^{+\infty} \Psi^{T}(AB)\partial_x \Psi dx \tag{71}$$

9. Breaking of Discrete Symmetries

Non-local properties of fractal operators prevent invariance under discrete symmetries: in general, fractional Dirac equation (67) is not invariant under space-time transformations [29].

This property is consistent with the well-known breaking of P and CP symmetries in weak interactions [1-3].

Consider for simplicity the Galileian transformation of space-time coordinates

$$t' = t, x' = x + \text{v}t \tag{72}$$

This transformation can be explicitly formulated as

$$\Psi(t,x') = W\Psi[t, x(x',t)] \tag{73}$$

where W is a 2×2 operator,

$$W = \begin{pmatrix} \text{w}_{11} & \text{w}_{12} \\ \text{w}_{21} & \text{w}_{22} \end{pmatrix} \tag{74}$$

Invariance of (67) to (72) is preserved if and only if [45]

$$\psi_L(t,x) = -(\frac{\text{w}_{22}}{\text{w}_{21}})\psi_R(t,x) + c(t) \tag{75}$$

in which $c(t)$ represents a constant function of x.

As it is known, the free Dirac equation for massive fermions is given by [1-3]

$$(\gamma^{\mu} p_{\mu} - m)\Psi = 0 \tag{76}$$

or, in terms of chiral components

$$\begin{pmatrix} -m & E + \vec{\sigma} \cdot \vec{p} \\ E - \vec{\sigma} \cdot \vec{p} & -m \end{pmatrix} \begin{pmatrix} \psi_R \\ \psi_L \end{pmatrix} = 0 \tag{77}$$

(77) implies that the mass parameter m induces a linear mixing between left and right spinors and a corresponding violation of chiral symmetry, that is,

$$\psi_R = \left(\frac{E + \vec{\sigma} \cdot \vec{p}}{m} \right) \psi_L \tag{78a}$$

$$\psi_L = \left(\frac{E - \vec{\sigma} \cdot \vec{p}}{m} \right) \psi_R \tag{78b}$$

Comparing (78b) with (75) for the trivial case $c(t) = 0$ leads to the identification

$$\frac{\mathrm{w}_{22}}{\mathrm{w}_{21}} = -\frac{E - \vec{\sigma} \cdot \vec{p}}{m} \tag{79}$$

We conclude from (75) to (79) that, imposing Galileian invariance of Dirac equation on fractal space-time (67), yields a *massive* Dirac equation in standard space-time. This finding is consistent with (62) and (64) which show that *massivation* is a direct consequence of placing massless field theories on fractal space-time. As stated, the underlying cause of massivation is that fractals tend to polarize classical space-time vacuum and convert it into an effective medium.

We close this section by recalling that breaking of parity and chiral symmetry violation are related to each other [46]. Consider the operation of parity in ordinary three-dimensional space,

$$x = (t, \mathbf{x}) \rightarrow x_P = (t, -\mathbf{x}) \tag{80}$$

Parity violation is seen to be closely related to breaking of chiral symmetry since

$$\psi_{L,R} \rightarrow P \psi_{L,R} P^{-1} = \gamma^0 \psi_{R,L}(x_P) \tag{81}$$

10. Effective Charges on Fractals and Anomalous Properties

Section (9) has built upon the idea that all physical quantities become effective on fractal space-time. This includes not only the electric charge but also $SU(2)$ and $SU(3)$ charges associated with gauge field theories. The conserved fermion current on fractals can be written as [28]

$$j_{eff}^{\mu}(D,\mathbf{r}) = c^{\mu}(D,\mathbf{r})\overline{\psi}\gamma^{\mu}\psi \tag{82}$$

where

$$c^{0}(D,\mathbf{r}) = c_3(D,\mathbf{r})\,, c^{i}(D,\mathbf{r}) = c_2(D,\mathbf{r})\,, i = 1,2,3 \tag{83}$$

(82) leads to a conserved charge

$$\boxed{g_{eff}(D,\mathbf{r}) = \int d^3x\, c_3(D,\mathbf{r})\overline{\psi}\gamma^{0}\psi} \tag{84}$$

Effective charge (84) has a different magnitude than its value on ordinary space-time $(D=4)$. Since the square of gauge charge gives the probability for emission or absorption of virtual particles, it follows that g_{eff}^2 may be able to explain, at least in principle, the *excess* of observed leptons, photons, jets that are anticipated to surface in some combination at LHC and other detector sites.

(84) may be also able to account for the source of anomalous magnetic moment of massive leptons (AMM). It is known that the magnetic moment $\boldsymbol{\mu}$ of a particle with mass m and charge e is related to the particle spin $\mathbf{S}$ by the gyro-magnetic ratio g [12]:

$$\boldsymbol{\mu} = g(\frac{e}{2m})\mathbf{S} \tag{85}$$

At the tree level, QED predicts the result $g = 2$ for all elementary fermions. Quantum effects produced by QED loop diagrams, from strong and weak interactions or from contributions arising above the electroweak scale lead to a deviation

$$a = \frac{1}{2}(g-2) \tag{86}$$

which measures the magnitude of AMM. Loop corrections from heavy particles with mass M are generally suppressed by a factor $(m/M)^2$. Therefore the effect of quantum corrections to AMM scales quadratically with the mass of charged leptons. The SM prediction for the muon anomaly, for example, is typically factored into a QED, EW and hadronic (leading and higher order) contributions [47]

$$a_\mu^{SM} = a_\mu^{QED} + a_\mu^{EW} + (a_\mu^{HLO} + a_\mu^{HHO}) \quad (87)$$

The difference between a_μ^{SM} computed with (87) and the most updated experimental value amounts to $\Delta a_\mu = +302(88)\times 10^{-11}$, which is on the order of 3, 4 standard deviations with all errors added in the quadrature. This numerical discrepancy can be attributed to two main sources: an erroneous determination of leading-order hadronic contributions (a_μ^{HLO}) or possible corrections induced by physics beyond SM. It is plausible, in this context, that the effective charge (84) may provide an appealing explanation for the muon anomaly.

Along the same line of arguments, (84) may alter conventional cross sections and favor new phase transitions beyond SM predictions. It appears likely that this scenario plays an important role in the phenomenology of QGP and color condensate (GLASMA), the multi-muon CDF anomaly, PAMELA excess of positrons, formation of hadronic and leptonic jets in relativistic proton-proton (pp) collisions, deviation from charge form-factors in studies of muonic hydrogen [48], deviations from branching ratios of B-mesons [49] and so on.

11. Generation Structure of SM Parameters

Consider the nonlinear RG flow equations (25a) and (25b). They define trajectories of fermion masses and Yang-Mills couplings in the space generated by $\varepsilon = 4 - D$. Suppose the flow depends on a single and generic control parameter λ. For example, λ may embody the perturbing effect of heavy fields near or above the characteristic scale of the system (Λ) [50]. Flow equations for masses and effective couplings become

$$\frac{dm}{d\varepsilon} = \beta_m(m, \lambda) \quad (88a)$$

$$\frac{dg_{eff}}{d\varepsilon} = \beta_g(g_{eff}, \lambda)$$

or, in condensed form,

$$\frac{d\sigma}{d\varepsilon} = \beta_\sigma(\sigma, \lambda), \sigma = \{m, g_{eff}\} \quad (88b)$$

We study the behavior of (88) using the following set of assumptions

- The analysis of (88) is carried out near the EW scale and in the neighborhood of $D = 4$, $\varepsilon = 0$.

- m and g_{eff} are considered *independent* parameters. As a result, (88) represents a system of autonomous and independent ODE equations.
- Gauge coupling g_{eff} is a vector whose components $(g_{eff})_r$ $r = 3,2,1$ reflect the $SU(3)\times SU(2)\times U(1)$ group structure of SM.
- Up to a first order approximation we take $(g_{eff})_r \approx g_r$ close to $\varepsilon = 0$.
- All functions are analytic in λ.
- System (88) has at least one limit cycle solution $\sigma_0(\varepsilon,\lambda)$.
- The limit cycle $\sigma_0(\varepsilon,\lambda)$ is stable for $\lambda < 0$ and it becomes unstable at $\lambda = 0$ after a period-doubling bifurcation created as a result of crossing the imaginary axis by one of the Floquet exponents.

Note that, since vector boson masses M depend on g_{eff}, no independent flow equation for M is considered in (88). According to [50-51], the first stage of the transition to chaos driven by the continuous variation of $\lambda > 0$ represents a Feigenbaum cascade of period-doubling bifurcations for $\sigma_0(\varepsilon,\lambda)$. Numerous examples of this scenario [51] show that the sequence of critical values λ_n, $n \in \mathrm{N}$, leading to the onset of super-stable orbits, satisfies the geometric progression

$$\lambda_n - \lambda_\infty \approx K\,\overline{\delta}^{-n} \tag{89}$$

Here, K is a multiplicative factor and $\overline{\delta}$ a scaling constant that is, in general, different than the standard $\delta = 4.669...$ for quadratic maps. Expanding $\sigma_0(\varepsilon,\lambda)$ around the critical value $\lambda = \lambda_\infty$ corresponding to the onset of fully developed chaos leads to

$$\sigma_0(\varepsilon,\lambda_n) = \sigma_0(\varepsilon,\lambda_\infty) + (\overline{\delta})^{-n}\left.\frac{\partial\sigma_0(\varepsilon,\lambda)}{\partial\lambda_n}\right|_{\lambda_\infty} + \frac{(\overline{\delta})^{-2n}}{2}\left.\frac{\partial^2\sigma_0(\varepsilon,\lambda)}{\partial\lambda_n^2}\right|_{\lambda_\infty} + ... \tag{90}$$

For $n = 2^p$, $p \geq 0$ the ratio of two consecutive terms in (90) is given by

$$\frac{\Delta\sigma_{0,n}}{\Delta\sigma_{0,n+1}} = \frac{\sigma_0(\varepsilon,\lambda_n)-\sigma_0(\varepsilon,\lambda_\infty)}{\sigma_0(\varepsilon,\lambda_{n+1})-\sigma_0(\varepsilon,\lambda_\infty)} = \frac{\sum_k c_k[K\overline{\delta}^{-n}]^k}{\sum_k c_k[K\overline{\delta}^{-(n+1)}]^k} \tag{91}$$

Under the assumption $c_1 \neq 0$ and $\overline{\delta}^{-n} \propto O(\varepsilon)$ for $p \gg 1$, we obtain

$$\boxed{\frac{\Delta\sigma_{0,2^{p+1}}}{\Delta\sigma_{0,2^p}} \approx \overline{\delta}^{-(2^p)}} \tag{92}$$

The table shown below contains a side-by-side comparison of estimated versus actual mass ratios for charged leptons and quarks, massive gauge bosons and ratios of interaction strengths [37, 50]. All masses are reported in MeV and evaluated at the energy scale set by the top quark mass (m_t). Using recent results issued by the Particle Data Group [52], we take

$$m_u = 2.12, m_d = 4.22, m_s = 80.9$$

$$m_c = 630, m_b = 2847, m_t = 170{,}800$$

Coupling strengths are evaluated at the scale set by the mass of the "Z" boson, namely

$$\alpha_{EM} = e^2/4\pi = 1/128\ , \alpha_W = g_w^2/4\pi = 0.0338\ , \alpha_s = g_3^2/4\pi = 0.123$$

Here, *u*, *d*, *s*, *c*, *b* and *t* stand for the six quark flavors, e, μ and τ represent the three flavors of charged leptons, W and Z the two flavors of EW gauge bosons and α_{EM}, α_W, α_s the coupling strengths associated with the electromagnetic, weak and strong interactions, respectively. Gauge bosons are spin-one self-interacting objects and the contribution of self-interacting energy needs to be accounted for when computing their masses [37]. Following the rationale of [53], the mass of the gauge boson scales as reciprocal of its coupling strength. In general, for two consecutive flavors of gauge bosons we expect

$$\frac{M_\kappa}{M_{\kappa+1}} = \left(\frac{g_{\kappa+1}}{g_\kappa}\right)^2 \tag{93}$$

where $\kappa = 1, 2, 3\ldots$. Since W carries both weak isospin and electric charges whereas Z is neutral, the first ratio in (93) corresponds to the EW sector and leads to [37, 53, 59]

$$\left(\frac{M_W}{M_Z}\right)^2 \approx \frac{1}{1+\left(\frac{e}{g_2}\right)^2} \approx 1-\frac{1}{\overline{\delta}} \tag{94}$$

Two points are worth mentioning at the end of this section, namely,

- A scaling law similar to (92) was recently linked to the infrared limit of QCD and the spectrum of hadron masses [64].
- (92) along with Tab. 1 can naturally recover Koide's mass formula

$$3(m_e + m_\mu + m_\tau) = 2(\sqrt{m_e} + \sqrt{m_\mu} + \sqrt{m_\tau})^2 \qquad (95)$$

Indeed, replacing $m_e = m_\mu \bar{\delta}^{-4}$ and $m_\mu = m_\tau \bar{\delta}^{-2}$ in (95) leads to the polynomial equation

$$(1 + \bar{\delta}^{-2} + \bar{\delta}^{-6}) = 4(\bar{\delta}^{-1} + \bar{\delta}^{-3} + \bar{\delta}^{-4}) \qquad (96)$$

(96) is solved by a Feigenbaum "delta" whose value matches closely the constant attributed to hydrodynamic flows ($\bar{\delta} \approx 3.9$).

Table 1. Actual versus predicted ratios of SM parameters

Parameter ratio	Behavior	Actual	Predicted
m_u / m_c	$\bar{\delta}^{-4}$	3.365×10^{-3}	4.323×10^{-3}
m_c / m_t	$\bar{\delta}^{-4}$	3.689×10^{-3}	4.323×10^{-3}
m_d / m_s	$\bar{\delta}^{-2}$	0.052	0.066
m_s / m_b	$\bar{\delta}^{-2}$	0.028	0.066
m_e / m_μ	$\bar{\delta}^{-4}$	4.745×10^{-3}	4.323×10^{-3}
m_μ / m_τ	$\bar{\delta}^{-2}$	0.061	0.066
M_W / M_Z	$(1-\frac{1}{\delta})^{1/2}$	0.8823	0.8623
$(\alpha_{EM} / \alpha_W)^2$	$\bar{\delta}^{-2}$	0.053	0.066
$(\alpha_{EM} / \alpha_s)^2$	$\bar{\delta}^{-4}$	4.034×10^{-3}	4.323×10^{-3}

12. Gravitation as Analog of Field Theory on Fractals

Besides its self-polarization property discussed in section 8, fractal space-time bears an intriguing resemblance to non-Euclidean geometry of classical gravitation [54]. We briefly review this connection below.

Consider the fractional analog of a free non-relativistic Hamiltonian system [54]

$$H(\alpha,\beta)=\frac{\alpha}{2(\alpha+\beta)}\eta_{\mu\nu}(p^{\mu})^{\frac{\alpha+\beta}{2}}(p^{\nu})^{\frac{\alpha+\beta}{2}} \tag{97}$$

A typical embodiment of (97) is the Hamiltonian describing the dynamics of classical free Dirac or Yang-Mills fields on fractal space-time. The case $\beta=\alpha=1$ recovers the familiar expression for kinetic energy density, namely,

$$T_{1,1}=\frac{1}{2}p^{\mu}p^{\nu} \tag{98}$$

Hamiltonian (97) may be cast in the equivalent form

$$H(\alpha,\beta)=g_{\mu\nu}(\alpha,\beta)p^{\mu}p^{\nu} \tag{99}$$

in which

$$g_{\mu\nu}(\alpha,\beta)=\frac{\alpha}{2(\alpha+\beta)}(p^{\mu})^{\frac{\alpha+\beta}{2}-1}(p^{\nu})^{\frac{\alpha+\beta}{2}-1} \tag{100}$$

The action of a minimally coupled classical field in curved space-time is defined as

$$S=\frac{1}{2}\int\sqrt{-g}\,d^4x\,g^{\mu\nu}p^{\mu}p^{\nu} \tag{101}$$

Direct comparison of (97-100) to (101) yields the straightforward identification

$$\boxed{(\sqrt{-g})g^{\mu\nu}\rightarrow 2g^{\mu\nu}(\alpha,\beta)} \tag{102}$$

Same results can be obtained by unveiling the formal analogy between Hausdorff measure and non-Euclidean metric in the action functional (101) [36]. It is instructive to point out that (102) may be interpreted as an extension in fractal space-time ($D=$ non-integer) of the AdS/CFT conjecture applied to conventional space-time ($D=$integer) [60].

13. Exotic Phases of Matter

A key hypothesis of sections 1 and 5 is that, at high energies, decoherence sets in and triggers the unavoidable transition from quantum to classical behavior of fields. There may be exceptions to this scenario provided, for example, by QGP near the transition temperature [55]. Fast thermalization occurs there since, the larger the temperature and collision frequency, the faster equilibration process is. This section explores what happens when decoherence is inhibited and fields preserve their quantum nature. We hold onto the assumption that non-perturbative effects become predominant and fractal operators are justified near full-scale invariance, that is, near $\varepsilon \approx 0$ (according to (14)).

Start from (3b) and consider for simplicity a free space-independent scalar field $\varphi(t)$ .Its classical Lagrangian in four space-time dimensions reads

$$L = \partial^{\mu}\varphi\,\partial_{\mu}\varphi - m^2\varphi^2 \tag{103}$$

and yields the following expression for the field momentum

$$\pi = \frac{\partial L}{\partial(\frac{\partial \varphi}{\partial t})} = \frac{\partial \varphi}{\partial t} \tag{104}$$

It is known that the standard technique of canonical quantization promotes a classical field theory to a quantum field theory by converting field and momentum variables into operators. To gain full physical insight with minimal complications in formalism, we work below in 0 + 1 dimensions. Ignoring the left/right labels, we define the field and momentum operators as

$$\varphi \to \varphi = \varphi \tag{105}$$

$$\pi \to \pi^{\alpha} = -i\frac{\partial^{\alpha}}{\partial|\varphi|^{\alpha}} \equiv -i\,D^{\alpha}$$

Without loss of generality, we set $m = 1$ in (103). The Hamiltonian becomes

$$H \to H^{\alpha} = -\frac{1}{2}D^{2\alpha} + \frac{1}{2}\varphi^2 = \frac{1}{2}(\pi^{2\alpha} + \varphi^2) \tag{106}$$

By analogy with the standard treatment of harmonic oscillator in quantum mechanics, it is convenient to work with the destruction and creation operators defined through

$$a^{\alpha} \,\square\, \frac{1}{\sqrt{2}}[\varphi + i\,\pi^{\alpha}] \tag{107}$$

$$a^{+\alpha} \,\square\, \frac{1}{\sqrt{2}}[\varphi - i\,\pi^{\alpha}]$$

Straightforward algebra shows that the following commutation rules are satisfied

$$[a,a]=[a^{+\alpha},a^{+\alpha}]=0 \tag{108}$$

$$[a^{+\alpha},a^{\alpha}]=i[\varphi,\pi^{\alpha}]=-\alpha\pi^{(\alpha-1)}$$

The second relation in (108) leads to

$$H^{\alpha}=a^{+\alpha}a^{\alpha}+\frac{1}{2}\alpha\pi^{(\alpha-1)} \tag{109}$$

The limit $\alpha=1$ recovers the quantum mechanics of harmonic oscillator, namely

$$H=a^{+}a+\frac{1}{2} \tag{110}$$

It was shown in [36, 56] that the fractional Hamiltonian (110) yields a continuous spectrum of states having non-integer numbers of quanta per state. These bizarre flavors of particles and antiparticles emerging as fractional objects were named "*complexons*". Similar conclusions have surfaced in a number of recent papers, where the possibility of a scale-invariant "hidden" sector of particle physics extending beyond SM has been investigated. A direct consequence of this setting is a continuous spectrum of massless fields having non-integral scaling dimensions called '*unparticles*'. The reader is directed to [56] for a brief discussion of '*unparticle*' and "*unmatter*' physics.

It can be shown that these exotic phases of matter, if they exist, display properties that substantially depart from SM [36, 56]. Aside from emerging as clusters of fractional objects, they may

- Show up as delocalized structures with long-range correlations in space-time,
- Show up as missing transverse energy in the deep inelastic regime of TeV collisions,
- Span multiple observation scales,
- Represent arbitrary mixtures of particles and antiparticles,
- Carry indefinite spin in four-dimensional space-time, at variance with the spin-statistics theorem of QFT.
- Show up as electrically neutral and ultra-weakly coupled to ordinary SM states. In particular, they may be classified as *sterile*, by analogy with the "would-be" sterile neutrinos,
- Surface as strongly-coupled fluids with anisotropic flow properties (analogous to QGP)

It would be interesting to investigate if these speculative forms of Terascale matter are in fact manifestations of non-baryonic *dark matter* [36].

Conclusion

LHC has opened up an unprecedented opportunity to detect signatures of new physics above the EW scale and shed light on some of the key unresolved issues of SM. There are many mainstream models promoting various avenues towards physics beyond SM. Most theories expand SM in ways that do not alter the basis of QFT in any fundamental way. Inspired by the growing evidence for complex behavior in non-linear physical phenomena, we have explored here an alternative direction. Our preliminary inquiry has uncovered several tentative findings:

Table 2. Tentative signatures for physics beyond SM

Anticipated behavior	Tentative evidence
Non-locality	- Long-range angular correlations in proton-proton collisions at the LHC [63]
Space-time asymmetries	- Breaking of discrete symmetries in the physics of K and B mesons [3] - Chiral symmetry violation in EW interactions and massive QCD [3] - Elliptic flow of charged QGP particles [16] - Dijet asymmetry in heavy ion collisions [15]
Matter-antimatter asymmetry	- Neutrino-antineutrino asymmetry in MiniBooNE data [14] - Predominance of baryonic matter [33] - Top quark-antiquark asymmetry [17]
Replication of parameters	- The fermion family problem of SM [2-3] - Hints for replication of EW gauge bosons [62]
Anomalous properties	- Anomalous magnetic moment of leptons [12] - Anomalous Lamb shift in muonic hydrogen [48] - The CDF muon anomaly [18] - The PAMELA positron anomaly [19]
New structures ultra-weakly coupled to SM	- Absence of heavy resonance-like structures in dijet states at the LHC [57]

- Dynamics of the Terascale region is likely to slide outside equilibrium.
- As a result, fractional dynamics becomes an attractive tool for model building near the EW sector or beyond SM.
- Fractional dynamics is tied to the underlying fractal topology of space-time or phase-space.
- Fractal topology of space-time enables perturbative renormalization of field theories in a manner similar to dimensional regularization.
- Fractal topology generates an intrinsic polarization of space-time which imparts mass to both gauge boson and fermion fields.
- Fractal topology accounts naturally for breaking of discrete space-time symmetries.
- Fractal topology is able to account, at least in principle, for anomalous gauge charges, anomalous magnetic moment of massive leptons and enhanced cross-sections.
- Onset of fractal space-time near the EW scale explains the generation structure of SM parameters.
- Classical gravity emerges as analog of field theory on fractal space-time.

- Fractal operators in field theory lead to the potential for exotic phases of matter that may be dynamically related to the composition of non-baryonic dark matter.

We conclude our report by listing a series of partially confirmed observations that may provide early evidence for fractional dynamics in high-energy physics.

The reader is urged to keep in mind that published data are strictly preliminary and subject to further revisions. The fast pace of change on the theoretical front reflects our long-standing effort for understanding Nature beyond the boundaries of current models.

References

[1] Weinberg, S. *The Quantum Theory of Fields*; Cambridge University Press: Cambridge, UK, 1995; Vol.1, pp 298-314.

[2] Burgess, C. P.; Moore G. D. The Standard Model: A Primer; Cambridge University Press: Cambridge, UK, 2007; pp 3-46.

[3] Donoghue, J. F.; Golowich, E.; Holstein, B. R. Cambridge Monographs on Particle Physics, *Nuclear Physics and Cosmology* **2**; Cambridge University Press: Cambridge, UK, 1994; pp 69-95.

[4] Nath, P. *et al.* (2010). *The Hunt for New Physics at the Large Hadron Collider*. http://arxiv.org/abs/1001.2693.

[5] Morrissey, D. E. *et al.* (2010). *Physics Searches* at the LHC. http://arxiv.org/abs/0912.3259.

[6] Rattazzi, R. (2006). *Physics Beyond the Standard Model*. http://arxiv.org/abs/hep-ph/0607058.

[7] Altarelli, G. (2010). *The Electroweak Symmetry Breaking Riddle*. http://arxiv.org/PS_cache/arxiv/pdf/1003/1003.3180v2.pdf.

[8] Altarelli, G. (2010). *LHC Physics in the Standard Model and Beyond*. http://indico.cern.ch/getFile.py/access?contribId=4andsessionId=0andresId=0andmaterialId=slidesandconfId=89312.

[9] Klaus, M. (Ed.) *Non-Equilibrium Physics at Short Time Scales: Formation of Correlations*; Springer: Berlin, 2004; Vol. XVI, pp 111-195.

[10] Van Vliet, C. M. *Equilibrium and Non-Equilibrium Statistical Mechanics*; World Scientific: Singapore, 2008; pp 531-632.

[11] Klages, R. Microscopic Chaos, Fractals and Transport in Non-Equilibrium Statistical Mechanics; Advanced Series in Nonlinear Dynamics; *World Scientific Books*; Imperial College Press: London, UK, 2007; vol. 24, pp 367-380.

[12] Stockinger, D. *J. Phys. G: Nucl. Part. Phys*. 2007, vol. 34, 45-91.

[13] Rodejohann, W. *Phys. Lett. B* 2010, vol. 684, 40-47.

[14] Aguilar-Arevalo, A. *et al.* (2010).*Observed Event Excess in the MiniBooNE Search for Muon Antineutrino to Electron Antineutrino Oscillations*. http://arxiv.org/PS_cache/arxiv/pdf/1007/1007.1150v3.pdf.

[15] Aad, G. *et al.* (The ATLAS Collaboration). (2010) *Observation of a Centrality-Dependent Dijet Asymmetry in the Lead-Lead Collisions with the ATLAS Detector at the LHC*. http://arxiv.org/PS_cache/arxiv/pdf/1011/1011.6182v1.pdf.

[16] Aamodt, K. *et al.* (The ALICE Collaboration). (2010). *Elliptic flow of charged particles in Pb-Pb collisions at sqrt(sNN)= 2.76 TeV.* http://arxiv.org/PS_cache/arxiv /pdf/1011/1011.3914v1.pdf.
[17] FERMILAB (the D0 Collaboration). (2010). *Measurement of the forward-backward production asymmetry of t and tbar quarks in ppbar-ttbar events.* http://www-d0.fnal.gov/Run2Physics/WWW/results/prelim/ TOP/T90/T90.pdf.
[18] Giromini, P. *et al.* (2008).*Phenomenological interpretation of the multi-muon events reported by the CDF collaboration.* http://arxiv.org/PS_cache/arxiv/pdf /0810/0810.5730v1.pdf.
[19] Feldman, D. *et al.* (2009). *PAMELA Positron Excess as a Signal from the Hidden Sector.* http://arxiv.org/PS_cache/arxiv/pdf/0810/ 0810.5762v3.pdf.
[20] Thompson, J. M. T.; Stewart H. B. *Nonlinear Dynamics and Chaos*; John Wiley and Sons Ltd.: Chicester, UK, 2002; pp 1-7.
[21] Strogatz, S. H. Nonlinear Dynamics and Chaos; *Studies in Nonlinearity*; Westview Press: Cambridge, MA, 2000; pp 16-19.
[22] vanWezel, J.; Zaanen, J.; van der Brink, *J. Phys. Rev. B* 2006, vol. 74, 094430.
[23] Kunihiro, T. *et al.* (2010). Chaotic behavior in classical Yang-Mills dynamics. http://arxiv.org/PS_cache/arxiv/pdf/1008/1008.1156v1.pdf.
[24] Prigogine, I. (1977). *Time, Structure and Fluctuations.* http://nobelprize.org/ nobel_prizes/chemistry/laureates/1977/prigogine-lecture.pdf.
[25] Tarasov, V. E. (2010). *Fractional Dynamics of Open Quantum Systems.* http://qfthep.sinp.msu.ru/programme/100911.
[26] Baleanu, D.; Guvenc, Z. B.; Machado, J.A.T. (Eds.). *New Trends in Nanotechnology and Fractional Calculus Applications*; Springer: Heidelberg, 2010, pp 123-181.
[27] Tarasov, V. E. *J. Phys.: Conf. Ser.* **7** 2005, 17-33.
[28] Tarasov, V. E. *Mod. Phys. Lett. A* **21** 2006, vol. 20, 1587–1600.
[29] Goldfain, E. *Comm. Nonlin. Sci. Numer. Simul.* 132008, vol. 3, 666-676.
[30] Lepage, G. P. (2005) What is Renormalization?. http://arxiv.org/abs/hep-ph/0506330.
[31] van Baal, P. (2001). A Course in Field Theory. http://www.lorentz.leidenuniv.nl /~vanbaal/FT/extract.pdf.
[32] Maggiore, M. A Modern Introduction to Quantum Field Theory; Oxford Master Series in Statistical, Computational, and Theoretical Physics; Oxford University Press: Oxford, UK, 2005; pp 135-146.
[33] Bouzas, A. O.; Cheng W. Y.; Gelmini G. *Phys. Rev. D* 1997, vol. 55, 4663-4672.
[34] Pich, A. (1998). Effective Field Theory. http://arxiv.org/abs/hep-ph/9806303
[35] Ref. [3], p.12 and p.96.
[36] Goldfain, E. Chaos, *Solitons and Fractals* 282006, vol. 4, 913-922.
[37] Goldfain, E. Electron. *J. Theor. Phys*. 24 2010, vol. 7, 219-234.
[38] Zhu, B. *et al* (2010).Same Sign WW Scattering Process as a Probe of Higgs Boson in pp collisions at sqrt(s) = 10 TeV. http://arxiv.org/PS_cache/arxiv/pdf/1010 /1010.5848v3.pdf.
[39] Petrosky, T.; Prigogine, I. *Proc. Natl. Acad. Sci. USA*. 20 1993, vol. 90, 9393–9397.
[40] Svozil, K. *J. Phys. A: Math. Gen.* **20** 1987, 3861-3875.
[41] Itzykson, C.; Zuber, J. B. *Quantum Field Theory; International Series in Pure and Applied Physics*; McGraw-Hill Inc.: New York, 1980; pp 9-10.
[42] Ref. [32], p. 247.

[43] Ref. [41], p. 569.
[44] Ref [32], p. 246.
[45] Pierantozzi, T.; Vasquez L. *J. Math. Phys.* **46**, 2005, 113512.
[46] Ref. [3], p. 8.
[47] Passera, M. *et al. Phys. Rev. D* 2008, vol. 78, 013009.
[48] Tucker-Smith, D.; Yavin, I. (2010). *Muonic hydrogen and MeV forces.* http://arxiv.org/PS_cache/arxiv/pdf/1011/1011.4922v1.pdf.
[49] Bhattacherjee, B. *et al.* (2010).*Do new data on B decays point to an early discovery of supersymmetry at the LHC?.* http://arxiv.org/PS_cache/arxiv/pdf/1012/1012.1052v1.pdf.
[50] Goldfain, E. *Intl. J. Bif. Chaos* **3** 2008, vol. 18, 891-896.
[51] Magnitskii, N. *Comm. Nonlin. Sci. Num. Simul.* **2** 2008, vol. 13, 416-433.
[52] Nakamura, K. *et al.* (Particle Data Group) *J. Phys. G* **37** 2010, 075021.
[53] Goldfain, E. *Intl. J. Nonlin. Sci.* **3** 2007, vol. 3, 170-180.
[54] Tarasov, V. E. *Phys. Rev. E* 2005, vol. 71, 011102.
[55] Das, S. K. *et al.* (2009). Equilibration in Quark Gluon Plasma. http://arxiv.org /PS_cache/arxiv/pdf/0912/0912.4089v1.pdf.
[56] Goldfain, E.; Smarandache, F. *Prog. in Phys.* 2008, vol. 4., 10-15.
[57] Aad, G. *et al.* (The ATLAS Collaboration) *Phys. Rev. Lett.* **105**, 2010, 161801.
[58] Khanna, F.; Matrasulov, D. (Eds.) *Nonlinear Dynamics and Fundamental Interactions*; Springer: Dordrecht, the Netherlands, 2006, pp 243-341.
[59] Goldfain, E. *Europhys. Lett.* **1**, 2008, vol. 82, 11001.
[60] de Boer, J. (2002). *Introduction to the AdS/CFT Correspondence.* http://www-library.desy.de/preparch/desy/proc/proc02-02/Proceedings/ pl.6/deboer_pr.pdf.
[61] Elze, H. T. *Nucl. Phys. B* **1-2**, 1995, vol. 436, 213-261.
[62] Salvioni, E. *et al.* (2009) *Non-universal minimal Z' models: present bounds and early LHC reach.* http://arxiv.org/PS_cache/arxiv/pdf/0911/0911.1450v1.pdf.
[63] CERN (the CMS collaboration) (2010). *Observation of long-range near- side angular correlations in proton-proton collisions at the LHC.* http://cms.web.cern.ch/cms /News/2010/QCD-10-002/QCD-10-002.pdf.
[64] Goldfain, E. Electron. *J. Theor. Phys.* **23** 2010, vol. 7, 75-84.
[65] Tsallis, C. *Introduction to Nonextensive Statistical Mechanics: Approaching a Complex World; Complexity Series*; Springer: New York, 2009; pp 221-295.
[66] Lebon G.; Jou D.; Vasquez J. C. *Understanding Non-equilibrium Thermodynamics. Foundations*, Applications, Frontiers; Springer-Verlag: Berlin, 2008, pp 43-54.

In: New Developments in the Standard Model
Editor: Ryan J. Larsen

ISBN: 978-1-61209-989-7

Chapter 3

UNEXPLORED REGIONS IN QFT AND THE CONCEPTUAL FOUNDATIONS OF THE STANDARD MODEL

Bert Schroer
CBPF, Rua Dr. Xavier Sigaud
Rio de Janeiro, Brazil
and Institut fuer Theoretische Physik der FU Berlin, Germany

Abstract

Massive quantum matter of prescribed spin permits infinitel many possibilities of covariantization in terms of spinorial (undotted/dotted) pointlike fields whereas massless finit helicity representations lead to large gap in this spinorial spectrum which for s=1 includes vectorpotentials. Since the nonexistence of such pointlike generators is the result of a deep structural clash between modular localization and the Hilbert space setting of QT, there are two ways out: gauge theory which sacrifie the Hilbert space and keeps the pointlike formalism and the use of stringlike potentials which allows to preserve the Hilbert space. The latter setting contains also string-localized charge-carrying operators whereas the gauge theoretic formulation is limited to pointlike generated observables.

This description also gives a much better insight into the Higgs mechanism which leads to a revival of the more physical "Schwinger-Higgs" screening idea.

The new formalism is not limited to m=0, s=1, it leads to renormalizable interactions in the sense of power-counting for all s in massless representations.

The existence of stringlike vectorpotentials is preempted by the Aharonov-Bohm effect in QFT; it is well-known that the use of pointlike vectorpotentials in Stokes theorem would with lead to wrong results. Their use in Maxwell's equations is known to lead to zero Maxwell charge. The role of string-localization in the problem behind the observed invisibility and confinemen of gluons and quarks leads to new questions and problems.

PACS: 11.10.-z, 11.15-q, 11.10Gh, 12.20.-m, 12.38.-t

1. Introductory Remarks

Particle theory of the past century has led to a vast body of knowledge, but many theoretical problems about the conceptual foundations of these discoveries remained unresolved. Ideas as quark/gluon confinemen have not been understood in terms of quantum fiel theoretical interactions, and even less ambitious problems as why and how the QED description of electrically charged particles requires a noncompact localization whereas the Schwinger-Higgs screening counterpart of scalar quantum electrodynamics (QED) is pointlike-generated, still lacks good understanding.

To get a feeling for the dimension of the problem, compare this situation to the post Maxwell pre Einstein era of classical electrodynamics. Already 20 years after its discovery, *everything*, except the problem of the ether, was in place; and when Einstein removed the ether from its throne, Maxwell's equations and its covariance transformations, the Lorentz transformation did not have to be modifie since the relativity interpretation was primarily a conceptual step abstracted from the structure of the Maxwell equations. For the post relativity particle theory however this step was essential.

When one says that quantum fiel theory (QFT), in particular the standard model (SM), explain an unprecedented amount of data, one thinks of the power of prediction (sometimes" post-diction") which results by combining calculations based on perturbative renormalization theory, in some cases combined with phenomenologically motivated assumption. This is indeed impressive, but should not create the illusion that QED, not to mention the complete SM, after half a century has arrived at the conceptual maturity which Maxwell's classical theory attained already after only two decades. To say that quantum theories are inherently conceptually more opaque is not convincing since QM in its epistemological standing does not lag behind classical fiel theory. Rather we should admit that, although we have developed a detailed vocabulary within an impressive set of results obtained by a variety of computational tools, the main impact has been to impress ourselves and become complacent. From a comprehensive understanding of all the physical concepts of e.g. QED we are presently further away than the discoverers of renormalization theory considered themselves to be when they discovered the impressive renormalized perturbative formulation of QED after world war II. Being aware about this state of affairs even with respect to our oldest and best studied QFT, it is less of a surprise that the central problems behind what we were rather quick to call gluon/quark confinemen meanwhile persevered for almost half a century without any essential progress.

The present work is an attempt to bring a bit of movement into this stagnating theoretical situation. The starting point is the recognition that higher spin $s \geq 1$ representations exhibit a certain clash between localization and the Hilbert space setting of quantum theory. More precisely it is not a clash between the $(m = 0, s \geq 1)$ representations as such, but an incompatibility with the existence of pointlike generators of those representations with certain prescribed covariance behavior as for s=1 the important vectorpotentials associated with fiel strengths. In such a situation there are two ways out: cede on the side of Hilbert space or cede on the side of pointlike localization.

The firs path leads to the gauge-theoretic formulation and is inexorably related to indefinite metric spaces instead of Hilbert spaces, whereas the second setting deals with *string-localized* covariant potentials in a Hilbert space which interact with matter field and transfer their noncompact localization to the charged matter; in this case the setting never leaves the quantum realm of Hilbert spaces. The stringlike localization of matter field (Maxwell charges, Yang-Mills charges) belong to what one could call the *non-local gauge-invariants*. Our approach shares some of the motivations with an old gauge-free but path-dependent attempt to formulate QED by Mandelstam [1]. It has the advantage of incorporating objects as e.g. electrically charge-carrying operators into the perturbative setting which have been known to have no pointlike generation in a Hilbert space (ghost-free) setting. In contrast to Mandelstam's time-like "path-localization"[1] our idea requires the presence of semiinfinit spacelike strings as the optimally localized generating vectorpotentials (the "core" of the spacelike-cone localized operators in [2]); they are "irreducibly stringlike" i.e. they cannot be obtained from integrating over pointlike objects.

But these intrinsically noncompact localized operators cannot be found among those constructed in the gauge theoretic formulation since they cannot be constructed from integrating over pointlike gauge-invariant operators or by putting together infinitel many compact localized gauge invariant operators. Hence there exists a physically important aspect (the description of electrically charged matter) which is not part of the iterative gauge formulation of QED. Since there is no classical counterpart of this strong delocalization of charge a Lagrangian quantization approach is not the appropriate setting for solving this problem. Neither does it have a counterpart in the QM with a Coulomb interaction between particles. Rather one has to return to the foundational aspects of causal localization in quantum theory (QT) and understand the conceptual origin of this clash and try to fin an appropriate perturbative description. This is the main motivation underlying the present work.

The use of non-pointlike free field which have no Lagrangians[2] characterization places the localization property on which QFT is founded back into the center stage which is presently dominated by the Lagrangian/functional quantization parallelism to classical physics. Of course this is not a one to one correspondence but rather an artful game which only relates a subset of classical theories to a subset of QFTs of which a famous vernacular (by Edward Nelson) says: quantization is an art, but second quantization is a functor (see section 4) . Actually in QM it is almost a one-sided correspondence in the sense that anything which can be formulated in the setting of canonical mechanics has a (at least one) counterpart in QM. This is not any more true in relativistic domain including relativistic classical fiel theory. For instance the classical square root Lagrangian of $ds^2 = g_{\mu\nu}\dot{x}_\mu(\tau)\dot{x}^\mu(\tau)$ whose Euler-Lagrange equations describe a covariant classical particle trajectory (with τ related to the propertime) has no covariant quantum counterpart (contrary to what Polchinski writes in his book [3]) and more generally any relativistic theory in which a would be operator $X_\mu(\tau)$ appears, as in string theory, cannot be related to stringlike spacetime localization; in each case such ideas are related to a incorrect understanding what causal localized QT is about. In fact the main reason behind Wigner's

[1]The quotation mark is supposed to indicate that localization in its QFT meaning refers to the spacelike. A picture of timelike particle orbits can only be maintained in the sence of "effective" (center of mass motions) localization in wave-packets.

[2]Even among the infinitel many spinorial pointlike free field which result from covariantizing the unique (m,s) Wigner particle representation most of them are not "Euler-Lagrangian fields and have a too high short distance dimension which makes them useless in renormalizable perturbations.

1939 representation theoretical characterization of particles was precisely the realization that this cannot be done by quantization of classical particle degrees of freedom. He missed the (recent) modular localization of states (which would have led him straight into QFT via mapping the causal localization of states into that of operators, see section 4) and constructed instead the noncovariant (frame dependent) quantum mechanical Newton-Wigner localization which is nothing else than the adapted Born localization associated with the position operator. The latter is shared between QM and QFT and misunderstandings on these issues carry the danger to overlook the enormous conceptual-mathematical difference between these two localization concepts of which only the modular localization is genuinely intrinsic.

Elsewhere it was shown how the identificatio of QFT with the idea of "just a kind of relativistic QM" has led particle physics into its firs serious and still unfinishe crisis [4] [5]. The topic of the present paper is more upbeat: the modular localization setting will be used to shed new light on remaining problems of gauge theory whose solution have a good chance to take particle theory out of its apparent present impasse and its less than perfect accounting for some of the experimental finding at the LHC. It is hard to imagine that there can be further conceptual progress on SM without solving some of the conceptual problems presented in this paper.

More concrete: it will be shown that modular localization resolves the clash between pointlike localization and Hilbert space structure which affects vector/tensor potentials of zero mass $(m = 0, s \geq 0)$ Wigner representations and the resulting covariant semiinfinit string-localized potentials in a QED-like interactions delocalize charged particles without affecting the pointlike localization of fiel strength. It turns out that the stringlike vectorpotentials themselves even after having interacted stay very close to pointlike localization; under a simple linear operation of differentiation they go over into the pointlike fiel strengths. The delocalization of charge-carrying matter field is however much more serious; there is no linear operation which does this; only by passing to charge neutral composites one returns to pointlike objects.

Since the historical roots of these observations are not so well known anymore (even among the older generation), the next section will review some within modern hindsight. Often problems, which at the time were too difficul to be solved in the realistic context, are analyzed by looking at their analog in the "theoretical laboratory" of two-dimensional QFTs. For the problem at hand this will be reviewed in section 3. The following section 4 contains the representation theoretical setting which together with modular localization leads to the interaction-free string-localized (s.l.) covariant potentials. Section 5 shows that already in the absence of interactions the use of pointlike indefinit metric vectorpotentials is a risky business since in contrast to the Hilbert space compatible s.l. potentials they may lead to incorrect results of which the absence of the well established quantum Aharonov-Bohm effect (the violation of Haag duality for toroidal regions) is an illustration.

In section 6 the main issue is to give perturbative arguments about the transfer of delocalization from potentials to charged field which thereby loose their pointlike innocence. This section also presents the Higgs mechanism in a different more physical setting of Schwinger-Higgs screening which leads to the re-localization of the nonlocal aspects of Maxwell charges in analogy to Debeye's screening in QM which converts the Coulomb potential into an effective short range interaction of the Yukawa kind always with the emphasis on the fact the its particle spectrum changing effect in QFT is much more radical than the appearance of effective short range forces. In section 7 some new naturally more speculative points of view about confinement/i visibility of states in connection with nonabelian gauge theories are mentioned before reviewing some of the high points and presenting concluding remarks.

2. Charged Fields and the Problem of Infraparticles in a Historic Perspective

The gradual increase in the understanding of the local quantum physical aspects of electrically charged field and their associated particles is one of the most fascinating projects of QFT which, even after its resounding success (dating back more than 70 years) in describing observational data with unbelievable accuracy, is not anywhere near its closure. It leads to a particle-fiel relation which is far more subtle than the textbook case of a QFT with an energy-momentum spectrum with mass gaps for which the validity of the LSZ scattering theory leading to the reduction formula of scattering amplitudes in terms of mass shell restrictions of time-ordered correlation functions. In the latter case one obtains in a well-known manner (via large time asymptotes) free fields together with the concomitant assumption of asymptotic completeness[3] one find that the Hilbert space of such a theory has the form of a Wigner-Fock space, a significan simplificatio which does not hold in QED and more generally in all models with zero mass vectorpotentials.

The appearance of zero mass particles as such do not necessarily cause a breakdown of these results, as long as the particle states in the presence of interactions manage to survive as particle state i.e. are "not sucked into the continuum" of the mass spectrum (example zero mass pions on massive nucleons) the results of scattering theory continue to be valid even though there are no mass gaps.

But interactions of charged particles with photons are not covered by this kind of scattering theory. Already the quantum mechanical Coulomb scattering leads to problems with the large time behavior of amplitudes, although in this case the multiparticle tensor structure of the Hilbert space remains unaffected and the large time contribution converges after removing a logarithmic time dependent phase factor [18]. In QED the failure of time-dependent scattering theory is more severe, in fact it is inexorably related to a breakdown of Wigner particle states and the absence of a Wigner-Fock structure of the QED Hilbert space. Instead of the mass shell contribution in the Källen Lehmann two-point function, one find a cut which is necessarily less singular and leads to a vanishing LSZ limit, whereas in perturbation theory the restriction to the mass shell causes infrared divergencies which, unlike the Coulomb scattering in QM, cannot be repaired on the level of scattering amplitudes but only by passing to infinit soft photon-inclusive cross sections.

It is the principle aim of this paper to establish that these unusual momentum space properties are caused by the string-localization of vector potentials which figurat vely speaking feed their string localization to the charged fields Whereas for the potential the

[3] All presently known constructions are based on the two-dimensional "bootstrap-formfactor setting" (factorizing models). For those constructions which start from particles and introduce field through their formfactors [6], the completeness property is built into the construction.

effect of string localization remains mild since although the vectorpotentials are needed in the perturbative evaluation of the interaction, the Hilbert space and the operator content can be fully described in term of fiel strengths together with charge-carrying matter fields the s.ll localization of the letter is genuine and cannot be converted into pointlike behavior by any linear or nonlinear operation. This weakening of localization of quantum Maxwell charges is behind the infraparticle concept and is the culprit for a radical change of particle and scattering aspects. Before we get there we will follow briefl the historical path.

Problems with the application of standard scattering theory to QED were noticed quite early since scattering theory in one form or other is one of the oldest tools which relate the dynamical concepts of QT to experimental observations [4]; the research on infrared divergencies begun in a 1934 paper by Bloch and Nordsiek in which scattering of charged particles and photons was analyzed in a simplifie model [7]. The important conclusions, namely that although the number of photons in the infrared is infinite the emitted energy and angular momentum remains finite as well as the message that the summation over all infrared photons up to a certain resolution determined by the measuring device is required in order to obtain a non-vanishing scattering probability (inclusive cross section), can already be found in this early paper. Perturbative calculations in QED in a new covariant renormalization setting 20 years later succeeded to confir and improve these results [8].

If QFT would be limited to the findin of successful recipes, then this formalism, which proceeds "as if" charged particles would be Wigner particles (and the LSZ asymptotic behavior would be valid modulo some imperfections which can be repaired by compensating infrared divergencies on the level of soft photon inclusive cross sections) then this formulation via recipes would have been the end of the story. Fortunately theoretical physics was never pragmatic in this extreme sense since most physicists realized that suppressed conceptual and philosophical questions always return sooner or later with vengeance.

In case of charged field in QED the left aside conceptual problem was to understand why the LSZ scattering theory, whose derivation depends on the *locality principle* and some mild assumptions about the energy momentum spectrum, did not work for QED. There is no cheap general answer by merely referring to the presence of zero mass particles. Models as the mentioned meson-nucleon interactions with zero mass mesons, in which the discrete mass nucleon state continues to exist in the presence of interactions, pose no problem with the LSZ scattering theory, so zero mass by its own is not the full explanation.

In this paper it will be shown that the infrared divergences are the *result of the presence of string-localized vectorpotentials* in the zero order interaction density. Their most important role is to *delocalize charge fields* with which they interact and as a result to lead to the phenomenon of the charged "infraparticles" which results in a continuous mass spread accumulated at the lower edge of a continuum. The effect on the string-localized vectorpotentials on themselves on the other hand remains very mild, even in the presence of interaction the associated fiel strength stay pointlike and the scattering theory for photons only remains close to the conventional setting [2]. This behavior occurs for all zero mass, $s \geq 1$ potentials but not for $s \leq 1/2$ (which includes the aforementioned zero mass meson to nucleon coupling). The reason behind the appearance of infraparticles is the

[4] In fact the Born probability concept was firs introduced in the setting of Born's scattering approximation; onl< afterwards Pauli started to use it to relate the Schrödinger wave function to localization probability associated to the spectral decomposition of the position operator.

semiinfinit string localization of the vectorpotentials which the interaction transfers to the charged particles and a resulting dissolution of the mass shell of the charged particle into the photon continuum is the momentum space consequence. In the presence of selfinte - actions between vectorpotentials (e.g. Yang-Mills) there are no linearly related pointlike fiel strengths and the rules for change of the string localization becomes more complicated (interaction-dependent). In order not to be misunderstood on this subtle point, string-localized covariant free field can also be define for massive representations with pointlike generators. As for massless s.l. vectorpotentials, there are also massive s.l. vectorpotentials and in both cases the string-localization has the expected scale dimension reducing effect $dimA_\mu = 1$ instead of 2 for p.l. massive A_μ; the difference is that the Wigner Hilbert space representation for $(m > 0,\ s = 1)$ does not need this s.l. vectorpotential since there exists a $dimA_\mu = 2$ vectorpotential where in the m=0 case this is not the case (the clash between p.l. and the Hilbert space positivity).

According to the previous remarks the two-point function of a charged fiel [5] in interaction should reveal a behavior at the mass shell which is different from a delta function. Unitarity imposes a strong restriction on the two-point function. For example derivatives of delta functions are excluded, since they are not positive measures; a moment of thought suggests that near $p^2 \approx m^2$ the Kallen-Lehmann function should have an anomalous (coupling-dependent) power; in fact in d=1+1 infraparticle models (next section) it behaves as $\theta(p^2 - m^2)\left(p^2 - m^2\right)^{f(\alpha)} g(p^2, \alpha)$ which for vanishing coupling $\alpha \rightarrow 0$ approaches the mass shell delta function and for finit $\alpha \neq 0$, for reasons of representing a positive measure, is milder than the delta function. In d=1+3 the power law into which the delta function is dissolved is more complicated in that the $p^2 - m^2$, as a consequence of the spontaneous breakdown of the Lorentz group [2], should have a directional dependence on the string direction e; even the exponent may have a directional dependence in addition to the dependence on the feinstructure constant α. But since the perturbative resummations to extract this behavior are quite difficult the normal practitioner hardly gets into these problems. This state of affairs (a milder mass shell singularity than a delta function) would immediately lead to a vanishing LSZ limit, which accounts for the observation by Bloch and Nordsiek about the vanishing of photon-less (or finit number of photons) charged particle scattering.

Since QED is hard to control outside perturbation theory and even poses unsolved problems concerning the perturbative description of (off-shell) physically charged fields physicists began in the 60s to look at simple soluble two-dimensional *infraparticle models;* which exhibited the expected coupling-dependent power behavior similar to that which appeared in the scattering formulas of YFS [8] after summing over leading logarithms in the infrared photon cutoff. A common feature of those two-dimensional infraparticle models is the presence of an exponential zero mass two-dimensional Bose fiel factor in the generating fiel [9]. What was still missing was a structural argument that electrically charged particles are really infraparticles in this spectral sense, as well as a conceptual basis for a new scattering theory which does away with cutoff tricks and, as the standard LSZ or Haag-Ruelle scattering theory, only uses spacetime properties of correlation functions.

---

[5] With charged fiel we always mean the physical charged fiel in a Hilbert space. The Lagrangian matter fiel ψ in the indefint metric setting does neither carry a Maywell charge nor has its localization any physical significanc [23] [22].

In QFT the overriding principle is causal localization and, thinking of the relation between energy positivity and localization (sections 3,4), it is in a certain sense the only one. This principle has a variety of different physical manifestations and the main problem of understanding a model of QFT consists in *finding the correct structural arguments which reveals the connection between the perceived properties of a model and their explanation in terms of causal localization*.

The role of localization can be exemplifie in two physically very important cases (section 5), the string-localization of electric charges and the Schwinger-Higgs charge screening as a kind of "re-localization" process, leading to a vanishing charge (a loss of the charge superselection) causing a disappearance of the charge symmetry and a copious production of neutral particles associated with the real (Higgs) fiel which replaces the complex charged field

In the case of the before mentioned two-dimensional infraparticle models this consists in the realization that the complex zero mass Bose fiel is really a semiinfinit string-like localized fiel (next section) and it is this noncompact localization behavior which is at the root of the dissolution of the mass shell delta function into a cut type singularity. Its screened counterpart is the d=1+1 Schwinger model in which the ψ-fiel together with the massless potential of the chiral current are converted into an exponential of a massive free fiel [10]. Computations may be easier in momentum space, but the deeper conceptual insight is always in the spacetime setting.

Suggestions that electrical charge-carrying field as the interacting electron-positron fiel have necessarily a noncompact extension entered the discussion quite early; practically at the same time of the Bloch-Nordsiek work Pascual Jordan [11] started to use the string-like formal presentation of the physical charge fiel which, since this also was on Dirac's mind and is also often linked with the later work by Mandelstam, will be referred to as the DJM presentation of a charged field

$$\Psi(x;e) = \text{"}\psi(x)e^{\int_0^\infty ie_{el}A^\mu(x+\lambda e)e_\mu d\lambda}\text{"} \tag{1}$$

$$\Phi(x,y;e) = \text{"}\psi(x)e^{\int_0^1 ie_{el}A^\mu(x+\lambda(x-y))(x-y)_\mu d\lambda}\bar{\psi}(y)\text{"} \tag{2}$$

Gauge invariance not only suggested that physical[6] electrically charged field have a noncompact localization, with the semiinfinit spacelike string (1) being the tightest (least spread) possibility, but also that overall charge-neutral opposite charged pairs are necessarily interconnected by a "gauge bridge" (2). The methods of local quantum physics permit to show the infinit extension of charges on the basis of a rigorously formulated quantum Gauss law (next section), but the above formula has no conceptual or computational preferential status and is not part of renormalization theory; in fact *a physical (Maxwell) charged field nor a gauge-bridges pair is not an object within the standard (Gupta-Bleuler, BRST perturbative formalism* [23][22].

The rigorous definitio of these charged field in renormalized perturbation theory is a nontrivial problem (the reason for the quotation marks in (1)). Steinmann had to develop a separate perturbative formalism only for definin the renormalized physical DJM

[6]The pointlike formal matter field which enter the Lagrangian and fiel equation of gauge theories are auxiliary quantities which act neither in a Hilbert space nor does their localization have a physical significance

charged field [12]. The gauge setting leads to a nice picture suggesting the semiinfinit string-localization of charged fields but, just as the abstract argument based on the quantum Gauss law, it does not really explain the origin of this weaker localization in terms of the form of the interaction. In the gauge setting the latter is pointlike i.e. it looks like any other local interaction and although the BRST invariance does not permit to treat (off-shell) physically charged fields this very fact at least shows that objects with a nontrivial Maxwell charge are not belonging to the compact localizable operators generated by smearing pointlike charged fields This shifts the problem of physical localization of charges outside the standard (Gupta-Bleuler, BRST) setting of gauge theory and hence reduces the operators in the range of the quantization parallelism to classical Maxwell theory to subobjects leaving spacetime localized charge-carrying operators outside.

The perturbation theory of QED in the BRST setting follows standard rules, but the calculation of BRST invariant correlations is prohibitively difficul since the BRST formalism resembles that of a nonlinear acting symmetry. In fact most of the papers present the formalism but fall short of calculation BRST invariant correlations of charged objects. Needless to add that for Yang-Mills interactions even the firs step is beset by inexorably intertwined ultraviolet-infrared divergencies which impede the execution of the renormalization program in any covariant gauge. Ultraviolet divergencies without intermingled infrared problems can be renormalized which expresses the fact that in a more intrinsic approach which avoids the quantum mechanical aspects of the standard (cutoff, regularization) methods in terms of a more intrinsic (Epstein-Glaser) approach (which takes better care of the intrinsically singular structure of quantum field resulting from vacuum polarization). Infrared divergencies on the other hand hide physical localization problems which result from a clash of quantum theory with point localization which exists for all theories in which massless higher spin $s \geq 1$ potentials participate in the interaction. Their control is not a matter of some mathematical-technical adjustments but rather requires a major revision of the physical setting.

The fact that the fundamental electrically charged field cannot have a better localization as semiinfinit stringlike, has of course (at least implicitly) been known since the DJM formula (1), but the infrared problem showing up in scattering and the problem of semiinfinite string-localized charged field have not been linked together for a long time . It is one of the aims of this paper to show that they represent two sides of the same coin.

The standard gauge formalism is, apart from some conceptual problems[7], very efficien in setting up a perturbative formalism for *local* gauge invariants. As mentioned one of the reasons why this formalism is not suited to formulate the problems behind the infrared divergencies is that it does not give any clue how to deal with electrically charged operators; they are just not part of any existing perturbative formalism. Attempts to attribute physical significanc to the Dirac spinors in the indefinit metric formalism have ended in failure; there is simply no subspace on which the Maxwell equations can be define and on which states with a nontrivial electric charge can be introduced, the pointlike $\psi's$ carry no Maxwell charge and their pointlike localization is a fake [23][22]. In section 5 it will be shown that their use in the QFT Aharonov-Bohm effect leads to the incorrect conclusion of its absence.

If the underlying philosophy of local quantum physics (LQP), which led to the construc-

[7]Different from non-gauge QFT the Hilbert space for the local observables is not define at the outset but rather results from an auxiliary indefinit Hilbert space through a cohomological construction.

tion of charge superselection sectors solely from the structural data of observables, would also apply to Maxwellian charges, one should be able to reconstruct the charge neutral bilocals with gauge bridges (2) from the algebra of charge-neutral pointlike local bilinears. For globally charged field (in which case there are no connecting gauge lines) this has been shown [13] in a model, but for Maxwellian charges there are yet no mathematically decisive result [14]. But gauge bridges are much more removed from what can be reached on the collocational radar screen of standard gauge theory than string-like potentials. There is a yet unpublished somewhat radical approach by Buchholz, Doplicher and Roberts in which the Minkowski spacetime is replaced by the forward light cone (our true history) within which there is no infrared problem but it is a general experience that the use of radical proposals is severely limited by their lack of producing concrete experimentally verifiabl formulas.

We propose a new approach for theories in which string-localized potentials associated to zero mass finit spin representations interact with massive quantum matter. This includes in particular models of gauge theories. The new setting is designed to incorporate the physical charged field into the perturbative formalism. Figuratively speaking, the potentials which have a rather harmless string dependence pass the string-localization onto the massive fiel where it keeps piling up in perturbation theory and (in contrast to the potentials themselves) becomes *irremovable by any linear operation* already in the lowest nontrivial order (section 5).

The starting point is a combination of Wigner's representation theory for $(m = 0, s \geq 1)$ with modular localization. The result is that, whereas the various possible "fiel strengths" are pointlike covariant wave functions (or pointlike quantum field in the functorially associated "second quantization"), their "potentials" are semiinfinit stringlike localized objects. The stringlike potential setting solves also another problem. For $s \geq 1$ the use of fiel strength does not allow interactions which are renormalizable in the sense of power-counting, but this interdiction does not hold for couplings with stringlike potentials; for any spin there always exist polynomial interactions of maximal degree 4 in terms of stringlike potentials of short distance scaling dimension $d_{sc} = 1$ which are renormalizable by power-counting. Whether desired interactions (as the Einstein-Hilbert action for s=2) are among those is a separate question.

Textbooks and review articles on gauge theories in general (and on QED in particular) often create the impression that they represent the best understood QFTs. It is certainly true that they are the physically most important models and they lead to rich perturbative calculations. It is also true that in the semiclassical form of quantum theory in external gauge field they led to deep mathematics (fibr bundles, realizations of zero modes related to the Atiyah-Singer index theory) and broadened the level of mathematical knowledge of several generations of theoretical physicists. But those geometrical structure, contrary to a widespread opinion[8], are not the kind of structures which are important for the solution of the above local quantum problems. Nevertheless both the (semi)classical and the local quantum physical aspects of these models are both very rich in their own right. They are certainly the most interesting theories, especially in the setting which we are going to present.

As mentioned, the models which are conceptional well-understood are those which have

[8]They did however raise the level of mathematical sophistication of particle physicists.

a mass gap and hence fall into the range of applicability of the LSZ/Haag-Ruelle scattering theory; but after the acceptance of the important standard model based on gauge theories, those models have been of a lesser observational and conceptional interest.

There are also theories which, at least at the outset, have a somewhat more hidden mass gap than those which served as illustrations of the LSZ formalism. These are the models which owe their mass to the *Schwinger-Higgs screening mechanism* and play an important role in the standard model. They involve massive vectormesons, but more generally all models which rely on interactions with higher ($s \geq 1$) spin objects are interesting because of the challenge they pose to renormalization theory, as it will be shown in section 3.

Schwinger's contributions to the screening idea has apparently been forgotten, therefore some historical remarks are in order. At the end of the 60s he envisaged the possibility of a "screened phase" in actual (spinor) QED in which the photon becomes a massive vectormeson. Since he could not fin any convincing argument, he looked at a soluable d=1+1 massless QED (the Schwinger model) where this screening idea can be exemplifie [15]; the model led to a vanishing (totally screened) charge showing also that the screening mechanism is not related to the Goldstone spontaneous symmetry breaking since the latter has no realization in d=1+1. The screened version of scalar QED in d=1+3 is identical to the model proposed by Higgs [16]. A screening mechanism for s=1/2 quantum matter in d=1+3 does, contrary to Schwinger's hope, not exist, at least not in perturbation theory. Therefore one always needs to start with scalar QED or with couplings of Yang-Mills field to complex scalar field in order to obtain massive vectormesons interacting with Schwinger-Higgs screened (real) scalars. If other field as s=1/2 Dirac field are coupled to massive vectormesons obtained from screening, the only charge which survives is the global charge carried by Dirac spinors whereas the Maxwell charge and its nonabelian counterpart are removed (screened).

We will limit our main attention to two models namely scalar QED with its delocalized (string-localized) charged fields and associated infraparticles, and the Higgs model, which in some sense to be made precise may be viewed as the result of "charge screening" leading to mass gaps and a return to pointlike locality. With the exception of this "mass-generating" Schwinger-Higgs screening mechanism, all gauge theories contain strongly delocalized objects, namely visible Maxwell type charges or invisible gluons/quarks.

In the following it will be shown that the infrared divergence properties and their firs cure in the famous Bloch-Nordsiek [7] scattering model and Jordan's stringlike (1) formula which represents a physical electrically charged field are really two different sides of the same coin; they are both consequences of the fact that certain quantum objects by their intrinsic nature do not permit pointlike generators but rather possess only noncompact localized generators. As the family of arbitrarily small double cone[9] localization (the natural shape of a compact causally closed region) is pointlike generated, and the tightest causally complete noncompact localization, namely a thin a spacelike cone, has as its core a semiinfinit spacelike semiinfinit string.

It may be more than a curiosity that both observation, the one on infrared divergencies in scattering of charged particles, and the stringlike DJM formula were made at practically the same time. Couldn't it be that there is an aspect of the subconscious in the Zeitgeist?

[9]A spacetime double cone is the unique kind of spacetime region which is compact, simply connected and causally complete.

Hard to substantiate because Jordan used the DJM formula mainly for deriving an algebraic magnetic monopole quantization in the same year that Dirac presented his geometric derivation. Historical speculations aside, the understanding of why these two observations belong together is more recent and constitutes a strong motivation for the present work.

The content is organized as follows. In the next section we review the two-dimensional infraparticle models and show that all of them are string-localized (half-space-localized). The section also contains the known rigorous statements about higher dimensional infraparticles.

Section 3 presents the theory of string-localized potential field starting from Wigner's representation theory and illustrated in more detail for s=1, 2.

The last section presents some rudiments of a new formalism, in which instead of enforcing pointlike potentials and paying the prize of being temporarily thrown outside quantum physics (by the occurrence of indefinit metric), one rather sticks to a setting in which one confronts the true stringlike localization instead of working with a simpler and familiar pointlike formalism which, apart from local BRST invariants does not describe the true localization of some of the most important physical objects a charged fields

3. Two-Dimensional Infraparticle and Their Fields

The Bloch-Nordsiek treatment of the infrared aspects of scattering of electrically charged particles and its extension to full perturbative QED in the work of Yennie, Frautschi and Suura [8] led to the notion of infrared finit *inclusive cross sections* in which infinitel many "soft" photons are summed over. This successful recipe did however not answer certain important conceptual questions. The quantum mechanical treatment of Coulomb scattering also leads to an infrared problem, namely a scattering theory of charged particles in which for large times a logarithmic time-dependent phase factor prevents the asymptotic convergence for large times [18] and where the remedy is either the removal of this factor from the amplitudes or passing directly from amplitudes to probabilities. In this case one knows in spacetime dependent terms what one is doing, however a manipulation in momentum space as in YFS is a recipe and only gains a conceptual status one find a spacetime explanation for what one is doing.

So the question arose: is the QFT scattering of charged particles similar to the quantum mechanical Coulomb scattering in which one particle states and their n-particle tensorproducts continue to exist in the Hilbert space of the theory and only the large time asymptotic convergence towards n-particle states is modifie by infrared factors, or is there something more dramatically happening in the QFT scattering of charged particles?

The fiel theoretic phenomenon of vacuum polarization in the presence of interactions leads to the mutual coupling of all channels as long as they are not separated by superselection rules. Since relativistic scattering theory treats one-particle states and multiparticle states on the same footing, one would expect that, different from the Coulomb scattering, radical changes in the scattering concepts can not happen without a major modificatio of the particle concept, so that even a charged one-infraparticle state cannot be described as a kinematical object in terms of Wigner's irreducible representation theory. Since, as most QFTs in d=3+1 which have no spectral gaps, QED is still outside mathematical-conceptual control, it has been useful to look for analogs in the more accessible two-dimensional "theoretical laboratory". QED shares with all other renormalizable theories in any dimension the divergence of the perturbative series so that great care has to be applied in drawing structural conclusions.

The exponentiation of leading logs to a power behavior in the YFS work suggested to consider models which contain a zero mass exponential Bose fiel φ[10], which in d=1+1 has formal dimension zero and therefore leads (still formally) to logarithmic correlation functions of exponential operators with an anomalous operator dimension

$$\dim : e^{ig\varphi(x)} : \simeq g^2 \tag{3}$$

$$\mathcal{L}_{int} = g\partial_\mu\varphi\bar{\psi}\gamma^\mu\psi,\ \psi = \psi_0(x) : e^{ig\varphi(x)} : \tag{4}$$

where the double dots, as usually, denote Wick-ordering. Historically the firs Lagrangian model used as an "infrared lab" [9] was a zero mass scalar meson coupled to a massive nucleon via a derivative coupling of strength (4) but actually this exponential should be called the Jordan model [11] since Jordan invented it at the same time of the Bloch-Nordsiek papers and his line integral presentation of gauge invariant charged field (1) though for a very different purpose[11]. The resulting "interacting" Dirac spinor (4) leads to a Kallen-Lehmann representation in which, instead of the one-particle mass shell delta function, one encounters a cut which is starting at the position of the mass and for $g \to 0$ converges in the sense of distributions to the mass shell delta function. The strength of this cut (the power) is bounded by unitarity since the latter forces the K-L weight to be a measure. The free Dirac fiel ψ_0 is a formal auxiliary object since in the autonomous Hilbert space constructed via Wightman's adaptation of the GNS reconstruction from the vacuum expectation values one obtains a reducible unitary representation of the Poincare group but without a spectral weight that supports a one-particle state with well-define mass.

Consistent with this structure is the observation that the LSZ large time asymptote vanishes i.e. instead of the standard incoming/outgoing free field one obtains zero. This is so because the infraparticle singularity is too weak in order to match the dissipative behavior of particle wave function; in perturbation theory one encounters however the typical well-known logarithmic infrared divergencies. The nonperturbative model of Bloch and Nordsiek, as well a the summing up of leading terms in the YFS work, lead to vanishing emission amplitudes for the emission of a finit number of photons, which as mentioned previously is in agreement with the vanishing of the LSZ limit as a result of the softening of the mass-shell singularity mentioned before.

The fiel φ (4) has infrared properties which prevent it from being a Wightman field since it cannot be smeared with all Schwartz test functions but only with those whose total integral vanishes. However the exponential is again a bona fid Wightman fiel if one imposes on the Wick contraction rules the charge superselection rule i.e. if one assigns to

[10]Interestingly enough, this was also a model used by Jordan though not for the present purpose but rather for developing his ill-fated "Neutrino theory of light".

[11]At that time physicist thought that, as in QM, one can study particle phenomena in low dimensions and the simply generalize to higher ones. But from Wigner's work on particles we know that the representation theory of the Poincare group is very much dimension dependent and hence to infer from the chiral Bosonization-Fermionization relation a "neutrino theory of light" is really far-fetched [11].

the exponential the charge g and to its adjoint the charge -g, so that the only products of field with total charge zero have nonvanishing vacuum expectations.

The selection rules would follow from the exponential of a massive spinless fiel in the massless limit by imposing the condition that the field are renormalized with appropriate powers of the mass in such a way that none of the correlations becomes infinite This requirement is well-known to lead to the vanishing of all correlations of exponential field except those involving charge neutral products. In the above derivative model (4) this local charge g selection rule is masked as a result of the presence of a second *global* charge conservation of a complex Dirac spinor.

For our purposes another method, which shows that the exponential fiel is *string-localized,* leads to more physical insight. It starts from a chiral current which is a well-define quantum fiel and define the exponential fiel as an exponentiated integral of the current over a *finite* interval followed by the spacetime limit which takes one of the end-points to $+\infty$ infinit , whereas the unwanted compensating charge sitting at of the charge neutral exponential is disposed of at infinit . Again requiring finitenes of all correlations in the large distance limit only expectations which are charge-neutral in the remaining finit endpoints survive. This method reveals that the resulting operators are localized along a string, namely the semiinfinit interval $[x, \infty]$ where x is the endpoint which has been kept fi ed. So formally the φ in the exponential should be viewed as

$$\Phi^{(g)}(x) =: e^{ig\int_x^\infty j(y)dy} : \tag{5}$$

$$\Phi^{(g)}(x, x') =: e^{ig\int_x^{x'} j(y)dy} : \tag{6}$$

The main difference to (1), apart form the spacetime dimension is that the current is a physical massless fiel whereas the pointlike vectorpotential is not. It is quite easy to show that all chiral correlation functions, including the charge superselection rule, can be obtained from string-connected bilocals as define in the second line (5). The e_{el}-charge has similarities with a Maxwell charge in that it is locally generated. In spinor QED there is besides the Maxwell charge also the global charge[12]. But, as will be seen later, a Schwinger-Higgs screening where both the Maxwell charge also the notion of global charge as well as the charge which comes with a complex fiel disappear is only possible with scalar complex field i.e. scalar QED. Needless to add, the zero chiral fiel plays a special role; it does not really exist rather what is behind it is a string-localized object which only makes sense in the exponential form.

String localized fields as those exponentials, are outside the standard Wightman fiel theory in that they neither commute nor anticommute for spacelike separations between endpoints x. In chiral conformal fiel theory as (5) the string localization is visible in *plektonic* (in the above abelien case *anyonic*) commutation relations representing braid group statistics. Just looking at the problem as a formal massless limit of a two-dimensional exponential would not reveal that behind the infrared divergences there is a transition from point- to string-like localization[13]. The global Dirac charge and the local charge carried

[12]"Global" in this context does not mean that there are no local properties or consequences. In the DHR theory [2] of superselected charges and inner symmetries, the global spinor charge is reconstructed via the (representation-theoretical) "shadow" it imprints on the charge-neutral local observables.

[13]String theorists defin a scalar Φ by treating it as a conformal object living on a compactifie circle. This

by the exponential line-integral in the above derivative model (4) lead to identical selection rules and have their counterpart in QED where the local charge is referred to as the Maxwell charge. In higher dimensions the string-like localized nature is more easily perceived, since field $\Phi(x,e)$ where e is the direction (spacelike unit vector) of the semiinfinit spacelike string $x+\mathbb{R}_{+}e$ have a Lorentz transformation law in which the e participates, whereas in d=1+1 there is no such variable direction. As usual for spontaneouly broken symmetries, the Lorentz transformation on d=1+3 charged field exists only as an algebraic automorphism (which also acts on the string direction) which cannot be globally unitarily implemented. The analog of the d=1+1 infraparticle factors (5) in d=1+3 which multiply the Dirac spinors are formally written in terms of line integrals over string-localized vectorpotentials (next section)

$$\Phi_{in}^{(e_{el})}(x;e,e') =: e^{ie_{el}\int_0^\infty dse'_\mu A_{in}^\mu(x+se';e)} : \tag{7}$$

$$\Psi_{in}^{(e_{el})}(x;e,e') = \psi_0(x)\Phi_{in}^{(e_{el})}(x;e,e') \tag{8}$$

where e is the string direction, e' the direction of integration and e_{el} the electromagnetic coupling constant. Different from the d=1+1 case the $A_{in}^\mu(x,e)$ in the"infrared factor" Φ which carries the Maxwell charge is not a solution of the fiel equation of QED but rather denotes the potential which is associated to the incoming fiel strength which according to the (slightly modified LSZ scattering theory still exists [2] even though the incoming infraparticle fiel $\Psi_{in}^{(e_{el})}$ which describes the dissolved mass-shell of the charge particles and represents a s.l. fiel which lives in a Hilbert space which is not a Wigner-Fock space [14].

The mentioned two-dimensional infraparticle models of the 60s and 70s have been recently re-discovered in order to illustrate a proposal in the setting of effective QFT called "unparticles" which burdened them with other more metaphoric tasks which they most probably cannot fulfill Unfortunately the authors have only sketched in vague perturbative momentum space terms what they want without even indicating in spacetime concepts what they want. The illustration of an unclear idea with the same 2-dimensional models as those which have been used in the 60s and 70s for the well established infraparticle nature of electrically charged particles certainly raises eyebrows. Writing down something in momentum space and naming it effective QFT without explaining what one wants in terms of localization is not revealing much.

Of course methods of"effective QFT" may have their justificatio in models whose conceptual structure has already been understood but were efficien computational techniques are still missing. The concept of infraparticles is not explained by pointing to momentum space properties of Kallen-Lehmann spectral function; its conceptual pillar is rather the weakening of localization of charge-carrying field as a result of de-localization through interactions with string-localized vectorpotentials (see next section). The infrared anomalous power cut in the K-L spectral function starting at the mass of the charged object as well as other momentum space anomalies are consequences and not the cause of the unexpected deviation from particle behavior (unexpected at least from the Lagrangian gauge theory viewpoint which does not reveal any eye-catching particularity). Whereas in d=1+1

leads to quantum mechanical zero modes, and although their presence destroys the full conformal covariant transformation law, the exponetials are again covariant.

[14]Fore details we refer to a forthcoming paper by Jens Mund.

formal zero mass scalars play a prominent role, in d=1+3 conformal scalars with anomalous dimensions cannot generate the noncompact localization which one needs to get the typical power cuts in the K-L two-point function, one really need string-like potentials to start with (next section)

Since most of the perturbative arguments in d=1+3 about Maxwell charges inherit the mentioned problems of charged field in the gauge theoretic setting, it is useful to know that there exists a rigorous conceptual argument based on a quantum fiel adaptation of the Gauss law [19]. The arguments can be found in Haag's book, it may however be helpful to remind the reader of its main content. One starts from a $t-r$ smeared fiel strength

$$F^{\mu\nu}(f_R) = \int f_R(t,r)dtdr \int F^{\mu\nu}(t,r,\theta,\varphi)f_2(\theta,\varphi)d\theta d\varphi \tag{9}$$
$$f_R(t,r) = R^{-2}f(\frac{t}{R},\frac{r}{R})$$

The test function smearing of the singular pointlike field is necessary in order to have a well-define operator, for the classical fiel strength this would be superfluous The angular integral define an operator which represents the average flu through a sphere of radius r at time t. There are two physically motivated assumptions about the charged state of interest ω

$$\omega(F^{\mu\nu}(f_R)) \neq 0 \tag{10}$$
$$\omega(F^{\mu\nu}(f_R)^2) < \infty$$

which is interpreted as a consequence of a quantum adaptation of Gauss's law i.e. the expectation value of the flu in a charged state deviates significantl from the vacuum, but the fluctuatio should remain as bounded as they are in the vacuum since the correlation of the fiel strength for large distances should not be influence by the presence of a charge. Whereas the previous formulae specify the mathematical definitio of electromagnetic flu through a spatial surface and charged state, the following commutation relation between the mass operator and the averaged fiel strength is at the heart of the matter:

$$\left[M^2, F^{\mu\nu}(f_R)\right] = iR^{-1}(P^\sigma F^{\mu\nu}(f_\sigma)_R) + F^{\mu\nu}((f_\sigma)_R P^\sigma \tag{11}$$

It then follows that the state ω cannot have sharp mass i.e. an electrically charged particles is necessarily an infraparticle. Furthermore the algebraic Lorentz symmetry is not unitarily implemented in a charged state (spontaneous symmetry breaking) and even stronger: the momentum direction of a asymptotically removed charged particle is a superselected quantity, since an infraparticle is inexorably burdened by infinitel many infrared photons which only can be pushed further into the infrared, but cannot be eliminated.

That especially the last consequence appears strange to us is because our intuition about charged states has been formed by charged particle in QM where there is no superselection rule forbidding the coherent superposition of electrons with different momenta. But as in real life, we cannot accept the good parts of our most successful theory and reject structural consequences which we do not like. Of course part of this difficult lies in the discrepancy between our structural knowledge and the present state of art about perturbation theory. The following sections are dedicated to the problem how to overcome these problems through a radical reformulation of gauge theory.

As mentioned on several occasions, interacting QFT is in many aspects much more radical than QM. Using a picturesque metaphor one may say that it realizes a benevolent form of Murphy's law: *states which are not interdicted by superselection rules to mutually couple, do inevitably couple*. The culprit of this complication or, depending one one's viewpoint, this blessing, which contributes a fundamental aspect to QFT not shared by QM, is the inexorable occurrence of vacuum polarization resulting from the realization of the locality principle in the presence of an interaction; the sharing of $\hbar$ does not bring them closer. So the question arose whether the charged one-particle states can remain unaffected. To exemplify what may happen, the idea of infraparticles was proposed in the mathematically controllable context of soluble two-dimensional models for which the anomalous power cut in the variable $\kappa - m$ of the Kallen-Lehmann spectral function resembled the power behavior in the YFS work for the soft photon inclusive cross section.

In the 70s there appeared the firs n^{th} order perturbative calculation which addressed the infrared properties of charge-carrying fields they could be interpreted as confirmin the momentum space infraparticle structure of charged states [20]. But this did not close the issue since they had two drawbacks; firs they did not deal with gauge invariant charged field and second they had nothing to say about the possible weaker localization, which in those 2-dimensional infraparticle models was the root of the infrared problem. Some of the perturbative observation were subsequently derived in a context which does not directly refer to perturbation theory. For example the statement that there are no states with nontrivial charge nor states on which the Maxwell equations hold on indefinit metric spaces (as the Gupta-Bleuler or BRST settings) showed that the covariant perturbation theory involving charged field has very serious physical defects [21][22]. At best one can use the pointlike fiel formalism for the calculation of gauge invariant vacuum expectation values and with the help of the Wightman reconstruction theorem arrive at a new Hilbert physical Hilbert space and physical (gauge-invariant) operators acting in it. Focussing attention on the local observables, the problem of infraparticles and spontaneous breaking of the Lorentz symmetry in charged states was taken up in [23] Further insights came from comparing the quantum problem of localization of charges with semiclassical arguments [24].

The most important structural (nonperturbative) enrichments, which came out of the infrared problem and its roots in noncompact localization, are certainly the aforementioned conclusions drawn from an appropriate formulation of the quantum Gauss law in conjunction with a nontrivial charge which led to charged states with an infinit extension. In the next section we will start to close the large gap between structural insight and computational implementation.

4. Zero Mass Wigner Representations and Localization Peculiarities

In this section we will for the firs time address some recent theoretical observations.

There is a very subtle aspect of modular localization which one encounters in the second Wigner representation class of *massless finite helicity representations*[15] (the photon-graviton class) which recently attracted some attention [26]. Whereas in the massive case all spinorial field $\Psi^{(A,\dot{B})}$ the relation of the physical spin s with the two spinorial indices follows the naive angular momentum composition rules [25]

$$\left|A-\dot{B}\right| \leq s \leq \left|A+\dot{B}\right|,\ m>0 \tag{12}$$

$$s=\left|A-\dot{B}\right|,\ m=0 \tag{13}$$

the "covariantization" of the zero mass finit spin representations leads to a much stricter relation between the given physical helicity and the possible dotted/undotted components of the covariant spinorial formalism for which the Poincaré covariance can be extended to conformal invariance. For helicity $s=1$ the best one can do is to work with covariant fiel strength $F_{\mu\nu}$ which in the spinorial formalism correspond to wave functions $\Psi^{(1,0)}, \Psi^{(0,1)}$; the desired vectorpotential representation $\Psi^{(\frac{1}{2},\frac{1}{2})}$ corresponding to the classical vectorpotential $A_\mu(x)$ simply does not exist[16].

Most particle theorists who access QFT by quantizing classical fiel structures do not think much about why such a procedure leads inevitably to a loss of Hilbert space (Gupta-Bleuler, BRST formalism) which is the basic pillar of QT; they only take an indirect notice and often are not aware that they have left the quantum physical terrain to which they must return in order to physically interpret results. The reason is that typical quantum phenomena which have no counterpart on the classical side are easily overlooked in a classical parallelism as quantization; but they stick out very clearly in an intrinsic approach to QFT as Wigner's representation theoretical setting.

At the root of this problem is a *fundamental clash* between the existence of pointlike covariant generating field (or corresponding generating wave functions in the Wigner representation space) and the existence of spinorial field which do not satisfy the restricted relation (13). Let us for brevity introduce the following terminology: all field fulfillin this restricted relation will be called "fiel strength" and the ones which cause that clash are "tensorpotentials". In the case of s=1 the fiel strength $F_{\mu\nu}$ would be the fiel strength in Maxwell's theory whereas the quantum theoretically incriminated $A_\mu(x)$ is the vectorpotential; for s=2 the fiel strength is a 4-index tensor whose symmetry properties are those of the Riemann tensor etc. The setting can be generalized to half-integer s in which case s=3/2 (Rarita-Schwinger field is the lowest spin for which the clash of spinor potentials with the Hilbert space structure occurs. It will be seen below that the tensor representation of the representation type $D^{(\frac{s}{2},\frac{s}{2})}$ are the most interesting ones.

The resolution of the clash between the existence of pointlike potentials and the Hilbert space of QT is very interesting. Instead of ceding on the side of the Hilbert space, the clear message is to keep the latter but do not insist in pointlike covariant localization but rather allow semiinfinit stringlike generators[17]. In order to not be misunderstood, all the $(m=$

[15]There are 3 positive energy classes: massive, massless with finit helicity and massless with infinit spin. The firs ans the third have the largest cardinality.

[16]The relation between Wigner's one-particle representation theory and free field will be explained in more detail below. In fact there is a one-to-one correspondence which permits to use the same letter for both (see below).

[17]As a pointlike fiel is a generator (the distribution-theoretical limit) of compact localized operator algebras,

$0, s \geq 1)$ Wigner representations possess pointlike generators but their covariantization does not produce those covariant field which one needs e.g. for formulating renormalizable interactions.

As mentioned the standard path is follow "quantum gauge theory" namely to keep pointlike field and the entire well-studied perturbative formalism at the prize of an indefinit metric formalism which usually comes with the introduction of additional unphysical degrees of freedom (ghosts). Since in perturbation theory one does not use the Hilbert space norm for control of convergence (Schwartz inequality) there is no problem in doing that. At the end of the calculation one has to reconvert the calculated correlation functions (e.g. by cohomological arguments) into a Hilbert space setting; this is the step for gauge variant to gauge invariant objects. To avoid any misunderstanding we are not proposing a new string localized approach because of any suspicion that the gauge theoretical setting may be incorrect. Rather the reason is that the latter masks important physical properties for example the noncompact localization of charges, not to mention even more serious problems in case of Yang-Mills interactions (just those problems which are connected to confinemen and invisibility of states). All these nonlocal (more precisely stringlike) effects must happen in the last step namely the cohomological descent; this radical change from pointlike gauge variant to stringlike gauge invariant overburdens the cohomological step apart from those objects which remain pointlike which are the pointlike gauge invariants. The most important physical objects as the charged field remain outside the range of this gauge formalism.

As will be seen keeping the Hilbert space and not imposing impossible requirements on localization leads automatically to string-localized covariant potentials $\Psi^{(A,\dot{B})}(x,e)$ which contain a spacelike string direction and are localized on $x+\mathbb{R}_{+}e$ for all $(A,\dot{B})$ as in (12) which includes the covariant vectorpotential $A_{\mu}(x,e)$ for s=1 and the tensorpotential $g_{\mu\nu}(x,e)$. The fiel strength for spin s have scale dimension which s+1 (the ones with higher dimensions can be written as derivatives) whereas the potentials have scale dimensions which fill the space between 1and s+1, more precisely $1 \leq dim\Psi^{(A,\dot{B})}(x,e) \leq s$. Of particular importance are the potentials $\Psi^{(\frac{s}{2},\frac{s}{2})}(x,e)$ because their scale dimension is $\dim \Psi^{(\frac{s}{2},\frac{s}{2})} = 1$ independent of s; the previous field $A_{\mu}(x,e)$ and $g_{\mu\nu}(x,e)$ are examples. They lead to interactions which are renormalizable in the sense of power counting. For pointlike field renormalizability in the sense of power counting is synonymous with renormalizability whereas for string-localized field this still must be established.

Since this clash between quantum theoretical positivity/unitarity and the existence of pointlike generators with a prescribed covariant transformation property is central to our new proposal, some more remarks are appropriate. The problem of constructing covariant free field from Wigner's unitary representation theory of the Poincaré can be systematically solved in terms of intertwiners $u(p,s)$ which are $(2A+1)\left(2\dot{B}+1\right)$ component functions on the mass shell (which then may be rewritten into the tensor calculus). These intertwiners (between the unitary and the covariant representation) and their adjoints can be systematically computed, either by using group theoretic methods [25] or "modular localization" [26]. For the massless finit helicity case with its different "little group" most of the intertwiners do not exist (among them the vectorpotential); only those for which the spinorial

the string-like generator serves the same purpose for noncompact regions in case the pointlike generation is not possible-.

indices are related to the physical spin in the more restrictive manner as in the second line (12) remain available. All the missing covariant realizations can however be recovered if one allows the intertwiners to be dependent on a string direction $u(p,e)$. This construction is most conveniently done in the modular localization setting [26].

This construction has no counterpart in the classical Lagrangian setting and hence cannot be formulated in the functional integral setting (perhaps one reason why the string-localized potentials have only been noticed recently). But fortunately perturbation is intrinsically define in QFT and does not need any crutches from the Lagrangian formalism; whether free field are Euler-Lagrange field or not is of no relevance for the working of perturbation theory [30]. However the importance of causal locality which is most visible in the Epstein-Glaser approach becomes even mor important in passing from pointlike to stringlike fields

The remaining question is then why does one need vectorpotentials if the fiel strength wave functions generate already the whole Wigner space or, if in interacting QED the quantum fiel strength together with the charged matter field generate the full Hilbert space and form an irreducible set of operators in it ? The short answer is that one does not need it for the description of QED but the string-localized potentials play a crucial role in understanding the delocalization of the charges; without having the potentials transfer their stringlike localization to the matter field the nonlocality of the latter remains a mystery. But even staying in the free Maxwell theory the use of a pointlike vectorpotential together with Stokes law would give a zero result for the Aharonov-Bohm effect whereas the string-localized potentials leads to the correct effect.

Some of the answers can already be given in the free theory in terms of the generalized Aharonov-Bohm effect (next section). The implementation of renormalizable interactions adds additional reasons.

If one keeps the quantum (unitarity, Hilbert space, probabilities) aspects, then the way out is to relax the pointlike localization which underlies the Lagrangian quantization approach. Localization is important for the physical interpretation and the derivation of scattering theory, but the localization principle does not require to be realized in a pointlike generating manner, i.e. whereas there is no lee-way on the side of the quantum theory requirements, there is no principle in QFT which requires that a theory can be point-like generated. The mathematical question about the tightest localized generating wave function (technically: wave function-valued distribution) which by smearing with test functions generate the full Wigner representation space has been answered: all positive energy representations have semiinfinit stringlike distribution-valued generators and the massive as well as the massless finit helicity representations permit pointlike distribution-valued generating wave functions (generalized fiel strengths).

Only the zero mass *infinite* spin representation are intrinsically string-localized in the sense that there is no "fiel strength" which generates the same representation [29]. In [26] it was conjectured that also the associated QFT has no pointlike generated subalgebras based on the argument that their existence would lead to implausible consequences, but the issue is not completely settled on the level of mathematical physics [18]. there is also the aforementioned curious difference between the two pointlike generated representation,

[18] In particular since there has been a recent claim (without proof) to the contrary by Ch. K'öhler, Institut fuer Theoretische Physik, University Goettingen, work in progress.

which at closer inspection reveals a subtle distinction in localization aspects. In the massive case the little group is compact, whereas in the massless case it is the noncompact Euclidean stability group $E(2)$ of a lightlike direction. In the finit helicity case this representation is finit dimensional, hence necessarily a unfaithful (degenerate) representation. Only in Wigner's infinit spin family this noncompact stability group ("little group") is faithfully represented.

All three families of representations massive, massless finit helicity and massless infi nite spin are positive energy representations and there is a structural theorem [27], stating that all unitary positive energy representations of the Poincaré group (irreducible or not) can be generated by semiinfinit string-localized fields But only in case of the infinit spin family (infinit dimensional representation of the stability group) this is the best possible localization. For the other two families the best (sharpest localized) generators are pointlike field (operator valued distributions in the associated free fiel theory) which makes them accessible via the classical-quantum parallelism known as quantization.

The above observation about the existence of gaps in the spinorial covariance spectrum (12) means that even though both families of spinorial representations are finit dimensional, the noncompactness of the little group still makes its presence felt by not allowing most of the spinorial generators which occur in the massive case. Using a terminology which generalizes the (m=0, s=1) case of (noninteracting) electromagnetism, we may talk about pointlike "fiel strengths" and their associated string-like "potentials ", which taken together *reconstitute the full spinorial spectrum* in the firs line of (12). So from now on "fiel strengths" denote the covariant pointlike objects the second line of (12), whereas "potential" is the generic terminology for the string-localized remainder which recover the full spinorial spectrum of the firs line.

The main idea is of course that, although string localized vectorpotentials do not fi into the standard formalism, it is better to face the problem on its physical side; always with the increased awareness that localization is the dominant principle, and in order to uphold it rather change the formalism than to compromise on physical principles. In order to avoid being misunderstood, we are not criticizing the gauge theory formalism in its efficien y to deal with local observables and we even have some sympathy for the temporary trespassing of the most cherished principles of QT by ghosts in the name of computational efficien y. Without the contributions of Feldman and t'Hooft, Faddeev and Popov and the BRST formalism of Stora et al., a consistent extension of the renormalization setting of QED to the standard model would not have been possible. But meanwhile almost 40 years have past, and although most people agree that the theory is nowhere near its closure, nothing of conceptual significanc has happened. It is natural in such a situation to search for unexplored corners of QFT and the issue of localization, which is central to the present work, is certainly a rather dark corner even in QED. The attempt to complete a theory may still turn into a unexpected radical change.

This paper is a plea to follow the localization principles and develop a new string-localization compatible formalism. We will present some of the firs steps in this direction. It is worthwhile to mention already here, that the origin of the string-localized electric charge-carrying field in QED, including their infrared aspects, is the result of the interaction of the matter current with the stringlike vectorpotentials. Whereas the influenc of the stringlike localization on their own physical properties in selfinteractin Yang-Mills models is hard to foresee without understanding their infrared aspects, it is clear that the localization of the linearly related fiel strength in abelian gauge theories remain pointlike and that although the physical content of QED can be described in terms of physical string-localized matter field $\psi(x,e)$ and pointlike $F_{\mu\nu}(x)$, the stringlike vectorpotentials remain indispensable in the formulation of the interaction and the perturbative calculations.

Since such perturbative calculations are fairly involved and the cohomological descent from gauge-variant to gauge-invariant correlation is only simple for correlations involving $F_{\mu\nu}(x)$ but not $\psi(x,e)$, it is not without interest that the quantum version of the Aharonov-Bohm like (next section)[19] permits to see the important role of the de-localized vectorpotential.

It is an ineradicable prejudice to believe that perturbation theory has to be set up in terms of quantized Lagrangians and functional integrals. The conceptually most pleasing perturbative approach consists in coupling covariant free field (obtained as above by covariantizing the 3 classes of positive energy Wigner representations) together in form of invariant polynomial interactions[20] ("causal perturbation theory") which only contain pointlike field and stringlike potentials of short distance scaling dimension $d_{sca} = 1$, since any higher dimensional fields/potential would violate the power-counting requirement. Whether the coupled free field obey an Euler-Lagrange equation is irrelevant for this perturbation theory, a fact which was already known to Weinberg [30].

It is another equally ineradicable misconception (usually related to the previous one), that the perturbative treatment of QFT contains intrinsic infinitie which are "renormalized away". Fact is that certain implementations of perturbation theory, especially those which treat the problem in the quantum mechanical spirit of bringing certain operator functions of free fiel under control without paying much attention to the fact that fields even in the absence of interactions, are rather singular objects which want their mathematical role as operator-valued distributions to be taken serious. If one does not do this, one still has a chance, but there is a prize to be paid: the removal of cutoff infinitie (or ad hoc regularization parameters). But there is always a finit way to arrive at those results, and if this would not be so, the whole renormalization approach would not have credibility.

For $s \geq 1$ such potentials with $d_{sca} = 1$ exist for all helicities. The property which is crucial in this approach is the localization structure, in fact the intertwiner formalism which leads to the spinorial field (12) can be solely based on modular localization instead of group theory [26]. It is therefore not surprising that also the renormalization procedure can be built on the iterative fulfillmen of the localization principle combined with a requirement of keeping the scaling degree of the counter-terms at its minimal possible value: the string-extended Epstein-Glaser approach.

For completeness and also for making some subsequent speculative remarks more comprehensible, it is important to say something about the large[21] third family of *infinite spin*

---

[19]There is a curious analogy between the abandonement of the ether, which was mainly important for the post Maxwell-Lorentz development of particle physics, and the string-localization of potentials which only has severe consequences for the charged sector in the interacting theory.

[20]One may call it the interaction Lagrangian but the free fields/potential used may and generally will not have a Euler-Lagrange structure. That perturbation theory does not require the existence of a Lagrangian for a free fiel was already known to Weinberg [30].

[21]Although it is a zero mass representation, the faithfulness of the representation of the little group brings a continuous parameter into the game which leads to a higher cardinality of representations than in the finit spin

representations. These are irreducible massless representations in which the euclidean E(2) stability subgroup is faithfully represented. In this case the representation theory of the little group leads to an infinit dimensional Hilbert space [26]. The Casimir invariant of the little group (i.e. the E(2) analog of the mass operator) takes on continuous values. In order to avoid the somewhat misleading terminology "continuous spin" in the older literature associated with a continuous representation of the "little" Hilbert space, it may be more appropriate to follow the recent terminology and refer to the "infinit spin representations".

Before we return to the discussion of consequences of stringlike localization, it is helpful to formalize the covariant field for all three families. Some of these formulas can be found in the firs volume of Weinberg's book [25] e.g. the following formula for massive free field

$$\Psi^{(A,\dot{B})}(x) = \frac{1}{(2\pi)^{\frac{3}{2}}} \int (e^{-ipx} \sum_{s_3} u^{(A,\dot{B})}(p,s_3) a(p,s_3) + e^{ipx} \sum_{s_3} v^{(A,\dot{B})}(p,s_3) b^*(p,s_3)) \frac{d^3p}{2\omega} \tag{14}$$

where the sum goes over 2s+1 spin values. The operators $a^{\#}$ and $b^{\#}$ are the momentum space annihilation/creation operators which transform according to the Wigner representation. The intertwiner $u^{(A,\dot{B})}(p,s_3)$ and their charge conjugate counterpart v convert the Wigner representation into the covariant representation. They are rectangular matrices transforming a $2s+1$ component Wigner spin into a $(2A+1)(2\dot{B}+1)$ component covariant space. For a given physical spin s there is an infinit of possibilities of which one only uses the ones with low $A, \dot{B}$ values which happen to have the lowest scaling degrees. The generating wave functions of the beginning of this section are obtained by replacing the Wigner creation/annihilation operators by the function $f(p) = 1$.

As mentioned, Weinberg's method to compute these intertwiners was group theoretical, but one can also base the computation on modular localization [26]; this is not surprising since covariance and locality of states are closely linked.

Practically the same formula, with the only change in the range of $s_3 = \pm s$ and different expression for the intertwiners, holds for the massless case. However there is an important caveat, the formula exists only for the restricted $A, \dot{B}$ values in the second line of (12) i.e. only for fiel strengths.

If one allows string-localization one can recover all the lost spinorial representations. These "potentials" are (by definition all string-localized and obey for $\left|A+\dot{B}\right| \geq s \geq \left|A-\dot{B}\right|$, (fiel strengths excluded i.e. $s \neq \left|A-\dot{B}\right|$) the following formula

case.

$$\Psi^{(A,\dot{B})}(x;e) = \frac{1}{(2\pi)^{\frac{3}{2}}} \int (e^{-ipx} \sum_{s_3=\pm s} u^{(A,\dot{B})}(p,s_3;e)a(p,s_3)+ \qquad (15)$$

$$+ e^{ipx} \sum_{s_3=\pm s} v^{(A,\dot{B})}(p,s_3;e)b^*(p,s_3))\frac{d^3p}{2\omega}$$

$$A^\mu(x,e) = \frac{1}{(2\pi)^{\frac{3}{2}}} \int (e^{-ipx} \sum_{s_3=\pm 1} u^\mu(p,s_3;e)a(p,s_3)+ \qquad (16)$$

$$+ e^{ipx} \sum_{s_3=\pm 1} \overline{u^\mu(p,s_3;e)}a^*(p,s_3))\frac{d^3p}{2\omega}$$

where the vectorpotential intertwiner, which has been written down separately, has the form

$$u^\mu(p,s_3;e)_\pm = \frac{i}{pe+i\varepsilon}\{(\hat{e}_\mp(p)e)p^\mu - (pe)\hat{e}^\mu_\mp(p)\} \qquad (17)$$

and the $\hat{e}_\pm$ denotes the two photon polarization vectors to be distinguished from the string direction. These operators transform covariantly and have stringlike commutation relations

$$U(\Lambda)\Psi^{(A,\dot{B})}(x,e)U^*(\Lambda) = D^{(A,\dot{B})}(\Lambda^{-1})\Psi^{(A,\dot{B})}(\Lambda x,\Lambda e) \qquad (18)$$

$$\left[\Psi^{(A,\dot{B})}(x,e),\Psi^{(A',\dot{B}')}(x',e')\right]_\pm = 0,\ x+\mathbb{R}_+e >< x'+\mathbb{R}_+e'$$

As expected, the scaling degree of the potential is $d_{sca}(A^\mu(x,e)) = 1$ i.e. better than that of the fiel strength. The resulting two-point function is of the form [28]

$$\langle A_\mu(x;e)A_v(x';e')\rangle = \int e^{-ip(x-x')}W_{\mu\nu}(p;e,e')\frac{d^3p}{2p_0} \qquad (19)$$

$$W_{\mu\nu}(p;e,e') = -g_{\mu\nu} - \frac{p_\mu p_\nu}{(p\cdot e-i\varepsilon)(p\cdot e'+i\varepsilon)}(e\cdot e')+$$

$$+\frac{p_\mu e_\nu}{(e\cdot p-i\varepsilon)} + \frac{p_\nu e'_\mu}{(e'\cdot p+i\varepsilon)}$$

The presence of the last 3 terms is crucial for the Hilbert space structure; without them one would fall back to the indefinit metric and negative probabilities.

Either from the two-point function or more directly from the form of the intertwiners one reads off the following two relations:

$$\partial_\mu A^\mu(x,e) = 0 = e_\mu A^\mu(x,e) \qquad (20)$$

These formulas are not imposed but are consequences of the requirement of having a covariant vectorpotential in the Wigner Hilbert space which, since it cannot be point-local, brings in an additional geometric parameter e.

Note that the string-localized potential looks like the axial gauge potential in the gauge theoretical setting, where e is a gauge parameter which, different from the above definition is inert against Lorentz transformations.

The difference between the gauge interpretation and string-localization is more conceptual than formal. The fluctuatio in both parameters x, e (e a point in 3-dim. de Sitter space) is very important for the lowering of the x-dimension, namely $d_x = 1$, at the cost of fluctuatio in e which manifest themselves as string-caused infrared divergences; the latter explain why the axial gauge, which overlooks the fluctuation in e is intractable (it was never of any computational use); despite of its welcome Hilbert space structure, the fluctuation in e prevented to control these divergencies, in fact it was not even possible to understand their physical origin. The localization based approach on the other hand treats the string potentials as distributions in x *and* e and addresses the question if and how the different e_i have to coalesce at the end. The axial gauge is besides the Coulomb gauge the only one which can be accommodated in a physical Hilbert space.

The interpretation of the e as a fluctuatin covariant string direction rather than as fi ed gauge parameter in the interpretation as a string-localized potential is the only meaningful interpretation. Hence the standard argument in favor of gauge invariance based on returning to a physical space from the unphysical indefinit metric formulation has lost its conceptual basis. But a new problem has emerged, namely how to treat fluctuatin string directions. If one has grown up with a "gauge principle", this may seem surprising, but the surprise should not be new, one could have asked the crucial question "does the axial gauge with its threatening infrared divergencies fi into the standard gauge-ideology ?" already a long time ago.

In order to obtain a theory in which the interaction between the vectorpotentials and matter leads to a subalgebra of local observables, one needs a relation which connects the potential for two different directions

$$A^\mu(x,e) \to A^\mu(x,e') + \partial^\mu \Phi(x;e,e') \tag{21}$$

$$\Phi(x,e,e') = \int e_\mu A^\mu(x+te',e)dt$$

The proof of pointlike locality of certain field amounts to the e-indepence; this is not different from the proof of independence of gauge parameters in the standard gauge theoretical setting. But the main purpose of this formalism is not the identificatio of local observables and the calculation of their correlation functions but rather to incorporate the string-localized charged field and their infraparticles into the perturbative formalism.

Another special case of significan interest is the case of s=2 whose fiel strength with the lowest scale dimension is an object $R_{\mu\nu\kappa\lambda}$ with the linear properties of the Riemann tensor and an $d_{sca} = 3$ and a string-like potential $g_{\mu\nu}(x,e)$

$$\langle g_{\mu\nu}(x,e) g_{\kappa\lambda}(x',e')\rangle = \int \frac{d^3p}{2p_0} e^{-ip(x-x')} W_{\mu\nu\kappa\lambda}(p;e,e') \tag{22}$$

$$W_{\mu\nu\kappa\lambda}(p;e,e') = W^R_{\mu\alpha\nu\beta\kappa\rho\lambda\sigma} \frac{e^\alpha e^\beta e'^\rho e'^\sigma}{(e\cdot p - i\varepsilon)^2 (e'\cdot p + i\varepsilon)^2}$$

where the superscript R refers to the fiel strength 2-point function whose 8 tensor indices correspond to the 4 tensor indices of the independent fiel strength and reflec the fact that the relation between the potential and the fiel strength is the linearized version of that between the metric tensor and the Riemann tensor [28]. The e-dependent factor obviously improves the short distance properties. As expected the $g_{\mu\nu}$-potential has $d_{sca} = 1$.

String-localized potentials with $d_{sca} = 1$ can also be constructed for massive theories, even though there is no compelling reason from the viewpoint of the Wigner representation theory for doing this since pointlike field with a fi ed physical spin exist for all spinorial pairs fulfillin (12).

In this case the *only reason would be the power counting requirement*. Since the increase of the short distance dimension with spin happens independent of the presence of a mass, there would be no renormalizable interaction of a spin one massive $A_\mu(x)$ fiel with other s=0 or s=$\frac{1}{2}$ matter fields the only way out is to take a string-localized massive $A_\mu(x,e)$ with $d_{sca} = 1$ instead of $d_{sca} = 2$ for $A_\mu(x)$. This enlarges the number of candidates for renormalizable interactions from a finit number to infinitel many. But even if some interactions which are power counting renormalizable turn out to lead to mathematical consistent theories, unless they have local observables in the form of pointlike generated subtheories, they are physically unattractive.

For infinit spin one find [26][28]

$$\Psi(x;e) = \frac{1}{(2\pi)^{\frac{3}{2}}} \int (e^{-ipx} \circ a(p) + e^{ipx} \sum_{s_3=\pm s} v(p;e) \circ b^*(p;e)) \frac{d^3p}{2\omega} \tag{23}$$

$$u(p;e)(\kappa) = \int_{\mathbb{R}^2} d^2 z e^{ikz} (\xi(z) B_p^{-1} e)^{-1+i\alpha}$$

where B_p is the p-dependent family of Lorentz-transformation, selected in such a way that they transform the reference $\bar{p}$ on the irreducible orbit into the generic p, and the circle product stands for the inner product in the little Hilbert space which consists of functions of a two-dimensional Euclidean space $\bar{f} \circ g = \int \bar{f}(\kappa) g(\kappa) d^2\kappa$. So the Wigner $a(p)^{\#}$ operators and the intertwiners depend on the euclidean κ variables make the dependence on e much more involved than in the finit helicity case. Nevertheless they are string-localized for all values of α. The κ dependence of the intertwiners results in a stronger form of string localization than for zero mass potentials.

This stronger form of delocalization shows up in the fact that there are no pointlike fiel strengths [29]. In fact the operator algebra generated by these field apparently has no compactly generated observable (pointlike generated) subalgebras[22]. In general stringlike generators of algebras which cannot be assembled as a union from compact parts (as our favorite example of operators carrying a nontrivial electric charge) do generate reducible states under the action of the Poincaré group if applied to the vacuum which are pointlike generated; the localization of operator algebras and the localization of states are two different pairs of shoes. But representations of the Poincaré group which have infinit spin components in their reduction can only appear in operator algebraic structures which were string-localized.

The only states which are intrinsically string-localized are those associated to the infi nite spin representation. Theories in which they occur would have serious problems with being accessible to observations[23]. This is because "counters" in QFT are compactly local-

[22] The pedestrian argument for bilinear operators in [26] can be generalized to monomials of arbitrary orders, but an elegant proof based on modular methods is still missing.

[23] This may have been the reason why Weinberg [25] dismissed them as unphysical, despite their fulfillmen of the positive energy requirement.

izable or, in order to avoid vacuum polarization problems [2], they should be at least localizable in the sense of quasilocality. Such a counter cannot register an intrinsically string-localized state, so that quantum matter related to the third Wigner class remains "invisible" despite the fact that it carries nonvanishing energy-momentum (and hence susceptible to gravity). Theories containing such representations are candidates for "invisible" quantum matter. So maybe Weinberg's "no" at the time of writing his book should be weakened to "not yet".

5. The Generalized Aharonov-Bohm Effect for ($m = 0, s \geqslant 1$) and the Violation of Haag Duality

Suppose one generates the ($m = 0, s \geqslant 1$) Wigner representation space (or the associated net of local algebras in the Wigner-Fock space) with pointlike fiel strength wave functions. Does the theory let us know that we forgot that there are string-localized potential which want to play an important physical role? In this case the Wigner representation should signal by its localization properties that there is a difference in localization between the massive to the massless representation, but does it really do this? Is there an intrinsic difference within the Wigner representation with respect to the modular localization structure which goes beyond the covariantization formula (12)?

There is indeed a subtle representation theoretical distinction which is connected with *Haag duality*. Whereas for simply connected convex double cone regions the localization spaces (real subspaces of the Wigner representation space, which are define in terms of modular localization[24] (29)) one find Haag duality for both the massive as well as the massless finit spin case

$$K(\mathcal{O}') = K(\mathcal{O})' \text{ or } K(\mathcal{O}) = K(\mathcal{O}')' \tag{24}$$

$$\mathcal{A}(\mathcal{O}) = \mathcal{A}(\mathcal{O}')', \quad \mathcal{O} \text{ double cone} \tag{25}$$

but there are interesting differences for *non simply connected regions* as relative causal complements of a smaller spacetime double cone within a larger one (the causal completion of a torus) where (24) refers to the Wigner representation theory and (25) is the second quantized algebraic version. The interpretation of Haag duality in the standard (von Neumann) quantum theoretical setting of measurements is that any measurement which is compatible with all measurement inside a causally complete spacetime region, must be associated with the causally disjoint observables (25); this rule can only be broken for very special and important reasons.

One knows from some old (but unfortunately unpublished) work [32] that for free QED, i.e. for the (m=0,s=1) Wigner representation, the Haag duality breaks down for a spacetime region $\mathcal{T}$ which results from the causal complement of a double cone inside a larger double cone or by sweeping an $x-t$ two-dimensional double cone subtended from a spatial interval $x \in [a, b], 0 < a < b$, by higher dimensional rotation around the origin. In d=1+2 the result would be topologically equivalent to the inside of a torus, whereas in d=1+3 it is the doubly

[24] The K-spaces are real subspaces of the complex Wigner space which are define as eigenspaces of the involutive Tomita S-operator. For a presentation of these spaces which is close to the spirit of the present paper we refer to [31].

connected 4-dimensional analog namely the causal completion of a 3-dimensional torus region at a fi ed time. One then find [32]

$$K(T') \subsetneq K(T)' \tag{26}$$

$$or\ K(T) \subsetneq K(T')',\ \ \mathcal{A}(T) \subsetneq \mathcal{A}(T')' \tag{27}$$

where in the second line we also wrote the violation of Haag duality relation in the interaction-free algebraic setting which are functorially related to the one formulated in terms of modular localized real particle subspaces K of the complex Wigner representation spaces.

These proper containment relations in the special case of the free electromagnetic fiel are quantum fiel theoretic analogs of the quasiclassical Aharonov-Bohm (also Ehrenberg-Siday) effect. This is an effect of a localized classical magnetic flu in an infinitel long solenoid exerts on a quantum mechanical electron which scatters on the solenoid even though it stays outside its thin magnetic flu tube. Its quantum fiel theoretic counterpart (A-B effect in QFT[25]) is more stringent; it states that despite the continued validity of Stokes theorem for the quantum magnetic field the quantum magnetic potential has modular localization properties which differ from the classical intuition since classically it should be localized on (or at least near) the Stokes boundary circumference. In other words the effect would disappear if the magnetic potential would be a pointlike fiel as in gauge theory.

This does of course not mean that gauge theory is misleading, it is only a warning against its unconstrained use. Matters of localization should only be discussed in a Hilbert space i.e. *after* having implemented the invariance under BRST transformations. But since the BRST "symmetry" acts in a nonlinear way, the construction of BRST invariant local correlations is very difficul and for this reason rarely done; it is especially the case if nontrivial Maxwell or Yang Mills charges lead to stringlike localizations after having done the cohomological BRST descent.

There are two remedies for the A-B effect at hand, either one introduces the notion of quantum cohomology of fiel strength [32], or one uses the *string-localized quantum vectorpotentials*. For free electromagnetic fiel the firs choice is completely adequate, but in the presence of interaction only the formulation based on string-localized potentials has a good chance to permit the perturbative construction of physical charged field and to really explain the origin of their de-localized nature.

In the cited work [32] the connection with the A-B effect was not mentioned and the representation theoretical basis of modular localization via intersection of wedges was not yet available. The calculation was done in the covariant fiel strength formalism of $\vec{E}, \vec{H}$. The main purpose was to formulate a warning against the use of pointlike vectorpotentials in QFT which leads to a contradiction with the A-B effect and its extension i.e. the violation of Haag duality (26). This, and the remark that string-localized magnetic potentials avoid this contradiction, is the precise reason why we revisited this age old problem. Since in the massive case the equality sign in (26) continues to hold, the toroidal Haag duality violation is the looked-for *intrinsic representation theoretic distinction* between massive and finit helicity massless representations.

[25] Usually the A-B effect refers to the semiclassical scattering of charged pstzivles on a thin solonoid whose magnectic fiel remains inside.

Perhaps the best way to present this result is to say that the pointlike localization for massless vectorpotentials clashes with the Hilbert space positivity and since the latter is the essence of quantum theory, it is the pointlike localization which has to cede. The only generalization of pointlike localization turns out to be semiinfinit stringlike; the positive energy condition of unitary representations of the Poincare group does not require to introduce *generating* wave functions (or free fields which are weaker localized than a semiinfinit string.

Following LTR one looks at a situation of two spatially separated, but interlocking regions $\mathcal{T}_1$ and $\mathcal{T}_2$ in which one represents as the smoothened boundary of two orthogonal unit discs D_1 and D_2 which intersect in such a way that the boundary of one passes through the center of the other. The delta function flu es through the D_i are smoothened by convoluting $\star$ with a smooth function $\rho_i(\mathbf{x})$ supported in an ε-ball B_ε; the interlocking $\mathcal{T}_i$ are then simply obtained as $\mathcal{T}_i = \partial D_i + B_\varepsilon$ $i = 1, 2$. One computes the following objects

$$Im(e(\vec{g}_1), h(\vec{g}_2)) \simeq [\vec{E}(\vec{g}_1)\vec{H}(\vec{g}_2)] = \int \vec{g}_1(x) rot \vec{g}_2(x) d^3x = \tag{28}$$

$$= \int \rho_1(x) d^3x \int \rho_2(y) d^3y, \quad \vec{g}_i = \vec{\Phi}_i \star \rho, \; \vec{\Phi}_i(\vec{f}) = \int_{D_i} \vec{f} d\vec{D}_i$$

where we have written the result in two different ways, on the right hand side the algebraic (commutator) expression and on the left hand side in terms of the associated Wigner wavefunction. The Φ_i is the functional which describe the flu through D_i, a kind of surface delta function.

The calculation of wavefunction inner product and its associated symplectic form define by its Imaginary part is more lengthy but straightforward. It is needed because it contains the information of the modular localization. We will omit it here and simply state the result. It confirm relation (26) since $e(\vec{g}_1) \in K(\mathcal{T}_1')'$, $h(\vec{g}_2) \in K(\mathcal{T}_2')'$ but none of these two wave functions are in the smaller spaces $K(\mathcal{T}_i)$ since the algebraic right hand side (28) is definitel nonvanishing. The QFT A-B effect is the only known violation of Haag duality for which the duality violating operators cannot be used for a "Haag dualization" i.e. an extension process by which Haag duality can be recovered for the extended not of local algebras.

The calculation can be done entirely in terms of fiel strengths, there is no need to use potentials and their two point functions (19). The firs term in (19) which contains only $g_{\mu\nu}$ and no string dependence e would lead to an indefinit inner product if taken for the two-point function of vector potentials; in fact this would describe the indefinit two-point function of pointlike vector potentials in the covariant Feynman gauge; but restricted to the fiel strength it is perfectly positive. On the other hand the full two-point function (19) is positive, this was the main achievement of reconciling modular localization and positivity via string-localization. For the cohomological argument supporting the QFT A-B effect or breakdown of Haag duality, one does not need the potentials. It is only if one wants to have a more operational argument for the discrepancy between Stokes theorem and modular localization than that based on cohomology that one needs the free stringlike potentials.

However the operational formulation in terms of *string-localized potential become absolutely crucial in the presence of interactions* for the understanding of the properties of physical charges. I know of no cohomological argument in terms of which one can understand the localization properties of interacting Maxwell charges. The fact that the A-B effect disappears if one uses the pointlike (and hence non Hilbert space) magnetic potential $\vec{A}(x)$ in Stokes theorem shows that one has to be very careful in drawing physical conclusions from the standard gauge formalism. Only after the imposition of BRST invariance and the cohomological descent (too difficul for correlation of charged field only stated but never performed) has one left the slippery ground.

There is no all-clear in the context of interacting gauge theories. It is well-kown that there is no electric charge in a formulation QED in which a pointlike vectorpotential is still present [21][23]. In such a case there exists only the nontrivial kinematical Dirac charge, but the Maxwell charge vanishes. As in the case of the above Aharonov-Bohm effect, this unphysical aspect disappears upon using the string-localized vector potentials in the Hilbert space formulation. Although we will not go into higher spin problems it may be interesting to remark on the side that the string-localization in Hilbert space has extensions to higher spin potentials whose scale dimension does not increase with s (with renormalizable interactions in the sense of power counting) whereas an extension of the gauge formulation beyond s=1 is not known.

The violation of Haag duality for conformal QFT on multifold connected spacetime regions is part of modular theory and this raises the question whether one can compute the modular group. The answer is positive and quite interesting; it will be deferred to an appendix.

Some more remarks about (the algebraic) Haag duality and its breaking are in order. Its validity for simply connected regions $\mathcal{O}$ (25) define a "perfect" world in which the quantum counterpart of the classical Cauchy propagation holds i.e. a local algebra is equal to that of its causal completion and the commutant of the algebra localized in the causal disjoint $\mathcal{O}'$ is equal to the original algebra.

As mentioned, interesting situations arise when the world of local quantum physics is not perfect and the Haag duality is violated i.e. the right hand algebra is genuinely bigger than the left hand side. The most common violation results from an observable algebra which is localized in several disconnected spacelike separated double cones (separated intervals in the chiral conformal case [33]). In case the observable algebra possess localizable superselected charges, the right hand side for such a multi-disconnected region $\mathcal{A}(\mathcal{O}')'$ is genuinely bigger because the *charge transporters* which carry the charge from one to the other region are in $\mathcal{A}(\mathcal{O}')'$ but not in $\mathcal{A}(\mathcal{O})$; the charge transporters are globally neutral, but they change the localization of charges between the localization regions.

Such models fall into the range of the DHR superselection theory [2]. The fina result of this theory is the (unique) existence if a "fiel algebra", which contains all superselected charges and a compact symmetry group[26], which acts on the fiel algebra in such a way that the observable algebra re-emerges as the fi ed point subalgebra. In the chiral case one can even compute geometric modular groups for such situations. They are associated with higher diffeomorphism groups beyond the Moebius group [34] and they require to trade the standard vacuum with the so-called split vacuum. In all those cases the violation of Haag-duality is an indicator of the presence of charge superselection sectors and a global symmetry.

[26]The appearance of compact group theory via the localization properties of observable algebras is perhaps the most surprising aspect of the power of quantum localization [35].

The idea underlying the relation between the charge neutral observables and the charge carrying field can be best by borrowing a famous phrase from Marc Kac in conncetion with Hermann Weyl's inversion problem namely "how to hear the shape of a drum?" If one substitutes drum by the full-fledge QFT containing globally charged field and the perceived sound by the observable charge neutral fields the existence of the superselection theory stands for the reconstruction of the full theory (containing all charges) from the observable "shadow". The existence and uniqueness of this inverse problem has given a significan insight into the inner workings of QFT; in particular it demystifie the origin of inner symmetries, a concept which started in the 30s with Heisenberg's isospin. Morally something like this should also hold for local (Maxwell) charges, if one allows fro the possibility that certain charges cannot be registered and probably not even produced in collisions of neutral particles (see section 7). However investigations in this direction did not lead to anything tangible. We view our approach as yet another attempt in this direction.

The violation of the Haag duality for the above doubly connected T has a quite different physical message. First it can be detected already in the Wigner setting, so it has nothing to do with superselection sectors and vacuum polarization. To appreciate its message, it is indicative to imagine that the fiel strength can be derived from a pointlike potential $A_\mu(x)$ which is the standard starting point of the indefinit metric gauge setting . In that case the flu es will be supported inside the spacelike separated T. The vanishing of the resulting expression is in in flagran violation of the above calculations. This shows that the the standard indefinit metric gauge formalism is unreliable. As we pointed out before it is not wrong, but one has to carefully distinguish situations where it can be applied from those where it leads to incorrect conclusions[27]. For the present purpose it is the *strongest support for the introduction of stringlike potentials* in the absence of interactions which would not create any contradiction in the above calculation.

Both, the charge superselection problem and the problem of multi-connectedness are intimately related to the way in which models of local quantum physics realizes the localization principle. The Aharonov Bohm effect is perhaps the most direct and simple illustration since it does not require composite field and vacuum polarization.

To generalize this subtle violation of Haag duality to arbitrary $(m = 0, s > 1)$ representations one needs a more adequate modular setting than the pointlike covariant fiel strength formalism used by LTR; the latter becomes increasingly complicated with the increasing number of tensor/spinor indices. A structural method which avoids the use of covariant fiel coordinatizations is the use of the unique representation theory combined with modular localized subspaces. The start for constructing such subspaces K is always the localization in wedge regions W since in that case the modular objects associated with the Tomita S-operator $S = J\Delta^{\frac{1}{2}}$ are part of Wigner's representation theory, namely the antiunitary J and modular unitary Δ^{it} coalesce with the spacetime reflectio on the edge of the wedge and with the wedge-preserving boost respectively [2]. The real localization subspace $K(W)$ is just the closed $+1$ eigenspace of S and the associated dense complex

[27]The point here is that in gauge theoretic calculations one computes numbers and is normally one is not interested in localization. On the other hand the commitment to gauge invariant results is mainly a lip-service, physical correlations are easily characterized in terms of BRST invariance, but the computation of charge neutral composites is quite a different story.

standard space is its complexificatio

$$K(W) = \overline{\{\varphi|S\varphi = \varphi\}} \qquad (29)$$
$$K(W) + iK(W)\ is\ standard$$

where standardness (K and iK have trivial intersection and taken together form a dense set) of spaces and operator algebras (where it is intimately related to the Reeh-Schlieder property) is one of the most important concepts of modular localization theory.

The modular objects for subwedge regions are determined by representing them in terms of intersections of wedges and showing the standardness of the associated subspaces. This is fairly easy for noncompact regions as spacelike cones C, whose core is a semiinfinit spacelike string with an arbitrary small opening angle. Evidently they can be represented by intersections of wedges which a shared origin which then becomes the apex of the spacelike cone C. In this case the standardness follows from the energy positivity [27] i.e. it is shared by all three Wigner representation. All 3 families have stringlike wave-function valued generators $\Psi(x, e)$, but only in the infinit spin case there are no better localized generators.

A systematic investigations along these purely modular lines would start with showing the standardness of compact double cone $\mathcal{D}$ localized subspaces $K(\mathcal{D})$ avoiding the use of covariant wavefunctions/field and relying entirely on the Wigner representation theory. The double cones are the causally closed regions in terms of which the setting of algebraic QFT (spacetime-indexed nets) is defined In that case the origin of the wedges cannot be fi ed (to the origin of the Minkowski spacetime coordinatization) but have to be passed around a circle on the two-dimensional spatial boundary if a symmetrically chosen (around the coordinate origin) double cone. From concrete calculations with pointlike generating field we know that generating wave functions $\Psi^{(A,\dot{B})}(x)$ obtained from covariantization of the Wigner wave function do generate standard compactly localized subspaces. But only an abstract version of this proof will reveal the importance of the nature of the little group and its impact on the localization problem, i.e. why unitary representations of compact- and finit dimensional representations of noncompact little groups are compactly localizable, whereas the infinit spin representation is not ($K(\mathcal{D}) = 0$).

Having prepared oneself in this way, there remains the structural understanding of the generalized Aharonov-Bohm effect namely for multiply connected regions there exist observables which, although they commute with the fiel strength in the causal disjoint, are not expressible in terms of fiel strength inside the multiply connected spacetime region. Since the fiel strength determine the global properties, the generalized A-B effect is another manifestation of the holistic aspects of QFT in this case one which distinguishes massive from massless representation.

Although the QFT A-B effect has only been established for (m=0, s=1), the role of the little group in localization leaves little doubt that there exists a generalization for s $>$1. Invoking a metaphoric principle namely that nature may have only few principles but an enormous variety of different manifestations, one is inclined to speculate that the increasing number of potentials with increasing s is associated with higher than double connectivity generalizations of the QFT A-B effect (and not only with an increase of the number of A-B operators in a geometric situation of n separated $\mathcal{T}$ regions. This would shed light into the dark corners of higher spin quantum matter which has been closed to gauge theory inspired ideas. Finally there is the question of the action of the modular group in massless theories, especially in cases where one expects this to be geometric, as in the case of $\mathcal{T}$. The answer to any of these questions would require more mathematics and would lead too far away from the spirits of this paper for which this section only serves to illustrate that there are indication for the role of string-localized potentials already in the free fiel theories. But the importance of this unexplored suggests to return to it in a more specifi future context.

The zero mass higher spin fiel strengths exhibits the above increase of scale dimension of pointlike generated fiel strength with spin and therefore shows a worsening of fiel strength associated short distance singularities. The bosonic potentials, namely n=s string-localized generators with increasing short distance dimensions fil the gap between d_{sca}=1 up to d=s where s+1 is the d_{sca} of the lowest dimensional fiel strength (the lowest dimension consistent with the second line in (12)). We already emphasized that the localization in a indefinit metric setting has no relation to the physical localization; this is the main message of the A-B effect and the violation of Haag duality for QFT.

The same second line (12) contains the considerably reduced number of spinorial descriptions for zero mass and finit helicity, although in both cases the number of pointlike generators which are linear in the Wigner creation and annihilation operators [26].

By using the recourse of string-localized generators $\Psi^{(A,\dot{B})}(x,e)$ one can *restore the full spinorial spectrum* for a given s, i. e. one can move from the second line to the firs line in (12) by relaxing the localization. Even in the massive situation where pointlike generators exist but have short distance singularities which increase with spin. there may be good reasons (lowering of short distance dimension down to d_{sca}=1) to use string-like generators. In all cases these generators are covariant and "string-local"

The explicit verificatio of stringlike locality is cumbersome because there are no simple x-space formulae for stringlike Pauli-Jordan functions. It is easier to avoid manifest x-space localization formulas and work instead with intrinsically define modular localization subspaces. In fact the construction of the singular stringlike generators are not based on any gauge theory argument but rather a consequence of the availability of stringlike intertwiners for all unitary positive energy representations of the Poincare group. In the present setting the equations

$$\partial_\mu A^\mu(x,e) = 0 = e_\mu A^\mu(x,e) \tag{30}$$

have nothing to do with a gauge condition but rather are a consequence of constructing intertwiners which localize on a string $x + \mathbb{R}_+ e$ which is the next best possibility in cases where the compact localized subspaces are empty and their pointlike generators nonexistent. The pointlike aspect of the gauge formalism is only physically relevant in case of gauge invariant operators i.e. the pointlike generated e-independent subalgebra coalesces with the gauge invariant subalgebra. So the stringlike approach complements the gauge invariant construction by incorporating the charged sector of QED with its infraparticle aspects and hopefully also the nonlocal aspects of gluons and quarks which are the key to their "invisibility".

Whereas free vectorpotentials have a harmless string localization since by applying a differential operator one can get rid of the semiinfinit string and return to the pointlike fiel strength, we will see in the next section that the interaction furnishes the charge carrying operators with a much more autonomous stringlike localization which cannot be removed by differential operators and in fact is intrinsic to the concept of electric charge.

The noninteracting $(0, s = 2)$ representation is usually described in terms of pointlike fiel strength in form of a 4-degree tensor which has the same permutation symmetries as the Riemann tensor (often referred to as the *linearized Riemann tensor*) with $d_{sca} = 3$ whereas its string localized covariant potential $g_{\mu\nu}(x, e)$ has the best possible dimension $d_{sca}\ g = 1$. By allowing string localized potential one can for all $(m = 0, s \geq 1)$ representations avoid the increase in the dimensions with growing spin in favor of $d_{sca} = 1$ (independent of spin) stringlike potentials from which one may return to the pointlike fiel strengths by applying suitable differential operators. In the massive case there is no reason for doing this from the point of localization rather the only physical reason for using the string like counterparts for the pointlike field is their lower short distance dimensions; again the optimal value is d_{sca}=1 for all spins. Hence candidates for renormalizable interactions in the sense of power counting exist for all spins.

In order to be able to continue with the standard pointlike perturbative formalism one took recourse to the Gupta-Bleuler or BRST gauge formalism. At the end one has to extract from the results of the pointlike indefinit metric calculations the physical data i.e. perturbative expressions in a Hilbert space

In this respect there is a significan conceptual distinction between e.g. classical ED and QED which is masked by the joint use of the same terminology "gauge". Whereas in classical theory the use of the gauge potential simplifie calculations and leads to interesting connections with the geometry, in particular with the mathematics of fibre- undles, the quasiclassical treatment of quantum mechanics in a classical external electromagnetic environment leads to the Aharonov-Bohm effect which is usually considered as the physical manifestation of the vectorpotential.

Finally in the quantum fiel theoretic setting of QED it becomes indispensable since without the minimal coupling of quantum matter to the potential it would be impossible to formulate QED. In this case the pragmatic meaning of the terminology "gauge principle" stands for the continued use of the standard pointlike fiel formalism of QFT within an indefinit metric setting and the return via gauge invariance to a restricted Hilbert space setting in which the formal pointlike localization is the same as the physical localization. The string-localized approach is strictly speaking not a gauge theoretic formulation in this sense. But neither is the closely related "axial gauge" formulation since the axial potential already lives in a Hilbert space and hence its localization is already physical. Although the clash between pointlike localization and Hilbert space representation continues to hold for the "potentials" of all $(m = 0, s \geq 1)$ representations, the analog of gauge theory does not exist or is not known. It seems that in those cases there is no "fake" pointlike formalism which can be corrected by a "gauge principle" which then selects the genuinely pointlike observables from the fake objects; in those cases one has to face the issue of string-localized field right from the beginning.

The next interesting case beyond $s = 1$[28] is $(m = 0, s = 2)$; in that case the "fiel strength" is a fourth degree tensor which has the symmetry properties of the Riemann tensor; in fact it is often referred to as the linearized Riemann tensor. In this case the

[28] We omit spinor fields as the zero mass Rarita-Schwinger representation (m=0,s=3/2).

string-localized potential is of the form $g_{\mu\nu}(x,e)$ i.e. resembles the metric tensor of general relativity. The consequences of this localization for a reformulation of gauge theory will be mentioned in a separate subsection.

6. String-Localization of Charged States in QED and Schwinger-Higgs Screening

In this section some consequences of working with physical [29] i.e. string-localized vector potentials in perturbatively interacting models will be considered. Whereas all charge neutral objects in QED are pointlike generated, this cannot be true for physical charge-carrying operators. From the previous sections we know that the best noncompact localized charged generators are semiinfinit spacelike strings which, as a result of their simultaneous fluctu ations in the Minkowski spacetime x and the spacelike direction e (3-dim. de Sitter) have improved short distance behavior in x, namely there always exists a potential with $d_{sca} = 1$ which is the best short distance behavior which the Hilbert space positivity allows for a vectorpotential associated with a fiel strength of scaling dimension $d_{sca} = 2$. The prize to pay for part of the fiel strength fluctuation having gone into the fluctuatio of the string direction e is the appearance of infrared divergences which require the distribution theoretical treatment of the variable e; this problem must be taken care of with special care in perturbative calculations.

In the previous section we learned that the full covariance spectrum (12) for zero mass finit helicity representation can be regained by admitting stringlike fields The pointlike *field strength* [30] is then connected with the *stringlike potentials* by covariant differential operators. We have presented structural arguments in favor of using stringlike potentials over fiel strength even in the absence of interactions when stokes argument is used to rewrite a quantum magnetic flu integral over a surface into an integral over its boundary. However the most forceful argument is that for each spin $s \geq 1$ there exists always a potential of lowest possible dimension namely $d_{sca}(\Psi^{(\frac{s}{2},\frac{\dot{s}}{2})}(x,e)) = 1$ which is the power-counting prerequisite for constructing renormalizable interactions.

This holds also in the massive case where the covariance for pointlike field covers the whole spinorial spectrum (12). Whereas the pointlike field have an $d_{sca} \geq 1$ which increases with s, there also exist stringlike field with $d_{sca} = 1$ for any s. The simplest example would be a massive pointlike vector fiel $A_\mu(x)$ with short distance dimension $d_{ssd} = 2$ and a stringlike potential $A_\mu(x,e)$ with dimension $d_{ssd} = 1$. It is only the stringlike potential which has a massless limit.

In this case there is no representation theoretical reason to introduce them (no clash of localization and positivity), rather the only reason for doing this is to meet the power-counting preconditions for renormalizability. Whereas with pointlike field the power-counting short distance restriction of maximal $d_{sd}(interaction) = 4$ only allows a fi

[29] Here we do not distinguish between "physical" and "operator in a Hilbert space" i.e. "unphysical" refers to an object in an indefinit metric space. Of course they maybe good reasons to further restrict this terminology within a Hilbert space setting in a more contextual way.

[30] We use this terminology in a generalized sense; all the pointlike generators (the only ones considered in [25]) are called fiel strength (generalizing the $F_{\mu\nu}$) whereas the remaining string-localized generators are named potentials.

nite number of low spin models, the stringlike situation increases this number to infinite since now power-counting renormalizable interactions with string-localized potentials of arbitrary high spin exist. For example for the string-localized s=2 symmetric tensor potential $g_{\mu\nu}(x,e)$ there exist interactions which obey the power-counting condition, but this of course does not mean that specifi interesting models with pointlike observables as the Einstein-Hilbert action, are among this larger class of renormalizable candidates. Instead of searching for a gauge principle which singles out pointlike generated observables (e.g. the Riemann tensor $R_{\mu\nu\kappa\lambda}$), the problem one faces is to understand the relation between a coupling dependent law for the change of potentials under the change of string direction $e\rightarrow e'$ and the form of the pointlike composites.

It was already mentioned that the string-localization has hardly any physical consequences for photons, since even in the presence of interactions the content of the calculated theory can be fully described in terms of linearly related pointlike fiel strengths. Even the scattering theory of photons in the charge zero sector has no infrared problems. However the interaction-induced string-localization of the charged fiel which is transferred from the vectorpotentials[31] is a more serious matter; it is inexorably connected with the electric charge, and there is no linear operation nor any other manipulation which turns the noncompact localization of charged quantum matter into compact localization. The argument (10) based on the use of the quantum adaptation of Gauss's law shows that the noncompact (at best stringlike) localization nature of generating Maxwell charge-carrying field is not limited to perturbation theory.

Its most dramatic observable manifestation occurs in scattering of charged particles. As mentioned before, the infrared peculiarities of scattering of electrically charged particles were firs noted by Bloch and Nordsiek, but no connection was made with the string-localization which was suggested[32] at the same time by the formula (1) from gauge theory. One reason is certainly that the standard perturbative gauge formalism (which existed in its non-covariant unrenormalized form since the time of the B-N paper) was not capable to address the construction of string-localized physical fields This is particularly evident in renormalized perturbation theory which initially seemed to require just an adaptation of scattering theory [8], but whose long term consequences, namely a radical change of one-particle states and the spontaneous breaking of Lorentz invariance, were much more dramatic.

These phenomena were incompletely described in the standard perturbation theory of the gauge setting which had no convincing practicable way to extend the requirement of gauge invariance to the charged sectors. In particular the observable part of the scattering formalism culminated in a calculational momentum space recipe for inclusive cross sections; it was not derived in a spacetime setting as the LSZ scattering formalism for interactions of pointlike fields The spacetime setting in a theory as QFT, for which everything must be reduced to its localization principles, is much more important than in QM where stationary scattering formulations compete with time-dependent ones. As mentioned before Coulomb scattering in QM can be incorporated into any formulation of scattering theory by

[31]Localization of the free fields in terms of which the interaction is define in the perturbative setting, is not individually preserved in the presence of interactions; the would be charged field are not immune against delocalization from interactions with stringlike vectorpotentials.

[32]Localization properties in terms of gauge dependent field are not necessarily physical.

extracting a diverging phase factor which results from the long range. Noncompact string-localization is a more violent change from pointlike generated QFT than long- versus short range quantum mechanical interactions.

Perturbative scattering (on-shell) processes represented by graphs which do not contain inner photon lines turn out to be independent of the string direction e i.e. they appear as if they would come from a pointlike interaction[33]. This includes the lowest order Møller- and Bhaba scattering. The mechanism consists in the application of the momentum space fiel equation to the u,v spinor wave functions so that from (19) only the $g_{\mu\nu}$ term in the photon propagator survives. The terms involving photon lines attached to external charge lines do however depend on the string directions, and if the scattering amplitudes would exist (they are infrared divergent), they would be e-independent. The on-shell infrared divergence and the e-dependence are two sides of the same coin. One expects the photon inclusive cross section to be finit and e-dependent (at least in the low energy domain). By using the additional resource of e-smearing one expects for the firs time the possible formulation of a large time convergence aiming directly at inclusive cross sections.

In the sequel some remarks on the perturbative use of stringlike vectorpotentials for scalar QED are presented which is formally define in terms of the interaction density [34]

$$g\varphi(x)^*(\partial_\mu\varphi(x))A^\mu(x,e) - g(\partial_\mu\varphi(x)^*)\varphi(x)A^\mu(x,e) \tag{31}$$

It is also the simplest interaction which permits to explain the Higgs mechanism as a QED charge-screening. The use of string-localized vectorpotentials as compared to the standard gauge formalism deflect the formal problems of extracting quantum data from an unphysical indefinit metric setting to the ambitious problem of extending perturbation theory to the realm of string-localized fields This is not the place to enter a presentation of (yet incomplete) results of a string-extended Epstein-Glaser approach. Fortunately this is not necessary if one only wants to raise awareness about some differences to the standard gauge approach.

It has been known for a long time that the lowest nontrivial order for the Kallen-Lehmann spectral function can be calculated without the full renormalization technology of definin time-ordered functions. With the fiel equation

$$(\partial^\mu\partial_\mu + m^2)\varphi(x) = gA_\mu(x,e)\partial^\mu\varphi(x) \tag{32}$$

the two-point function of the right hand side in lowest order is of the form of a product of two Wightman-functions namely the point-localized $\langle\varphi(x)\varphi^*(y)\rangle = i\Delta^{(+)}(x-y)$ and that of the string-localized vectorpotential (19)

$$\left\langle A^\mu(x,e)A^\nu(x',e')\right\rangle \left\langle\partial_\mu\varphi(x)\partial_\nu\varphi^*(x')\right\rangle \tag{33}$$

leading to the two-point function in lowest (second) order

$$(\partial_x^2 + m^2)(\partial_{x'}^2 + m^2)\left\langle\varphi(x)\varphi^*(x')\right\rangle_{e,e'}^{(2)} \sim g^2\left\langle A^\mu(x,e)A^\nu(x',e')\right\rangle\left\langle\partial_\mu\varphi(x)\partial_\nu\varphi^*(x')\right\rangle \tag{34}$$

[33]The time-ordered correlation functions, of which they are the on-shell restriction, are however string-dependent.

[34]The integral over the interaction density is formally e-independent.

which is manifestly e-dependent in a way which cannot be removed by linear operations as in passing from potentials to fiel strength. One can simplify the e dependence by choosing collinear strings $e = e'$, but the vectorpotential propagator develops an infrared singularity and in general such coincidence limits (composites in d=2+1 de Sitter space) have to be handled with care (although these objects are always distributions in the string direction i.e. can be smeared with localizing testfunctions in de Sitter space); just as the problem of definin interacting composites of pointlike field through coincidence limits. The infrared divergence can be studied in momentum space; a more precise method uses the mathematics of wave front sets[35]. This simple perturbative argument works for the second order two-point function, the higher orders cannot be expressed in terms of products of Wightman function but require time ordering and the Epstein-Glaser iteration.

Not all functions of the matter fiel φ are e-dependent; charge neutral composites, as e.g. normal products $N(\varphi\varphi^*)(x)$ or the charge density are e-independent. On a formal level this can be seen from the graphical representation since a change of the string direction $e \to e'$ (21) corresponds to an abelian gauge transformation. The divergence form of the change of localization directions together with the current-vectorpotential form of the interaction reduces the e-dependence of graphs to *vectorpotentials propagators attached to external charged* lines while all e-dependence in loops cancels by partial integration and current conservation. This is in complete analogy to the standard statement that the violation of gauge invariance and the cause of on-shell infrared divergencies on charged lines result from precisely those external charge graphs; external string-localized vectorpotential lines cause no problems since they loose their e-dependence upon differentiation. A *neutral external composite* as $\varphi\varphi^*$ on the other hand does not generate an external charge line; again the gauge invariance argument parallels the statement that such an external vertex does not contribute to the string-localization.

Hence both the gauge invariance in the pointlike indefinit metric formulation and the e-independence in the string-like potential formulation both lead to pointlike localized sub-theories[36]. But whereas the embedding theory (Gupta-Bleuler, BRST) in the firs case is unphysical[37], the string-like approach uses Hilbert space formulations throughout. The pointlike localization in an indefinit metric description is a fake. Its technical advantage is that pointlike interactions, whether in Hilbert space or in a indefinit metric setting, are treatable with the same well known formalism. The gauge invariant correlation defin (via the GNS construction) a new Hilbert space which coalesces with the subspace obtained by application of the pointlike generated subalgebra of the physical string-like formulation to the vacuum.

[35]Technical details as renormalization, which are necessary to explore these unexplored regions, will be deferred to seperate work.

[36]Note however that the spacetime interpretation of the e is not imposed. The proponents of the axial gauge could have seen in in the free two-pointfunction of vectorpotentials and in all charge correlators if they would have looked at the commutators inside their perturbative correlation functions. The axial "gauge" is not a gauge in the usual understanding of this terminology.

[37]The pointlike localization in an indefinit metric description is a fake. Its technical use is that pointlike interactions, whether in Hilbert space or in a indefinit matric setting, come with a well known formalism. The gauge invariant correlation defin (via the KMS construction) a new Hilbert space which coalesces with the subspace obtained by application of the pointlike generated subalgebra of the physical string-like formulation to the vacuum.

But whereas the noncompact localized charge-carrying field are objects of a physical theory it has not been possible to construct physical charged operators through Gupta-Bleuler of BRST cohomological descent. The difficult here is that one has to construct non-local invariants under the nonlinear acting formal BRST symmetry. So the simplicity of the gauge formalism has to be paid for when it comes to the construction of genuinely nonlocal objects as charged fields

This leaves the globally charge neutral *bilocals* in the visor. Their description is expected to be given in terms of formal *bilocals which have a stringlike "gauge bridge"* linking the end points of the formal bilocals (2). In contrast to the string-localized single operators it is difficul to construct them in perturbation theory starting from string-localized free fields they are too far removed from the form of the interaction (see also next section). In order to understand the relation between such neutral bilocals and infraparticles one should notice that in order to approximate a scattering situations, the "gauge-bridge" bilocals will have to be taken to the limiting situation of an infinit separation distance, so that the problem of the infinit stringlike localization cannot be avoided since it returns in the scattering situation. The only new aspect of the proposed approach based on string-localized potentials which requires attention is that the dependence on the individual string directions e is distributional i.e. must be controlled by (de Sitter) test function smearing and moreover that composite limits for coalescing $e's$ can be defined

Finally there is the problem of Schwinger-Higgs mechanism in terms of string localization. The standard recipe starts from scalar QED which has 3 parameters (mass of charged field electromagnetic coupling and quadrilinear selfcoupling required by renormalization theory). The QED model is then modifie by Schwinger-Higgs screening in such a way that the Maxwell structure remains and the total number of degrees of freedom are preserved. The standard way to do this is to introduce an additional parameter via the vacuumexpectation value of the alias charged fiel and allow only manipulations which do not alter the degrees of freedom. We follow Steinmann [12], who find that the screened version consists of a selfcoupled real fiel R of mass M coupled to a vectormeson A^μ of mass m with the following interaction

$$L_{int} = gmA^\mu A_\mu R - \frac{gM^2}{2m}R^3 + \frac{1}{2}g^2 A_\mu A^\mu R^2 - \frac{g^2M^2}{8m^2}R^4 \tag{35}$$

$$\Psi = R + \frac{g}{2m}R^2 \tag{36}$$

The formula in the second line is obtained by applying the prescription $\varphi \to \langle\varphi\rangle + R + iI$ to the complex fiel within the neutral (and therefore point-local) composite $\varphi\varphi^*$ and subsequently formally eliminating the I fiel by a gauge transformation. The result is the above interaction where A_μ and R are now massive fields Since the fiel Ψ is the image of a pointlike $\varphi\varphi^*$ under the Higgs prescription, the real matter fiel Ψ is point-local.

The important point which formalizes the meaning of "screening" is that the algebraic Maxwell structure as well as the degrees of freedom remain preserved[38] even though the interaction in terms of the new field R and the massive vectorpotential $A_\mu(x,e)$ breaks the charge symmetry (by "screening" i.e. trivializing the charge, see below) and the even-odd

[38]The degrees of freedom of the real massive fiel I(x) went into the conversion of a photon into a massive vectormeson.

symmetry $R \to -R$ of the remaining R-interaction. It is this discrete symmetry breaking which renders the even-odd selection rule ineffective and by preventing that R can have a different localization from R^2 the pointlike localization of the quadratic terms is transferred to the linear R. The stringlike $d_{sca} = 1$ massive vectormeson $A_\mu(x,e)$ played the important technical role in the renormalizability of the theory but is not needed to describe the constructed theory in terms of generating fields a pointlike $F_{\mu\nu}$ (or an associated pointlike $A_\mu(x)$) and a pointlike $R(x)$.

Hence in the present context the string-localized potentials, as well as the BRST formalism, behave as a "catalyzer" which makes a theory amenable to renormalization. The former have the additional advantage over the latter that the Hilbert space is present throughout the calculation.

One has to be careful in order not to confuse computational recipes with physical concepts. Nonvanishing vacuum expectations (one-point functions) are part of a recipe and should not be directly physically interpreted, rather one should look at the intrinsic observable consequences[39] before doing the physical mooring. The same vacuum expectation trick applied to the Goldstone model of spontaneous symmetry breaking has totally different consequences from its application in the Higgs-Kibble (Brout-Englert, Guralnik-Hagen) symmetry breaking.

In the case of spontaneous symmetry breaking (Goldstone) the *charge associated with the conserved current diverges* as a result of the presence of a zero mass Boson which couples to this current. On the other hand in the Schwinger-Higgs screening situation *the charge of the conserved current vanishes* (i.e. is completely screened) and hence there are no charged objects which would have to obey a charge symmetry with the result that the lack of charge resulting from a screened Maxwell charge looks like a symmetry breaking.

$$Q_{R,\Delta R} = \int d^3x j_0(x) f_{R,\Delta R}(x), \;\; f_{R,\Delta R}(x) = \begin{matrix} 1 \; for \; |\mathbf{x}| < R \\ 0 \; for \; |\mathbf{x}| \geq R + \Delta R \end{matrix}$$

$$\lim_{R\to\infty} Q_{R,\Delta R}^{spon} |0\rangle = \infty, \; m_{Goldst} = 0; \;\; \lim_{R\to\infty} Q_{R,\Delta R}^{screen} \psi = 0, \; all \; m > 0$$

That the recipe for both uses a shift in fiel space by a constant does not mean that the physical content is related. The result of screening is the vanishing of a Maxwell charge which (as a result of the charge superselection) allows a copious production of the remaining R-matter. Successful recipes are often placeholders for problems whose better understanding needs additional conceptual considerations. In both cases one can easily see that the incriminated one-point vacuum expectation has no intrinsic physical meaning, i.e. there is nothing in the intrinsic properties of the observables of the two theories which reveals that a nonvanishing one-point function was used in the recipe for its construction. For a detailed discussion of these issues see [37].

The premature interpretation in terms of objects which appear in calculational recipes tends to lead to mystification in particle theory; in the present context the screened charged particle has been called the "God particle". As mentioned before the Schwinger-Higgs screening is analog to the quantum mechanical Debeye screening in which the elementary Coulomb interaction passes to the screened large distance effective interaction which has

[39]These are properties which can be recovered from the observables of the model i.e. they do not depend on the particular method of construction.

the form of a short range Yukawa potential. The Schwinger-Higgs screening does not work (against the original idea of Schwinger) directly with spinor- instead of scalar matter. If one enriches the above model by starting from QED which contains in addition to the charged scalar field also charged Dirac spinors then the screening mechanism takes place as above via the scalar fiel which leads to a loss (screening, bleaching) of the Maxwell charge while the usual charge superselection property of complex Dirac field remains unaffected.

The Schwinger-Higgs mechanism has also a scalar fiel multiplet generalization to Yang-Mills models; in this case the resulting multicomponent pointlike localized massive model is much easier to comprehend than its "charged" string-localized origin. As the result of screening there is no unsolved confinement/i visibility problem resulting from nonabelian string-localization.

The Schwinger Higgs screening suggests an important general idea about renormalizable interactions involving massive $s \geq 1$ fields namely that formal power-counting renormalizability ($d_{sdd} = 1$) is not enough. For example a pure Yang-Mills interaction with massive gluons (without an accompanying massive real scalar multiplet) could be an incorrect idea because the string-localization of the Hilbert space compatible gluons could spread all over spacetime or there may exist other reasons why the suspicion that such theories are not viable may be correct. Such a situation would than be taken as an indication that a higher spin massive theory would always need associated lower spin massive particles in order to be localizable; in the s=1 case this would be the s=0 particle resulting from Schwinger-Higgs screening. Before one tries to understand such a structural mechanism which requires the presence of localizing lower spin particles it would be interesting to see whether these new ideas allow any renormalizable s=$\frac{3}{2}$ (Rarita-Schwinger) theories. Even though there may be many formal power-counting renormalizable massive $s \geq 1$ interactions only a few are expected to be pointlike localized.

It is interesting to mention some mathematical theorems which support the connection between localization and mass spectrum. The support for placing more emphasis on localization in trying to conquer the unknown corners of the standard model comes also from mathematical physics. According to Swieca's theorem [36][37] one expects that the screened realization of the Maxwellian structure is local i.e. the process of screening is one of reverting from the electromagnetic string-localization back to point locality together with passing from a gap-less situation to one with a mass gap. Last not least the charge screening leads to a Maxwell current with a vanishing charge[40] and the ensuing copious production of alias charged particles. The loss of the charge superselection rule in the above formulas (35) is quite extreme, in fact even the $R \leftrightarrow -R$ selection rule has been broken (35) in the above Schwinger-Higgs screening phase associated with scalar QED. The general idea for constructing renormalizable couplings of massive higher spin potentials interacting with themselves or with normal s=0,1/2 matter cannot rely on a Schwinger-Higgs screening picture because without having a pointlike charge neutral subalgebra for zero mass potentials as in QED, which is the starting point of gauge theory, there is no screening metaphor which could preselect those couplings which have a chance of leading to a fully pointlike localized theory, even though renormalizability demands to treat all $s \geq 1$ as stringlike objects with d_{sca}=1. Of course at the end of the day one has to be able to fin the renormalizabe models

[40]Swieca does not directly argue in terms of localization but rather uses the closely related analyticity properties of formfactors.

which maintain locality of observables either in the zero mass setting as (charge-neutral) subalgebras (QED,Yang-Mills) or the massive theories obtained from the former with the help of the screening idea. gauge theory is a crutch whose magic power is limited to s=1, for s> 1 it lost its power and one has to approach the localization problem directly.

The existence of a gauge theory counterpart, namely the generalization of the BRST indefinit metric formalism to higher spins, is unknown. So it seems that with higher spin one is running out of tricks, hence one cannot avoid confront the localization problem of separating theories involving string-localized potentials which have pointlike generated subalgebras from those which are totally nonlocal and therefore unphysical. This opens a new chapter in renormalization theory and its presentation would, even with more results than are presently available, go much beyond what was intended under the modest title of this paper.

An understanding of the Schwinger-Higgs screening prescription in terms of localization properties should also eliminate a very unpleasant previously mentioned problem which forces one to pass in a nonrigorous way between the renormalizable gauge (were the perturbative computations take place) and the "unitary gauge" which is used for the physical interpretation. The relation between the two remains somewhat metaphoric.

The screened interaction between a string-localized massive vectorpotential and a real fiel (35) remains pointlike because the string localization of the massive vectorpotential only serves to get below the power counting limit but does not de-localize the real matter field since the pointlike fiel strength together with the real scalar fiel generate the theory, the local generating property holds. In an approach based on string-localization there is only one description which achieves its renormalizability by string-localized potentials.

The BRST technology is highly developed, as a glance into the present literature [39] shows. It certainly has its merits to work with a renormalization formalism which starts directly with massive vectormesons [40] instead of the metaphoric "photon fattened on the Higgs one point function". It is hard to think how the BRST technology for the presentation of the Schwinger-Higgs screening model which starts with a massive vectormeson in [39] can be improved. For appreciating this work it is however not necessary to elevate "quantum gauge symmetry" (which is used as a technical trick to make the Schwinger-Higgs mechanism compliant with renormalizability of massive s=1 fields from a useful technical tool to the level of a new principle.

Besides the Schwinger-Higgs screening mechanism which leads to renormalizable interactions of massive vectormesons with low spin matter, there is also the possibility of renormalizable "massive QED" which in the old days [41] was treated within a (indefinit metric) gauge setting in order to lower the short distance dimension of a massive vectormeson from $d_{sca} = 2$ to 1, and in this way stay below the powercounting limit. Such a construction only works in the abelian case; for nonabelian interactions the only way to describe interacting massive vectormesons coupled to other massive s=0,1/2 quantum matter is via Higgs scalars in their Schwinger-Higgs screening role. Whereas the local Maxwell charge is screened, the global charges of the non-Higgs complex matter field are preserved. It seems that Schwingers original idea of a screened phase of spinor QED cannot be realized, at least not outside the two-dimensional Schwinger model (two-dimensional massless QED).

But the educated conjectures in this section should not create the impression that the role of the Schwinger-Higgs screening in the renormalizability of interactions involving selfcoupled massive vectormesons has been completely clarified if anything positive has been achieved, it is the demystificatio of the metaphor of a spontaneous symmetry breaking through the vacuumexpectation of a complex gauge dependent fiel and the tale of "God's particle" which creates the masses of s=1/2 quantum matter. Actually part of this demystificatio has already been achieved in [39].

This leads to the interesting question whether, apart from the presence of the Higgs particle (the real fiel as the remnant of the Schwinger-Higgs screening), there could be an intrinsic difference in the structure of the vectormeson. Such a difference could come from the fact that the screening mechanism does not destroy the algebraic structure of the Maxwell equation, whereas an interaction involving a massive vectormeson coming in the indicated way from a S-H screening mechanism and interacting with spinorial matter field maintains the Maxwell structure. In the nonabelian case this problem does not arise since apparently the Schwinger-Higgs screening mechanism is the only way to reconcile renormalizability with localizability (or a return to physics from an indefinit metric setting).

This raises the interesting question whether renormalizability and pointlike locality of interactions with massive higher spin $s > 1$ potentials is always related to an associated zero mass problem via an analog of a screening mechanism in which a lower spin fiel plays the analog of the Higgs field

Whereas for interactions between spin one and lower spin field the physical mechanism behind the delocalization of matter (or rather its noncompact re-localization) is to some degree understood, this is not the case for interacting higher spin matter. Stringlike interactions enlarge the chance of potentially renormalizable (passing the power counting test) theories, in fact stringlike potentials with dimension $d_{sca} = 1$ exist for any spin (hence infinitel many) whereas the borderline for pointlike interaction is $s = 1/2$ and with the help of the gauge setting $s = 1$. Certain interactions, as the Einstein-Hilbert equation of classical gravity probably remain outside the power-counting limit even in the stringlike potential setting, but certain polynomial selfinteraction between the $g_{\mu\nu}(x,e)$ with $dimg_{\mu\nu}(x,e) = 1$ may be renormalizable. The existence of free pointlike fiel strength (in this case the linearized Riemann tensor) indicates that there may be renormalizable interactions which lead to pointlike subalgebras, but the presence of self-couplings modifie the transformation law under a change of e (21) which now depends on the interaction as it is well-known from the gauge theoretical formulation for Yang-Mills couplings.

One of course does not know whether QFT is capable to describe quantum gravity, but if it does in a manner which is compatible with renormalized perturbation theory, there will be no way to avoid string-localized tensorpotentials even if the theory contains linear or nonlinear related pointlike localized fiel strength. The trick of gauge theory, by which one can extract pointlike localized generators without being required to construct firs the string-localized ones, is a resource which does not seem to exist for higher spins, not even if one is willing to cope with unphysical ghosts in intermediate steps. The most interesting interactions are of course the selfinteraction between $(m = 0, s > 1)$. Here one runs into similar problems as with Yang-Mills models (next section). The independence on $e's$ of the local observables leads to nonlinear transformation laws which extend that of free stringlike potentials and the non-existence of linear local observables. Although saying this does not solve any such problem, the lack of an extension of the gauge idea to higher spin makes one at least appreciative of a new view based on localization.

There is one important case which we have left out, namely that of massless Yang-Mills theories interaction with massive matter. This will be discussed in the next section.

There are 2 different categories of delocalization: string-localization with nontrivial pointlike-generated subalgebras. Generically the coupling of string-localized field leads to a theory with no local observables. The models of physical interest are those which contain eindependent subfields For the case at hand the crucial relation is that the change in the string direction can be written as a derivative as in (21).

7. A Perturbative Signal of "Invisibility" and "Confinement"?

The reader may wonder why a concept as confinement which for more than 4 decades has been with us and entered almost every discourse on strong interactions appears in the title of this section in quotation marks. The truth is that these issues are the least understood aspects of gauge theory, and there are even reasons why this terminology may be questionable.

In QM a confinemen into a "cage" can be implemented by choosing a confinin potential; since the firs models for quark-confinemen were quantum mechanical, this explains the origin of the terminology. However the conceptual structure of QFT is radically different, and a spatial confinemen is not implementable in a setting in which causal localization is the main physical principle. In fact it was (and still is) one of the conjectures [37] of the 60s that compact localizability together with a limitation on the phasespace degree of freedom (in the old days it was describe by "compactness", nowadays it is "nuclearity"[41]) would result in "asymptotic completeness" i.e. the property that every state of the theory can be written as a superposition of (generically infinitel many) particle states. The LSZ scattering theory addresses precisely this case and there is a vast conceptionally not understood terrain outside these limitations. Inasmuch as already the notion of infraparticles can hardly be handles in the classical physics rooted Lagrangian quantization setting, one is well advised to be prepared that phenomena of invisibility/confinemen which one expects in Yang-Mills theories are even further removed from quantization parallelisms.

N-particle states, in a LQP scenario are simply the stable n-fold counter clicks in a coincidence/anticoincidence arrangement in which the spacetime continuum has been cobbled with counters measuring excitations of the vacuum. Here stable means that a counter-registered event of n–fold excitation at large times does not later turn any more into an m-state $m \neq n$ with changed velocities. Only in this case of asymptotic stability when the centers of wave-packets are far separated it makes sense to replace the word "excitation" (of the vacuum) by particle. Objects localized on semiinfinit spacelike strings blur this picture.

Part of this picture can be proven. It is true that a stable n-fold vacuum excitation state is a tensor product state of n Wigner particles [2][42]. The only known physical counter example, the electrically charged infraparticle states which are obtained by applying a smeared string-localized Maxwell-charged operator to the vacuum and studying its asymptotic be-

[41] The expression refers to the cardinality of states in a phase space cell being equal to states in the range of a compact or trace-class operator.

havior, contain the mass shell component with vanishing probability [42] and require a conceptually different scattering treatment. Infraparticles are most suitably described in terms of "weights" (a kind of singular state which cannot be associated to a state vector), which are directly related to probabilities. Fortunately the Wigner representation theory extends to weights. Such weights are define by expectations of suitably selected observables in plane wave infraparticle weights[43] .

No matter how much one compresses the momentum support against the lower value $p^2 = m^2$, one never arrives at a state which is not populated by infinitel many photons; one can only control the value of the measuring resolution Δ, but there will be always infinitel many photons with energies below Δ which escape detection; if we refin our registering precision i.e. $\Delta \to 0$, we do in the end not register anything since the charged particles with sharp mass are not states but "weights". In other words there is a certain intrinsic lack of precision in registering infraparticles in their states; assuming that the radiation of photons is the only way by which we can measure the presence of charged particles, the infraparticles with small Δ would not be perceived in agreement with the vanishing inclusive cross section for $\Delta \to 0$.

But what happens in case of pure Yang Mills interaction of string-localized gluons which live in a ghost-free physical space, and which by the very nonlinear structure of Yang-Mills theories have to play simultaneously the role of the charged particles and of the mediators of their interactions? It is not easy to propose even a metaphoric idea based on such a vague analogy.

The difference to the abelian situation discussed in the previous section becomes particularly evident on the issue of localization. Among the field in terms of which the (perturbative) interaction is defined there is none which is linearly related to pointlike generated basic fields all *local observables are generated by nonlinear composites,* or equivalently the interactions of the string-localized higher spin potentials involve self-interactions. This means that the independence under changes of string directions, i.e. the characterization of pointlike generated observables, is a more demanding issue since now the change of potentials under a change of string direction becomes entangled with the very nature of interaction and the composite nature of pointlike field which generate the observable subalgebras.

The simple rule (21) holds only in zero order; again the formal aspects of the axial gauge are of valuable help in findin the new dynamical law for the change of string localization which replaces the law (21). Related to this complication is the fact that, unlike in QED, there are now no identifiabl perturbative closed gluon loop structures *inside* Feynman diagrams which are e-independent rather one needs to form sums of loop structures. This complicates the case of Yang-Mills gluons from the viewpoint of localization and the ensuing infrared divergence behavior of its correlations [43].

Even if one succeeds to control all the notorious infrared problems of nonabelian gauge theories by e_i-smearing and/or compositeness, the nontrivial task of how to adapt perturbation theory to recover the pointlike generators of the composite local observables (squares of "fiel strengths") still remains. The BRST formalism delegates the problem of pointlike

[42] The Hilbert space of QED does not contain a Wigner particle with the mass of the electron.

[43] Infrared divergencies are the momentum space manifestations of noncompact localizations required by the principles in certain interacting higher spin QFTs.

gauge invariance to the problem of identifying BRST invariant polynomials in the gluon potentials and quark fields But since this "symmetry" acts in a nonlinear way, this is no trivial matter; in fact looking at the literature it seems to be impossible to directly construct perturbative renormalized gauge invariant correlation functions of F^2 (e.g. two-point function) without passing through the renormalization of s.l. correlations. In YM models there are no "good gauges" (as most gauges in QED except axial gauges) in which the gauge-variant pointlike correlations still exists and the infrared complications only appear in the on shell amplitudes (to be formally removed by gauge-invariant prescriptions for inclusive cross sections) one rather has to confront the problem of renormalization of s.l. off-shell correlations. The big gain of the stringlike formulation over the standard formulation is that the *physical origin of the infrared problem has been identified* which shows the possible cure with the help of split point limits in the de Sitter space of the $e's$, an additional handle on the problem which is not available in the standard gauge formulation.

The form of the interactions is often presented as resulting from the imposition of a gauge group symmetry in analogy to an internal symmetry group. As already noted by Stora [44] this is not correct since the form of the Yang Mills interaction follows from the number of selfinteractin string-localized potentials and the requirement for the existence of pointlike generated subalgebras whereas the in the gauge theoretic setting it becomes part of the technicalities of the consistency of the nonabelian BRST formalism. In this way one keeps in mind that the BRST symmetry is a formal device which has absolutely nothing to do with physical symmetries [44] but results from the requirement about the existence of *local* observables and their pointlike generators. Of course all symmetries are related to localization [2], but gauge "symmetries" are most directly related than standard inner symmetries for which one needs the DHR superselection theory to see there origin in the locality principle.

This closeness to the axial gauge should also dispels the impression that the subject of the present paper is something very speculative and distant. From a pragmatic viewpoint it is nothing else than the attempt to make sense of the axial gauge by understanding the origin of its apparent incurable infrared divergences and figurin out how they arise from overlooked localization problems which cause fluctuatin string directions and reduce those in x. There is certainly nothing more conservative than tracing the infrared divergences to their origin and taming them by controlling the fluctuatio which cause them.

There is of course the still open problem of generalizing the Epstein Glaser setting from pointlike to stringlike fields a formalism which practitioners of QFT hardly pay attention to since its impact on computation is insignificant but which in the new context gains in importance. A de-localization of quantum matter which cannot be controlled in terms of the correlated field (but worsens with the growing internal structure of Feynman graphs) would not be acceptable. It is actually easy to see that the Epstein-Glaser iteration works on restricted e-configuration for $e's$ on a hyperplane [38].

From the point of view of string-localized potentials the infrared divergencies of the axial gauge fi ed QED are to be expected, since the $e's$ are *not fixed gauge parameters* as in axial gauge theory, but belong to operator-valued distributions whose arguments fluctuat in x *and* e; so in order to get to coalescing $e's$ (points in de Sitter space), one has to use similar ideas as those by which one define pointlike *composites* (in Minkowski space) in standard ($s < 1$) non-gauge models. These infrared divergencies, which remain unexplained in the gauge approach, are the strongest signal yet that "gluons" and quarks are badly localized, since their infrared singularities are much worse than those of charged particles in fi ed gauges.

The idea which is most close to the phenomenological hadronization through "gauge bridge breaking" consists in assuming that, although a gluon/quarks may be color-charged states in the Hilbert space they may carry infinit energy so that the although the expectation of the conserved energy momentum tensor density in such a stated is well-defined its *total integral diverges* thus assigning an infinit energy to isolated gluon/quark states. This would look a bit as the spontaneous symmetry breaking except that the conserved current is not one which, as in Goldstone's case, belongs to an internal symmetry, and therefore the standard conclusions about the presence of a zero mass Goldstone boson do not apply. Such a situation would immediately explain the origin of the spontaneous broken Poincaré covariance which in the QED case was only indirectly inferred as a result of the quantum properties of the Gauss law which spontaneously breaks the Lorentz boosts.

The idea of breaking up local composites into their elementary pieces is already quite tricky in theories with $s \leq 1/2$ field as the following illustration shows. Let $A(x)$ be a free scalar fiel and ask the simple question whether the local composite $: A^2(x) :$ can be split into bilocals:

$$:A^2(x): \quad \stackrel{?}{\longrightarrow} \quad A(x)A(y) \tag{37}$$

This is indeed possible by applying a subtle lightlike limiting procedure [13] to the product $: A^2(x) :: A^2(y) :$ which makes use of the singularity appearing for lightlike separation. This idea was used in chiral models to show that currents determine bilocals [45]. But there is hardly any experience with this splitting in gauge theories [14]. A successful splitting would of course automatically generate a gauge invariant bridged bilocal. It would be a firs step in an extension of the DHR superselection theory to gauge theories

The infinit spin class of Wigner representations offers another somewhat more speculative explanation. This rather large class of representations lead to indecomposable semi-infinit strings; in contrast to infraparticle states they are not only indecomposable on the algebraic level but even remain so as states in fact the Wigner representation space is the only one which contains no pointlike generating wave function at all. A (necessarily) localized counter cannot register a piece of an irreducible semiinfinit string; in that case the energy is finit and the reason of invisibility is more directly related to localization than the infinit energy argument.

The appearance of string-localized representations[44] of the third Wigner class (massless infinit spin) in gauge theories is not very plausible, since in a perturbative setting the kind of irreducible representations of the Poincaré group which appear in an interacting theory is believed to be already decided by the zero order input. In other words, it is difficul to conceive of an mechanism whereby a free multiplet of gluon potential $A_\mu^a(x, e)$ (16) through interactions with itself passes to an object $A_\mu^a(x, e)_{i.s.}$ whose application to the vacuum contains infinit spin components (23). However outside of perturbation theory this may not be true; there exists presently no theorem which excludes the possibility that

[44] In section 3 we made a distinction between string-localized representations and zero mass string-localized covariant potentials in pointlike generated (by "fiel strengths") representations which do not exist as pointlike objects and whose only mark on the representation is the A-B effect.

the application of interacting gluons to the vacuum contains an irreducible infinit spin representation component since such a statement would also rule out bound states. Since infinit spin representations are inert, apart from their coupling to gravity (since they carry nontrivial energy-momentum), they fi better to dark matter than to gluons/quarks.

The aim of this paper is to recall unexplored (and not fully explored) regions in QFT and shed new light onto them from the principle of localization. All properties met in QFT models can an must be traced back to this principle, only then one can claim to have understood a problem. It has been shown elsewhere [31] that QFT allows a presentation solely in terms of the "modular positioning" of a finit number of "monads" where a monad stands for the algebraic structure (hyperfinit type III_1 von Neumann) which one meets in the form of localized algebras in QFT. The only reason why this is mentioned here is its Leibnizian philosophical content: the wealth of QFT can be encoded into the abstract positioning of a finit number of copies of one monad into a common Hilbert space. The encoding encloses even spacetime (the Poincare group representations) and the information about the kind of quantum matter. In other words *relative modular positions in Hilbert space have physical reality, the substrate*[45] *which is being positioned does not.* Modular positioning, modular localization and Poincaré symmetry are inexorably interwoven. What is important is not the individual localized operator but the ensemble of operators encoded into a localized algebra, a fact which is sufficien for the computation of the S-matrix (the difference between localized individual operators can be absorbed into normalization constants [2]). This viewpoint parallels that of an experimentalizer who's only means to enhance the precision of measurements is to reduce the size of the effective part of the counters and place them further away from the interaction regions where two beams meet. It is well-known that the vacuum restricted to a local measuring device is a thermal KMS state [4]. Such thermal states have an intrinsic natural concept of probability, namely that of statistical mechanics. This poses the question whether the Born probability of QM which refers to individual operators (example the position operator) should not be considered as an artificia extension to individual events of the natural intrinsic probability coming from the thermal aspects of modular localization. It seems that such an origin of probabilities in terms of thermal ensembles could have even been acceptable to Einstein who was a life-long adversary of Born's quantum mechanical probability. this points towards other consequences of modular localization than those used her in order to resolve the clash between localization and Hilbert space positivity which is resolved for all $(m = 0, s \geq 1)$ representations in terms of stringlike localization.

8. Resumé and Concluding Remarks

The main aim of this paper is the presentation of old important unsolved problems of gauge theory in a new light. The standard gauge approach to the renormalizable model of QED and its Yang-Mills generalization maintains the pointlike formalism for s=1 vectorpotentials and sacrifice the Hilbert space in intermediate computational steps; the advantage is that the classical formalism is quantized in the standard pointlike way. The new string-localized

[45] Modular positioning is the most radical form of relationalism since the local quantum matter arises together with internal and spacetime symmetries. In other words the concrete spacetime ordering is preempted in the abstract modular positioning of the monads in the joint Hilbert space.

setting on the other hand avoids the indefinit metric Gupta-Bleuler/BRST formalism and indicates how string-dependent charge-carrying operators in QED should be constructed.

In the case of the Higgs mechanism there is in addition the necessity to pass between the unitary (physical) and the renormalizable gauge which is absent in the s.l. setting. In terms of the non-covariant axial gauge which has the attractive property of coming with a Hilbert space representation but has an incurable infrared divergency the s.l. description can be seen as a radical re-interpretation: this gauge is not really a gauge in the standard use of the terminology, rather it is a semiinfinit string-localized vectorpotential with variable spacelike string directions which acts in a Hilbert space which transforms covariant (in the reinterpretation e plays an important role in the covariance law)

A more profound justificatio is of course to forget about gauges altogether and use the fact that, although certain covariant field cannot exist in the setting of the Wigner representation theory, the situation changes completely if one allows semiinfinit spacelike string-localized covariant field $\Psi(x, e)$ which for any spin always maintain the minimal scale dimension 1. These field we summarily called *potentials*, since the vectorpotential is the prime example (for higher s there are also tensorpotentials) whereas the pointlike generators whose scale dimension increases with spin are the (generalized) fiel strengths. The potential field fulfil the correct power counting prerequisite for renormalizability and do not need any power lowering BRST formalism, not even in the massive case.

It may be helpful to collect the arguments for the use of those noncompact localized potentials (instead of the pointlike indefinit metric potentials) presented in this paper:

- The gauge theoretic argument why electrically charged operators cannot be compactly localized remains obscure. Although the structural argument based on Gauss law is rigorous, it does not really explain the de-localization in terms of localization properties of the interaction itself.

- Rewriting the quantum magnetic flu through a surface via Stokes theorem into an integral over a pointlike vectorpotential leads to a contradiction with the QFT A-B effect, whereas the use of string-localized vectorpotential removes this discrepancy. Although this rather simple calculation does not instruct how to formulate interactions, it does show that in order to avoid incorrect conclusions about localizations, one should always work with s.l. instead of p.l. vectorpotentials outside the Hilbert space setting.

- In most perturbative calculations in the gauge theoretical formalism the condition of gauge invariance in terms of BRST invariance is clearly formulated, but gauge invariant correlation functions of composite operators (not to mention s.l. charged correlators) are, as a result of computational difficulties rarely calculated. In the new approach there is *no gauge conditions*, rather the perturbative results are already the physical one.

- The reformulation of the Higgs phenomenon in the Schwinger-Higgs screening setting removes some mysterious aspects of the former and brings it into closer physical analogy with the Debeye screening mechanism of QM. Whereas the latter explains how long range Coulomb interactions pass to effective Yukawa potentials, the former describes the more radical change from semiinfinit electrically charged strings with infrared photon clouds to massive Wigner particles associated with pointlike fields This more radical screening is accompanied by a breaking of the charge symmetry (vanishing charge) and the breaking of the even-oddness symmetry of the remaining real fiel which makes the screening contribution from the alias charge neutral $\varphi\varphi^*$ sector (after screening) indistinguishable from that of φ.

- The localization issue in case of Yang-Mills interactions and QCD (as well as for self-interacting $s \geq 2$ models) is more involved since the change under string direction is dynamical law instead of the kinematical law (21) which follows from Wigner's representation theory. This leads to a much stronger infrared behavior, in fact all spacetime correlators are infrared divergent and only some spacetime independent coupling-dependent functions as the dimensionally regularized beta function are infrared finite The new string-localization approach explains this and proposes to take care of the e-fluctuation which cause the infrared divergences and clarify their role in confinement/i visibility and gauge-bridge breaking (jet formation). This, as well as the still missing presentation of an Epstein-Glaser approach in the presence of string potentials, will be the topic of a separate work[46].
- The approach based on string-localized potentials does not only replace the gauge setting, which resulted from a resolution of the clash between pointlike localization and quantum positivity with the brute force method of indefinit metric in order to maintain quantization, but it is also meant to be useful for higher spins (example: $g_{\mu\nu}$ string tensorpotentials) where such a gauge trick is not known. In addition to the avoidance of indefinit metric it also lowers the short distance dimension of pointlike fiel strength s+1 (for spin s) to the lowest value $d_{sca} = 1$ allowed by unitarity which is the prerequisite of having renormalizable interactions for any spin.

Within the conventional standard terminology of QFT the present project to incorporate string-localized objects into already existing settings (standard model, s=2 "gravitons") would be considered as a kind of "nonlocal" QFT. To make QFT compatible with nonlocality is one of the oldest (but never suffientl understood) projects of relativistic QT. Apart from early (pre-renormalization) attempts to modify the quantum mechanical commutation relations to make them more quantum-gravity friendly, the more systematic investigations in QFT started in the 50s with attempts by Christensen and Møller to improve the behavior of interactions in the ultraviolet by spreading interaction vertices in a covariant manner. Later attempts included the Lee-Wick proposal to modify Feynman rules by pair of complex poles and their conjugates. All these models were eventually shown to contradict basic macro-causality properties which are indispensable for their interpretation[46]. There are of course relativistic quantum mechanical theories which lead to a Poincaré-invariant clustering S-matrix [31], but do not fi into the causal localization of the QFT setting.

The more recent interest in nonlocal aspects originated in ideas about algebraic structures (noncommutative QFT) which are supposed to replace classical spacetime as the firs step towards "quantum gravity". These attempts usually start from a modificatio

[46] Jens Mund and Bert Schroer, in progress.

of the quantum mechanical uncertainty relation which, since they involve position operators, strictly speaking does not exist in QFT[47]. The only analog of the uncertainty relation which comes to one's mind is the statement that one can associate to a localized algebra $\mathcal{A}(\mathcal{O})$ and a "collar" of size ε (the splitting distance) which separates $\mathcal{A}(\mathcal{O})$ from its causal disjoint [2], a localization entropy (or energy) $Ent(\varepsilon)$ which is proportional to the surface and diverges for $\varepsilon \to 0$ in a model-independent manner [48]. Whether such relations between the sharpness of localization and the increase of entropy/energy can be the start of a noncommutative/nonlocal project remains to be seen.

If one can consider this thermal relation as a QFT substitute of an uncertainty relation, it points into a quite different direction than the proposal for a more noncommutative modificatio of QFT [47]. Namely it looks like an invitation to explore connections between thermodynamics/statistical mechanics, a project which Ted Jacobson has pursued for some time [49]. Unlike algebraic modification for position operators it has the appealing feature of not having to struggle with problems of frame dependence.

The philosophy underlying the noncommutative approach has been nicely exposed in a recent essay by Sergio Doplicher [50]. His emphasis that a principle as causal localization can only be overcome by another principle which contains the known one in the limiting situation of large distances is certainly well taken, as in many cases, the devil is in the details.

In the present work, the nonlocal behavior remains part of QFT; it may go against certain formalisms as Lagrangian quantization or functional representations, but it certainly does not lead to reasons to change principles of QFT and it also is not "revolutionary", it only belongs to one of its unexplored corners. Even with respect to quantum gravity the two nonlocal approaches remain different. Within the nonlocality allowed by QFT it would be tempting to relate gravity with selfinteractin $d_{sca} = 1$ string-localized tensor potentials $g_{\mu\nu}(x, e)$. The hope is that one can access this problem (and if necessary dismiss it) with more conventional means.

The aim of this paper was to cast new light on unexplored regions of gauge theory based on recent progress in the understanding of modular localization. There was however no attempt to go into the important details of the new perturbation theory in terms of string-localized potentials. This will be the subject of forthcoming work [38].

Acknowledgements: I thank Jakob Yngvason for his invitation to visit the ESI in Vienna and for his hospitality. Part of the research which entered this paper was carried out at the ESI. I also acknowledge several discussions with Jakob Yngvason especially on matters related to section 5.

This work is in a way a continuation of a paper written several years ago together with Jens Mund and Jakob Yngvason. I am indebted to Jens Mund with whom I have maintained close scientifi contact and who has kept me informed about recent results about attempts to adapt the Epstein-Glaser renormalization and the idea of wave front sets to string-localized fields I owe Karl-Henning Rehren thanks for bringing me up to date on matters which were relevant for the appendix.

[47] As a result these attempts lead to problems with the principle of independence of the reference frame and in a certain sense open the backdoor for the return of the ether.

9. Appendix

9.0.1. Modular Group for Conformal Algebras Localized in Doubly-Connected Spacetime Rings

It has been known for a long time that the modular group for a conformal double cone which is placed symmetrically around the origin is related to that of its two-dimensional counterpart by rotational symmetry. In other words if

$$x_{\pm}(\tau) = \frac{(1+x_{\pm}) - e^{\tau}(1-x_{\pm})}{(1+x_{\pm}) + e^{\tau}(1-x_{\pm})}, \quad -1 < x_{\pm} < 1 \tag{38}$$

represents the two-dimensional conformal modular group in lightray coordinates for a two-dimensional double cone symmetrically around the origin, then the modular group of a symmetrically placed four-dimensional double cone which results from the two-dimensional region by rotational symmetry acts as above by simply replacing $x_{\pm}(\tau)$ in the above formula by their radial counterpart [51]

$$x_{\pm}(\tau) \rightarrow r_{\pm}(\tau),\ \varphi, \theta,\ \tau - independent \tag{39}$$

The generalization to two and more copies of double cones in two dimensions, symmetrically placed on both sides of the origin is obviously a group which in terms of $x_{\pm}$ has 4 or 2n fi ed points which are the endpoints of two separated intervals. The construction of explicit formulae for n intervals $E = I_1 \cup I_2 .. \cup I_n$ with 2n fi ed points is well-known; they are most conveniently obtained as Caley-transforms of one-parametric subgroups of $Diff(S^1)$

$$f_{\tau}^{(n)}(z) = \sqrt[n]{Dil(-2\pi\tau)z^n},\ Dil(-2\pi t)x = e^{2\pi t}x \tag{40}$$
$$x \rightarrow z = \frac{1+ix}{1-ix},\ Cayley\ transf.\ \dot{R} \rightarrow S^1$$

This diffeomorphism group in terms of x is infinity-preserving By applying further infinit preserving symmetry transformations (translations..) we may achieve the desired symmetric situation with respect to the origin.

For n=2 the two double cones are the x-t projections of 4-dimensional matter localized in $\mathcal{T}$ and not matter in a 2-dimensional conformal theory. This suggests that in looking for a geometric analog of (39) one should be aware that the e.g. the full diffeomorphism group $Diff(S^1)$ has no analog in 4 dimensions; in fact not even the Moebius subgroup associated to the Virasoro generator L_0 has a counterpart. Hence arguments which are based on properties of L_0 as the necessity to work with split vacua states [52] or with mixing [53][34] are not applicable here.

The use of the above formalism in connection with modular theory of multi-intervals and two-dimensional multi-double cones has been presented in great detail in [52]. In particular it was shown that in the presence of the L_0 in the Virasoro algebra there is no global representation of the $f_{\tau}^{(n)}(z)$ diffeomorphism. Rather the best one can do by choosing instead of the global vacuum the so-called split vacuum modular recent is to represent this diffeomophism group on E and have an non-geometric action on its complement E', or construct a "geometric state" (another split vacuum) for E' and then to put up with an nongeometric action on E.

In the special case of a chiral Fermion one can achieve a global quasi-geometric action in the vacuum at the expense of a mixing between the different intervals by a computable mixing matrix [53][34]. But only the projections of localized zero mass matter in d=1+3 are candidates for a pure geometric action in their standard vacuum state.

This difference extends to the explanation of violation of Haag duality for $(m = 0, s \geq 1)$. Whereas in the chiral case this is due to charge transporters whose construction requires the setting of fiel theory with its characteristic property of vacuum polarization, the Aharonov Bohm effect in QFT can be fully described in the Wigner one-particle representation and the crucial property is positivity in the sense that the use of covariant pointlike vectorpotentials which violates positivity would also violate the A-B effect in QFT.

The localization of n $\mathcal{T}$ symmetrically placed around the origin has a x-t projection which consists of n symmetrically arranged two-dimensional double cones. The diffeomorphism group which leaves this figur invariant is a particular diffeomorphism group which in lightray coordinates is a diffeomorphism with 2n fi ed points. The number of stringlike potential associated with a pointlike fiel strength increases with spin s; there is always one with the lowest possible dimension which is $d_{sca} = 1$ and the one with the highest dimension has a d_{sca} which is smaller than that of the lowest fiel strength. So the A-B flu es which account for the string-localized potentials are certainly expected to increase with s. But the situation of n disconnected $\mathcal{T}$ appears repetitive. It would be fascinating if the increase of s could be linked with the occurrence of a new type of A-B effect in higher genus (higher connectivity) analogs of $\mathcal{T}$ instead of being n-$\mathcal{T}$ repetitive.

References

[1] S. Mandelstam, Phys. Rev. **112**, (1958) 1344 and Phys. Rev. **115**, (1959) 1741

[2] R. Haag, *Local Quantum Physics*, Springer, second edition, 1996

[3] J. Polchinski, *String theory I*, Cambridge University Press 1998

[4] B. Schroer, *Causality and dispersion relations and the role of the S-matrix in the ongoing research*, arXiv:1107.1374

[5] B. Schroer, *Particle crossing versus field crossing; a corrective response to Duff's recent account of string theory*, arXiv:1201.6328, and prior work cited therein

[6] H. Babujian and M. Karowski, *Int. J. Mod. Phys.* **A1952**, (2004) 34, and references therein to the beginnings of the bootstrap-formfactor program

[7] F. Bloch and A.Nordsiek, *Phys. Rev.* **52**, (1937) 54

[8] D. Yenni, S. Frautschi and H. Suura, *The infrared divergence phenomena and high energy processes*, *Ann. of Phys.* **13**, (1961) 370-462

[9] B. Schroer, *Fortschr. Physik* **143**. (1963) 1526

[10] J. H. Lowenstein and J. A. Swieca, *Ann.Phys.* **68**, (1971) 172

[11] B. Schroer, *Pascual Jordan's legacy and the ongoing research in quantum field theory* , in preparation

[12] O. Steinmann, *A New Look at the Higgs-Kibble Model*, arXiv:0804.2989

[13] J. Langerholc and B. Schroer, *Can current operators determine a complete theory* ?, *Commununications in Mathematical Physics* **4**, (1967) 123

[14] S. Jacobs, *Eichbrücken in der klassischen Feldtheorie*, DESY-THESIS-2009-009

[15] J. Schwinger, *Phys. Rev.* **128** (1962), 2425; Theoretical Physics, Trieste Lectures, 1962 p- 89, I.A.E.A, Vienna , 1963

[16] W. Higgs, *Phys. Rev. Lett.* **12**, (1964) 132

[17] G. Lechner, *An Existence Proof for Interacting Quantum Field Theories with a Factorizing S-Matrix*, *Commun. Mat. Phys.* **227**, (2008) 821, arXiv.org/abs/math-ph/0601022

[18] Dollard, J. D. *J. Math. Phys.* **5**, (1964) 729

[19] D. Buchholz, *Phys. Lett. B*174, (1986) 331

[20] T. W. B. Kibble, *Phys. Rev.* **173**, (1968) 1527

[21] R. Ferrari, L. Picasso and F. Strocchi, Comm. Math. Phys. **35**, (1974) 25; *Nuovo Cim.* **39** A, (1977) 1

[22] G. Morchio and F. Strocchi, *Nucl. Phys.* **B211**, 471, (1983) 232

[23] J. Fröhlich, G. Morchio and F. Strocchi, *Ann. Phys.* **119**, (1979) 241

[24] D. Buchholz, S. Doplicher, G. Morchio, J. E. Roberts and F. Strocchi, *Annals Phys.* **290** (2001) 53

[25] S. Weinberg, *The Quantum Theory of Fields I*, Cambridge University Press 1995

[26] J. Mund, B. Schroer and J. Yngvason, *String-localized quantum fields and modular localization*, *CMP* **268** (2006) 621, math-ph/0511042

[27] R. Brunetti, D. Guido and R. Longo, *Modular localization and Wigner particles* , *Rev.Math.Phys.* **14**, (2002) 759

[28] J. Mund, *String–Localized Quantum Fields, Modular Localization, and Gauge Theories,* New Trends in Mathematical Physics (V. Sidoravicius, ed.), Selected contributions of the XVth Int. Congress on Math.Physics, Springer, Dordrecht, 2009, pp. 495-508.

[29] J. Yngvason, *Commun. Math. Phys.* **18** (1970), 195

[30] S. Weinberg, Feynman rules for any spin, *Phys. Rev.* **133B** (1964) 1318

[31] B. Schroer, *Studies in History and Philosophy of Modern Physics* **41** (2010) 104–127

[32] P. Leyland, J. Roberts and D. Testard, *Duality For Quantum Free Fields*, C.N.R.S. preprint July 1978, unpublished

[33] Y. Kawahigashi, R. Longo and M. Mueger, *Commun. Math. Phys.* **219** (2001) 631

[34] R. Longo, P. Martinetti and K-H Rehren, *Geometric modular action for disjoint intervals and boundary conformal field theory,* arXiv:0912.1106

[35] S. Doplicher and J.E. Roberts, *Commun. Math. Phys.* **131** (1990) 51

[36] J. A. Swieca, *Phys. Rev. D* **13**, (1976) 312

[37] B. Schroer, *Particle physics in the 60s and 70s and the legacy of contributions by J. A. Swieca*, arXiv:0712.0371

[38] J. Mund and B. Schroer, work in progress

[39] M. Dütsch, J. M. Gracia-Bondia, F. Scheck and J. C. Varilly, *Quantum gauge models without classical Higgs mechanism*, arXiv:1001.0932

[40] M. Duetsch and B. Schroer, *J.Phys.* **A33**, (2000) 4317

[41] J. H. Lowenstein and B. Schroer, *Phys. Rev. D* **6**, (1972) 1553

[42] V. Enss, *Commun. Math. Phys.* **45**, (1975) 35

[43] D. Buchholz and M. Porrmann, Ann. Inst. Henri Poincare - *Physique théorique* **52** (1990), 237

[44] R. Stora, *Local gauge groups in quantum eld theory: perturbative gauge theories* , talk given at the workshop: Local quantum physics", Erwin Schrodinger Institute, Vienna, 1997

[45] K-H Rehren, *Lett. Math. Phys.* **30**, (1994) 125

[46] B. Schroer, *Ann. Phys.* **319**, (2005) 49

[47] D. Bahns, S. Doplicher, K. Fredenhagen and G. Piaticelli, *Quantum Geometry on Quantum Spacetime: Distance, Area and Volume Operators*, arXiv:1005.2130

[48] B. Schroer, *BMS symmetry, holography on null-surfaces and area proportionality of "light-slice" entropy*, to be published in *Foundations of Physics,* arXiv:0905.4435

[49] T. Jacobson, *Phys.Rev.Lett.* **75** (1995) 1260

[50] S. Doplicher, *The principles of locality, Effectivness, fate and challenges,* arXiv:0911.5136

[51] P. Hislop and R. Longo, *Commun. Math. Phys.* **84**, (1982) 71

[52] Y. Kawahigashi and R. Longo, *Commun. Math. Phys.* **257**, (2005) 193

[53] H. Casini and M. Huerta, Reduced density matrices and internal dynamics for multi-component regions *Class. Quant. Grav.* **28**, (2009) 185005

In: New Developments in the Standard Model
Editor: Ryan J. Larsen
ISBN: 978-1-61209-989-7

Chapter 4

SUPERSYMMETRIC STANDARD MODEL, BRANES AND DEL PEZZO SURFACES

S.L. Cacciatori[1,*] ***and M. Compagnoni***[2,†]
[1] Dipartimento di Fisica e Matematica, Università dell'Insubria, Como, Italy,
and I.N.F.N., sezione di Milano, Italy,
[2] Dipartimento di Matematica, Politecnico di Milano, Milano, Italy

Abstract

Even though the Standard Model of particles has been confirme by several experiments, many questions require improvements. Beyond the problem of Grand Unification the mass gap problem, the question of hierarchies, low boson masses and dynamical soft supersymmetry breaking, there is the really hard difficult in including gravity in a full quantum paradigm of the Standard Model. The most famous scheme elaborated in order to solve the last and, possibly, all this points is String Theory. Dualities, mirror symmetry, M-theory and AdS/CFT are some of the powerful tools which permit to perform several progresses in all the mentioned directions, at least in principle. However, interactions of String Theory with phenomenology are really recent results. A way to get a contact between theory and phenomenology is the so called bottom-up approach. We will present here a possible String Theory approach to the (Minimal Supersymmetric) Standard Model based on the geometric engineering construction firs proposed in [H. Verlinde and M. Wijnholt, JHEP 0701, 106]. We will study the relevant geometry along the lines of [S.L. Cacciatori and M. Compagnoni, JHEP 1005:078,2010], and the related physics. We will study the singular orbifold $\mathbb{C}^3/\Delta_{27}$, with Δ_{27} a suitable non abelian group, its geometry and show how it can be desingularized. To render technical computations as simple as possible we will work also with a simplifie toric version, studing its main properties at K-theory level, and we will discuss how such calculations should be extended to the non abelian case. The associated relevant physics will be discussed.

*E-mail address: sergio.cacciatori@uninsubria.it
†E-mail address: marco.compagnoni@polimi.it

1. Introduction

The Standard Model of particles physics is one of the most important successful results of the work of the last century physicists. Even though it has not been fully understood, quantum fiel theory, together with a good number of relevant ingredients, has provided an elegant working synthesis of the physics of elementary particles. Already at classical level, fiel theory gave rise to the concept of unificatio as we can learn from the Maxwell theory of electromagnetic fields It has also given rise to a reanalysis of the concept of relativity and at the same time it has provided the seed of the corpuscle-wave duality. The firs success against the difficul conciliation between quantum physics and special relativity has been the construction of the theory of quantum electrodynamics (QED) which, despite the presence of divergences, resulted to be able to provide predictions with surprising precision. Renormalizability is one of the basic ingredients in formulating perturbative theories. The reinterpretation of electrodynamics in terms of a dynamical $U(1)$-local symmetry has led to the concept of gauge theories, with non abelian local symmetry groups (Yang-Mills theories) next providing an explanation of weak nuclear interactions and culminating in the electroweak fiel theory, which presents electromagnetic and weak interactions as aspects of a unique phenomenon. In conclusion, the Standard Model is able to take account of the known particle fields which amount to hadrons, constituted by quarks, leptons and gauge bosons, manifesting an $U(1)_y \otimes SU(2)_L \otimes SU(3)_c$ gauge group symmetry, which at low energies is broken down to a $U(1)_{em} \otimes SU(3)_c$ symmetry. The breaking is mediated by the presence of the Higgs scalar field which in this scheme is the only fiel whose existence has not yet been confirmed

Though this is a formidable result, not all are roses. There are many questions which require an answer to give a satisfactory model. Let us list some of them:

1. The gauge group $U(1)_y \otimes SU(2)_L \otimes SU(3)_c$ unifie electromagnetic and weak interactions, but includes separately the strong interactions described by the $SU(3)_c$ color group. Could it happen that all interactions are low energy manifestations of a unique interaction? Hints in this direction are given by the renormalization flu es;

2. The universal covering of the hypercharge $U(1)_y$ group is non compact. For this reason the quantization of the hypercharge y is not a prediction of the theory. However, it could happen that the gauge group is the subgroup of a unifying simple group with compact covering. Is this the case?;

3. If the unificatio is actual at high energies, in lowing energy it must be broken well above the electroweak breaking scale. Why such a hierarchy should arise?;

4. The $SU(2)_L$ chiral symmetry group assumes zero mass for neutrini. But we know that not all neurino families can be massless. So, why chiral symmetry is manifest?;

5. Assuming Grand Unificatio (GUT), why are there three families? Could they further unifie in a unique family in some way?

6. Chiral symmetry protects fermion masses from running to cutoff scales. But why boson masses are much smaller than the cutoff scale?;

7. Should and could one include gravity in such a scheme beyond a semiclassical level? Can gravity be unifie to the other interactions end/or included in a unique quantization scheme?

Answering to these and other related questions leads beyond the standard model. GUT theories with unifying $SU(5)$, $SO(10)$ or E_6 groups can partially answer the firs two questions, supersymmetric models can help in answering to question 6, and so on. No need to say, a complete theory should answer all question, including the Standard Model as a limit case. In this sense, string theory is one of the most promising theories, at least in principle. The supersymmetric version is able to take account to supersymmetry, and to include in a mathematically consistent way the gravity. At low energies it "lives" in ten spacetime dimensions. This is the dimension required to replicate families by means of Kaluza-Klein reduction down to four dimensions (thus answering to question 5). String theory gives place to a so large number of possibilities that it could contain a solution to all the problems.

Actually, history is not confirmin such a possible conclusion. Indeed, string theory is on the scenes from many years now. It has overtaken initial technical difficultie undermining its consistency, and undergone more then a revolution that have finall led to amazing interconnections between physics and mathematics. However, up to the time we are writing these notes, it has not yet given rise to a natural top-down physical prediction. String theory is not yet a testable theory: it needs to be connected to predictive models that can be tested via experiment. This has led many people to search for such a connection by following a bottom-up approach: starting from the main features which should characterize a working extension of the Standard Model, for example a GUT model, one tries to embed it in string theory scenario. As D-brane physics provides a full geometrical interpretation of gauge symmetries, one way to do it is by means of D-brane engineering. Recently, there are appeared several promising models reproducing almost realistic four dimensional supersymmetric gauge theories decoupled from gravity, see, for example, [BHV1], [BHV2], [BBGW], [DW1], [DW2].

Among the possible constructions, we decide to describe one of the simplest ones, proposed some time ago by H. Verlinde and M. Wijnholt [VW]. It consists in realizing a supersymmetric quiver version of the Standard Model as a world-volume theory on a D3-brane. The relevant geometry is that of a noncompact Calaby-Yau threefold realized starting from the non Abelian orbifold $\mathbb{C}^3/\Delta_{27}$, where Δ_{27} is the Heisenberg group of order 27 obtained by the central extension of the abelian group $\mathbb{Z}_3 \times \mathbb{Z}_3$ by $\mathbb{Z}_3$.

In section 2. we will review the construction of the Verlinde-Wijnholt model. In section 3. we analyze the "classical" geometry of the model. According to the correspondence conjectured by Verlinde and Wijnholt we will relate the resolution of $\mathbb{C}^3/\Delta_{27}$ to a quasi del Pezzo surface $q - d\mathbb{P}_8$. With "classical" geometry, we mean that we will only identify the nature of singularities and their resolution, and will not work out a basis for the K-theory or derived category of coherent sheaves, which should however necessary in order to describe the brane physics, in order to avoid excessive technical difficulties To bring the discussion to a simpler technical level, we will discuss a toric version of the Verlinde-Whinjholt model, based on the abelian quotient $\mathbb{C}^3/\mathbb{Z}_3 \times \mathbb{Z}_3$, which is the toric model nearest to the non abelian one. This will be presented in section 4.. This is a cone over a singular cu-

bic surface in $\mathbb{P}^3$ possessing several crepant resolutions, all related by flops which can be easily determined via toric methods. Among them, a special one can be identifie as the total space of the canonical bundle over a quasi del Pezzo surface of degree 3, that is the surface obtained by blowing up three points on $\mathbb{P}^2$ twice. The strategy used here is the same yet adopted for the non abelian case. However, in this case we will employ toric geometry methods in order to compute the K-theory generators, which are indeed generators of the derived category of coherent sheaves, supporting the possible brane configurations We will compute the prepotential for all the resolutions.
Section 2. is based on the original paper [VW]. Section 3. and the initial part of section 4. are based on our paper [CaCo-1]. Most of the material in section 4. is new.

2. The Verlinde-Wijnholt Model

In this section we will provide a short review of the paper [VW]. We will not enter the details, which can be found in the original paper.
D3-brane physics allows to realize effective (3+1)-dimensional supersymmetric fiel theories that, under suitable conditions, at low energy decouple from gravity and from higher dimensional dynamics. Moreover, D-branes provide a beautiful interpretation of internal gauge symmetries in terms of true geometrical external symmetries thus permitting a pure geometrical engineering. From a unifying theory, one should expect that the standard model will appear in a natural way as a low energy limit, in a top-down process. However, no signals of such naturalness appeared from string theory and a bottom-up procedure is often adopted in order to try to relate string theory to phenomenology. In such an approach it is not at all obvious to start from a GUT theory, which should be eventually suggested by the theory. Instead it appears more natural to start from the experimental data, thus from Standard Model. This is indeed the viewpoint expressed by the authors in [VW].
The main idea is to localize the Standard Model on a D3 brane sited at the apex of a highly warped throat of a Calabi Yau manifold. In the gravity decoupling limit, where the Planck scale is sent to infinit , the warped throat is replaced by a singular noncompact Calabi Yau with the brane placed on a singularity. If the singularity is a quotient one, the D3-brane can breaks in several fractional branes F_i, each one with multiplicity n_i. At low energies each fractional brane will give rise to a $U(n_i)$ vector multiplet V_i. Brane intersections (F_i, F_j) give rise to bifundamental matter multiplets. At high energies, string theory sees matter fields gauge modes and couplings on the same footing as dynamical degrees of freedom. Al low energies, the role of the field is distinguished. In particular, in the model we are considering, closed string dynamical freedoms are frozen thus giving rise to continuously tunable constant couplings. The resulting low energy effective theory is a “quiver gauge theory”: the dynamical degrees of freedom are described by quiver, where each node represents the fractional brane F_i with its gauge group $U(n_i)$ and oriented lines from i to j represent bifundamental matter transforming in the $(\bar{n}_i, n_j)$ representation. To give rise to phenomenological models, a quiver must respects some basic rules:

- The absence of non abelian gauge anomalies imposes that the number of incoming and of outgoing lines at each node must be equal;
- The absence of adjoint matter forbids lines starting and ending on the same node;

- True chirality of bifundamental matter requires that between any two nodes there must be only one type of oriented lines.

The phenomenological quivers must be built up using these rules and recalling that any triangle compatible with them selects the non vanishing Yukawa couplings.
The authors of [VW] proposed a quiver gauge theory associated to the gravity decoupling limit of a D3-brane placed on a dP_8 del Pezzo singularity, obtained by shrinking a degree 8 del Pezzo surface in a Calabi Yau threefold down to a point. This choice is dictated by the well known fact that, roughly speaking, the intersection structure of the exceptional curves obtained by blowing up eight points of $\mathbb{P}^2$ give origin to a E_8 Dynkin diagram structure, representing the possibility to develop ADE type singularities. If E_i, $i = 1, \ldots, 8$ are the exceptional curves and H is the hyperplane curve in dP_8, then the curves $\alpha_i = E_i - E_{i+1}$, $i = 1, \ldots, 7$, $\alpha_8 = H - E_1 - E_2 - E_3$, behave as the simple roots of E_8. Sheaves F_i over points, curves and the whole dP_8 surface, represent possible fractional brane configurations Their RR (D7,D5,D8)-charges $(Q_{D7}, Q_{D5}, Q_{D3})_i$ are given by the Chern character

$$(Q_{D7}, Q_{D5}, Q_{D3})_i = \mathrm{Ch}(F_i) \equiv (\mathrm{Rank}(F_i), c_1(F_i), \mathrm{Ch}_2(F_i)). \tag{1}$$

As the D3 brane, decomposed in fractional ones, corresponds to the sheaf $\mathcal{O}_p \simeq \sum_i n_i F_i$, the charges must satisfy the integrality condition

$$(0, 0, 1) = \mathrm{Ch}(\mathcal{O}_p) = \mathrm{Ch}(\sum_i n_i F_i) = \sum_i n_i (Q_{D7,i}, Q_{D5,i}, Q_{D3,i}). \tag{2}$$

The firs step in realizing the quiver gauge theory is given by a choice of a collection of curves allowing for a configuratio of satisfying the above rules. This is a so called "exceptional collection." A firs choice is given fractional branes F_i, $i = 1, \ldots, 11$ such that

$$\mathrm{Ch}(F_i) = \begin{cases} (1, H - E_i, 0), & i = 1, 2, 3, 4 \\ (1, E_i - K, 1), & i = 5, 6, 7, 8 \end{cases} \tag{3}$$

$$\mathrm{Ch}(F_{10}) = (3, -K + \sum_{i=5}^{8} E_i, -1/2), \tag{4}$$

$$\mathrm{Ch}(F_{11}) = (6, -3K + 2\sum_{i=5}^{8} E_i, 1/2), \tag{5}$$

where H is the hyperplane class and $K = -3H + \sum_{i=1}^{8} E_i$ is the canonical class. The integrality condition gives $n_i = 1$ for $i = 1, \ldots, 9$ and $n_{10} = -n_{11} = 3$. Moreover, it can be shown that the matter links between F_i and F_j are given by

$$\mathcal{X}(F_i, F_j) = \sum_k (-1)^k \mathrm{dim} Ext^k_{dP_8}(F_i, F_j) = \int_{dP_8} \mathrm{Ch}(F_i^*)\mathrm{Ch}(F_j)\mathrm{Td}(dP_8). \tag{6}$$

This gives $\mathcal{X}(F_{11}, F_i) = 1$, $\mathcal{X}(F_i, F_{10}) = 1$, and $\mathcal{X}(F_{10}, F_{11}) = 3$, $i = 1, \ldots, 9$. Thus, the corresponding gauge theory is characterized by two $U(3)$ vector multiplets, nine $U(1)$ vector multiplets, nine $(\bar{1}_i, 3_{11})$, nine $(\bar{3}_{10}, 1_i)$ and a $(\bar{3}_{11}, 3_{10})$ bifundamental matter multiplets, where the subscript numbers the fractional brane. Of course, this is not properly the right field content for the Standard Model. However, we have seen that the quiver depends on the choice of the exceptional basis. Geometrically, an exceptional basis can be changed into a new one by means of a mutation, which, physically, corresponds to a Seiberg duality: in [VW] it is considered the mutation at F_{10} (see [VW] for explanations: F_{11} is the only node whose links to F_{10} incomes to the latter)

$$F_{11} \mapsto \tilde{F}_{11}, \qquad \mathrm{Ch}(\tilde{F}_{11}) = -\mathrm{Ch}(F_{11}) + \mathcal{X}(F_{10}, F_{11})\mathrm{Ch}(F_{10}) = (3, \sum_{i=5}^{8} E_i, -2), \tag{7}$$

and the new gauge group at the corresponding node is $U(\tilde{n}_{11})$, $\tilde{n}_{11} = 6$. This gives a $U(6) \otimes U(3) \otimes U(1)^9$ gauge theory.
To get the Standard Model quiver one then needs to break the gauge group by assigning suitable expectation values to the Fayet-Iliopoulos couplings. We will not describe it here, but only comments that geometrically it correspond to partially resolve the singularity. In particular, by blowing up the A_2 type singularity given by the roots α_1 and α_2 provides an effective $U(3) \otimes U(2) \otimes U(1)^7$ group based Standard Model. As the total $U(1)$ charge vanishes, one of the $U(1)$ factors decouples. Other two, associated to non abelian anomalies, are reduced to global anomalies through the Green-Schwartz mechanism at the string theory level. The remaining ones, exception made for the $U(1)_{em}$ electromagnetic combination, can be ruled out by giving a mass to the corresponding superneutrini partners. More concretely, all this can be realized directly by tuning the closed string effective free parameters.
In this way, one obtains a string theory description of the $N = 1$ minimally supersymmetric Standard Model. However, we have not yet discussed the geometry of the noncompact Calabi Yau manifold corresponding to the gravity decoupling limit. In [VW] it is argued that the right limit is given by the orbifold $X = \mathbb{C}^3/\Delta_{27}$, where Δ_{27} is the finit Heisenberg group associated to the abelian group $\mathbb{Z}_3 \times \mathbb{Z}_3$. The firs observation is that it gives the same quiver gauge theory as the one discussed above, associated to the basis F_i, $i = 1, \ldots, 11$. Indeed, let us consider a D3-brane fillin $M^{1,3}$ in $M^{1,3} \times \mathbb{C}^3$. Its position in $\mathbb{C}^3$ will be determined by three complex chiral field X^i, $i = 1, 2, 3$. By orbifolding under the fundamental action R_3 of Δ_{27} over $\mathbb{C}^3$, the brane and its images under the group will split in in fractional branes S_i, each one with multiplicity n_i. The corresponding worldvolume theory will be a gauge theory with gauge group $G = \prod_a U(n_a)$. More precisely, strings living on the D3-brane will carry Chan-Paton indices, which carry a regular representation R of the group Δ_{27}, which will decompose under irreducible representations (irreps): $R = \oplus_a n_a R^a$, where $n_a = \mathrm{dim} R^a$ defin the gauge group G. The orbifold projection requires $R^{-1} X^i R = \sum_j (R_3)_{ij} X^j$ for the chiral fields and $R^{-1} V R = V$ for the vector multiplets. The group Δ_{27} has order 27 organized in 11 conjugacy classes corresponding to 11 irreps, two 3 dimensional and 9 one dimensional. The associated gauge group is then $G = U(3)^2 \otimes U(1)^9$. Moreover, strings stretched between fractional branes (F_a, F_b) will give rise to bifundamental field transforming under $(\bar{n}_a, n_b)$. The numbers n_{ab} of such field can be computed according to the relation $R_3 \otimes R^a = \oplus_b n_{ab} R^b$. These give rise to the same quiver described above.
To further corroborate their identificatio of X with $\mathbb{C}/\Delta_{27}$, Verlinde and Wijnholt noted that if one uses (X, Z, Z) as coordinates for $\mathbb{C}^3$, then there is a basis of four Δ_{27}-invariant polynomials

$$x = XYZ$$

$$\begin{aligned} z &= X^3+Y^3+Z^3 \\ y &= (X^3+\omega Y^3+\omega^2 Z^3)(X^3+\omega^2 Y^3+\omega Z^3) \\ w &= (X^3+\omega Y^3+\omega^2 Z^3)^3 \end{aligned} \tag{8}$$

which have relative weight $(1,1,2,3)$ thus giving a representation of X into the weighted projective space $\mathbb{P}^3_{1,1,2,3}$. In particular, restricted to the image of this map, the coordinates (x,z,y,w) satisfy the polynomial equation

$$w^2+y^3-27wx^3-3wyz+wz^3=0 \tag{9}$$

in $\mathbb{P}^3_{1,1,2,3}$. As dP_8 surfaces can in general be represented as the vanishing locus of degree 6 quasi-homogeneous polynomials in $\mathbb{P}^3_{1,1,2,3}$, they conjectured that this is indeed a singular dP_8 del Pezzo surface isomorphic to the $\mathbb{C}^3/\Delta_{27}$ orbifold.
We will put these considerations in a set of precise statements in the next section.

3. The Geometry of $\mathbb{C}^3/\Gamma$

In this section we will study the classical geometry of the Verlinde-Wijnholt model, relating the non Abelian orbifold $\mathbb{C}^3/\Gamma$ to a qualsi del Pezzo surface.
This section is based on notes provided to us by professor Bert van Geemen, to whom we are indebted.

The adopted strategy is the following one: The orbifold $\mathbb{C}^3/\Gamma$ is the cone over $\mathbb{P}^2/\tilde{\Gamma}$, where $\tilde{\Gamma}$ is the quotient of $\Gamma=\Delta_{27}$ by its center $\mathbb{Z}_3$. $\tilde{\Gamma}$ has four degenerate orbits on $\mathbb{P}^2$ with non trivial stabilizer. These identify the four singular point on $\mathbb{P}^2/\tilde{\Gamma}$. On the other side, a $d\mathbb{P}_8$ surface is obtained by blowing up eight points in general positions on $\mathbb{P}^2$. Now, all cubics on $\mathbb{P}^2$ passing through eight points also have in common a ninth point. When these nine lie in a square array, the corresponding set of cubics is called a Hesse pencil. An Hesse pencil contains exactly four degenerate cubics, each one being the union of three lines. By blowing up eight of the nine base points, we will obtain a surface S whose anticanonical divisor $-K_S$ is not ample. This is not a del Pezzo surface, but we can call it a quasi del Pezzo surface. The anticanonical maps, instead of giving a projective embedding of S in $\mathbb{P}_{1,1,2,3}$, will shrunk down to a point that components of the four singular cubics that do not pass through the ninth base point of the pencil. Thus, the resulting singular variety can be identifie with $\mathbb{P}^2/\tilde{\Gamma}$. The anticanonical maps then provide a desingularizing map from S to $\mathbb{P}^2/\tilde{\Gamma}$. We will prove this by showing that pinching out eight base points, the Hesse pencil will describe both $\mathbb{P}^2$ and $\mathbb{P}^2/\tilde{\Gamma}$ as elliptic fibration over $\mathbb{P}^1$. In particular, we will identify the Hessian group with the image of the normalizer of Γ in $PGL(3,\mathbb{C})$.

3.1. The Verlinde-Wijnholt Isomorphism

We will firs prove that the map $(X,Y,Z)\mapsto(x,y,z,w)$ define by

$$\begin{aligned} x &= XYZ \\ z &= X^3+Y^3+Z^3 \\ y &= (X^3+\omega Y^3+\omega^2 Z^3)(X^3+\omega^2 Y^3+\omega Z^3) \end{aligned} \tag{10}$$

$$w = (X^3 + \omega Y^3 + \omega^2 Z^3)^3$$

that is a map from $\mathbb{C}^3/\Delta_{27}$ to the tautological bundle over the weighted projective space $\mathbb{P}_{1,1,2,3}$, is indeed an isomorphism between algebraic varieties.
The Heisenberg group $\Gamma := \Delta_{27} \subset SL(3, \mathbb{C})$ is generated by the matrices

$$g_1 = \begin{pmatrix} 1 & 0 & 0 \\ 0 & \omega & 0 \\ 0 & 0 & \omega^2 \end{pmatrix}, \qquad g_2 = \begin{pmatrix} 0 & 0 & 1 \\ 1 & 0 & 0 \\ 0 & 1 & 0 \end{pmatrix}, \tag{11}$$

where $\omega = e^{\frac{2\pi i}{3}}$. They satisfy

$$g_1^3 = g_2^3 = I, \qquad g_2 g_1 = \omega^2 g_1 g_2, \tag{12}$$

and the center of Γ is

$$C = \{I, \omega I, \omega^2 I\}, \tag{13}$$

so that its abelianization is $\tilde{\Gamma} := \Gamma/C \simeq \mathbb{Z}_3 \times \mathbb{Z}_3$. This shows that Γ is the non abelian central extension of the group $\mathbb{Z}_3 \times \mathbb{Z}_3$ by $\mathbb{Z}_3$. We see that its generators commute in $\mathbb{P}^2$. In other words, the group $\tilde{\Gamma}$ is the image of Γ in $PGL(3, \mathbb{C})$, the group of automorphisms of the projective plane. It has four cyclic subgroups, each one isomorphic to $\mathbb{Z}_3$, which are

$$\begin{aligned}
&< g_1 > = \{g_1, g_1^2, I\} && g_1 = \begin{pmatrix} 1 & 0 & 0 \\ 0 & \omega & 0 \\ 0 & 0 & \omega^2 \end{pmatrix}, \\
&< g_2 > = \{g_2, g_2^2, I\} && g_2 = \begin{pmatrix} 0 & 0 & 1 \\ 1 & 0 & 0 \\ 0 & 1 & 0 \end{pmatrix}, \\
&< g_1 g_2 > = \{g_1 g_2, g_1^2 g_2^2, I\} && g_1 g_2 = \begin{pmatrix} 0 & 0 & 1 \\ \omega & 0 & 0 \\ 0 & \omega^2 & 0 \end{pmatrix}, \\
&< g_1^2 g_2 > = \{g_1^2 g_2, g_1 g_2^2, I\} && g_1^2 g_2 = \begin{pmatrix} 0 & 0 & 1 \\ \omega^2 & 0 & 0 \\ 0 & \omega & 0 \end{pmatrix}.
\end{aligned} \tag{14}$$

The fi ed point in $\mathbb{P}^2$ of these cyclic subgroup of $\tilde{\Gamma}$ are

$$\begin{aligned}
\text{Fix}(< g_1 >): &\quad (1:0:0), \quad (0:1:0), \quad (0:0:1), \\
\text{Fix}(< g_2 >): &\quad (1:1:1), \quad (1:\omega^2:\omega), \quad (1:\omega:\omega^2), \\
\text{Fix}(< g_1 g_2 >): &\quad (1:\omega:1), \quad (1:1:\omega), \quad (\omega:1:1), \\
\text{Fix}(< g_1^2 g_2 >): &\quad (1:\omega^2:1), \quad (\omega^2:1:1), \quad (1:1:\omega^2).
\end{aligned} \tag{15}$$

and can be easily determined as the common eigenspaces in $\mathbb{C}^3$ of the generators of each given subgroup. Any set of fi ed points of a given cyclic subgroup selects a degenerate orbit of order three of $\tilde{\Gamma}$ in $\mathbb{P}^2$, the only orbits in the projective plane having non-trivial stabilizer in $\tilde{\Gamma}$. The action of the stabilizer subgroup at each fi ed point is locally given by

$\mathrm{diag}(\omega, \omega^2)$. Thus, the singular locus of the quotient surface $\mathbb{P}^2/\tilde{\Gamma}$ consists of four singular points of type A_2, so that near each singular point the surface is isomorphic to the orbifold $\mathbb{C}^2/\mathbb{Z}_3$. $\mathbb{C}^3/\Gamma$ is then the cone over the singular surface $\mathbb{P}^2/\tilde{\Gamma}$ and the four singular points correspond to the lines

$$\begin{aligned} \mathcal{C}_1: &\quad (t,0,0) \sim (0,t,0) \sim (0,0,t), \\ \mathcal{C}_2: &\quad (t,t,t) \sim (t,\omega^2 t,\omega t) \sim (t,\omega t,\omega^2 t), \\ \mathcal{C}_3: &\quad (t,\omega t,t) \sim (t,t,\omega t) \sim (\omega t,t,t), \\ \mathcal{C}_4: &\quad (t,\omega^2 t,t) \sim (\omega^2 t,t,t) \sim (t,t,\omega^2 t). \end{aligned} \tag{16}$$

passing through the origin.
We can then state the following theorem

Theorem 1. *([CaCo-1], Theorem 3.1) The ring homomorphism associated to (10)*

$$\begin{aligned} \phi^*: \quad \mathbb{C}[x,z,y,w] &\longrightarrow \mathbb{C}[X,Y,Z]^\Gamma \\ (x,z,y,w) &\longrightarrow (f_0,f_1,f_2,f_3) \end{aligned} \tag{17}$$

with

$$\begin{aligned} f_0 &= XYZ, \\ f_1 &= X^3+Y^3+Z^3, \\ f_2 &= (X^3+\omega Y^3+\omega^2 Z^3)(X^3+\omega^2 Y^3+\omega Z^3), \\ f_3 &= (X^3+\omega Y^3+\omega^2 Z^3)^3, \end{aligned} \tag{18}$$

defines a ring isomorphism:

$$\mathbb{C}[X,Y,Z]^\Gamma \simeq \frac{\mathbb{C}[x,z,y,w]}{(w^2+y^3-27wx^3-3wyz+wz^3)}. \tag{19}$$

The proof of this theorem uses the following three technical lemmas, whose proof can be found in [CaCo-1].

Lemma 1. *The morphism*

$$\begin{aligned} \phi_R: \quad \mathbb{C}^3 &\longrightarrow \mathbb{C}^3 \\ (X,Y,Z) &\longmapsto (f_0,f_1,f_2) \end{aligned} \tag{20}$$

is surjective.

Lemma 2. *The graded ring*

$$\frac{\mathbb{C}[x,z,y,w]}{(w^2+y^3-27wx^3-3wyz+wz^3)} \tag{21}$$

with gradation

$$deg(x^a z^b y^c w^d) := 3a+3b+6c+9d \tag{22}$$

is the direct sum of vector spaces of dimension

$$dim\left(\frac{\mathbb{C}[x,z,y,w]}{(w^2+y^3-27wx^3-3wyz+wz^3)}\right)_n = \begin{cases} 0 & if\, n \neq 0 \bmod 3, \\ 1+\frac{d(d+1)}{2} & if\, n = 3d. \end{cases} \tag{23}$$

Lemma 3. *The graded ring* $\mathbb{C}[X,Y,Z]^\Gamma$ *of invariant polynomials under the action of* Γ, *with gradation*

$$deg(X^A Y^B Z^C) := A + B + C\,, \tag{24}$$

is the direct sum of vector spaces of dimension

$$dim\left(\mathbb{C}[X,Y,Z]^\Gamma\right)_n = \begin{cases} 0 & if\, n \neq 0 \; mod\, 3, \\ 1 + \frac{d(d+1)}{2} & if\, n = 3d. \end{cases} \tag{25}$$

We can prove the theorem as follows. As

$$(w^2 + y^3 - 27wx^3 - 3wyz + wz^3) \subset \mathrm{Ker}\,\phi^*, \tag{26}$$

we have the ring homomorphism (that we call again ϕ^*)

$$\phi^* : \quad \frac{\mathbb{C}[x,z,y,w]}{(w^2 + y^3 - 27wx^3 - 3wyz + wz^3)} \longmapsto \mathbb{C}[X,Y,Z]^\Gamma \tag{27}$$

Lemma 1 implies that the map ϕ_R is surjective. Then, the map

$$\begin{array}{rccc} \phi_R^* : & \mathbb{C}[x,z,y] & \longrightarrow & \mathbb{C}[X,Y,Z]^\Gamma \\ & (x,z,y) & \longrightarrow & (f_0, f_1, f_2) \end{array} \tag{28}$$

between the rings is injective and there are not relations involving x, z, y. Moreover, the image of $\mathbb{C}^3/\Gamma$ under the geometric map ϕ is a hypersurface in $\mathbb{C}^4$ contained in $(w^2 + y^3 - 27wx^3 - 3wyz + wz^3) = 0$, that is an irreducible hypersurface, so that the image of ϕ coincides with it. Hence the map ϕ^* in (27) is injective. In particular, it is factorizable in vector space morphisms

$$\phi_n^* : \left(\frac{\mathbb{C}[x,z,y,w]}{(w^2 + y^3 - 27wx^3 - 3wyz + wz^3)} \right)_n \longmapsto \left(\mathbb{C}[X,Y,Z]^\Gamma\right)_n . \tag{29}$$

The map ϕ^* is surjective if ϕ_n^* is surjective for any n. $\mathrm{Ker}(\phi^*) = \varnothing$ implies $\mathrm{Ker}(\phi_n^*) = \varnothing$ for any n, hence

$$\dim \mathrm{Im}(\phi_n^*) = \dim \left(\frac{\mathbb{C}[x,z,y,w]}{(w^2 + y^3 - 27wx^3 - 3wyz + wz^3)} \right)_n . \tag{30}$$

Therefore, ϕ^* is surjective if

$$\dim \left(\frac{\mathbb{C}[x,z,y,w]}{(w^2 + y^3 - 27wx^3 - 3wyz + wz^3)} \right)_n = \dim \left(\mathbb{C}[X,Y,Z]^\Gamma\right)_n \tag{31}$$

for any n. This is ensured by lemmas 2 and 3, which then imply the theorem.

Remark: The isomorphism ϕ^* determines the geometric isomorphism ϕ mapping the orbifold $\mathbb{C}^3/\Gamma$ in the hypersurface

$$w^2 + y^3 - 27wx^3 - 3wyz + wz^3 = 0 \;\subset\; \mathbb{C}^4, \tag{32}$$

and sending the singular curves of the orbifold to the singular locus of the hypersurface, given the union of four curves intersecting in the origin

$$\begin{aligned} C_1 &\longmapsto (0, t, t^2, t^3), \\ C_2 &\longmapsto (t, 3t, 0, 0), \\ C_3 &\longmapsto (t, 3\omega t, 0, 0), \\ C_4 &\longmapsto (t, 3\omega^2 t, 0, 0). \end{aligned} \tag{33}$$

This define an isomorphism of surfaces between the quotient $\mathbb{P}^2/\tilde{\Gamma}$ and a singular hypersurface in a weighted projective space

$$\begin{aligned} \tilde{\phi}: \quad \mathbb{P}^2/\tilde{\Gamma} &\longrightarrow \mathbb{P}_{1,1,2,3} \\ (X:Y:Z) &\longmapsto (x:z:y:w), \end{aligned} \tag{34}$$

sending the singular points to

$$\begin{aligned} q_1 &= (1:0:0) \sim (0:1:0) \sim (0:0:1) \longmapsto (0:1:1:1), \\ q_2 &= (1:1:1) \sim (1:\omega^2:\omega) \sim (1:\omega:\omega^2) \longmapsto (1:3:0:0), \\ q_3 &= (1:\omega:1) \sim (1:1:\omega) \sim (\omega:1:1) \longmapsto (1:3\omega:0:0), \\ q_4 &= (1:\omega^2:1) \sim (\omega^2:1:1) \sim (1:1:\omega^2) \longmapsto (1:3\omega^2:0:0). \end{aligned} \tag{35}$$

3.2. The Automorphisms Group of Γ and the Hessian Group

To get contact between the orbifold and the Hesse pencil we have to prove the following theorem:

Theorem 2. *([CaCo-1], Theorem 3.5) The largest subgroup of the automorphism group $PGL(3,\mathbb{C})$ of the projective plane $\mathbb{P}^2$ that respects the quotient $\mathbb{P}^2/\tilde{\Gamma}$ is the Hessian group.*

The Hessian group acts naturally on the Hesse pencil of plane cubic curves. The Hesse pencil behavior under the quotient will suggest the right desingularization for the singular surface $\mathbb{P}^2/\tilde{\Gamma}$ as a quasi del Pezzo surface. To this aim we firs need to deepen our knowledge of the properties of Γ. The normalizer of Γ, $N := \{n \in GL(3,\mathbb{C}) : n\Gamma n^{-1} \subset \Gamma\}$, acts naturally on $\mathbb{C}^3$ and it is the largest subgroup of $GL(3,\mathbb{C})$ which sends Γ orbits into Γ orbits. The image of N in $PGL(3,\mathbb{C})$ is the quotient $\tilde{N} := N/D$, where $D := \{zI : z \in \mathbb{C}^*\}$ is the normal subgroup of N of diagonal matrices in $GL(3,\mathbb{C})$. It acts naturally on $\mathbb{P}^2$ and is the largest subgroup of $PGL(3,\mathbb{C})$ which sends a $\tilde{\Gamma}$ orbit into another one. $\tilde{\Gamma}$ is a normal subgroup of $\tilde{N}$ and the quotient group $\tilde{N}/\tilde{\Gamma}$ acts naturally on the quotient surface $\mathbb{P}^2/\tilde{\Gamma}$.

Theorem 3. *([CaCo-1], Theorem 3.6) The group $\tilde{N}/\tilde{\Gamma}$ is isomorphic to $SL(2,\mathbb{Z}_3)$ and the group $\tilde{N}$ has order 216.*

The proof of this theorem is based on following two technical Lemmas, whose proof can be found in [CaCo-1].

Lemma 4. *The homomorphism χ is a surjection from N to $SL(2,\mathbb{Z}_3)$.*

Lemma 5. *The kernel of χ is exactly $D \cdot \Gamma$, i.e. any element $n \in \ker \chi$ can be seen as the product of a diagonal matrix times an element $g \in \Gamma$.*

The proof of the theorem then goes as follows.

Proof. Any element $n \in N$ define an automorphism

$$\phi_n: \quad \Gamma \xrightarrow{\simeq} \Gamma \qquad g \longmapsto ngn^{-1}. \tag{36}$$

These kind of automorphisms of Heisenberg groups are studied in great generality in [We] (see also [Lig, Yo, Ge1, Ge2, CMS]). An element in Γ can be written uniquely as $\omega^k g_1^a g_2^b$, with $k, a, b \in \{0, 1, 2\}$. For any $n \in N$ the automorphism ϕ_n is determined by its action on the generators g_1, g_2 of Γ. Therefore, any ϕ_n is determined by the elements $k, l, a, b, c, d \in \{0, 1, 2\}$:

$$\phi_n(g_1) = \omega^k g_1^a g_2^b, \qquad \phi_n(g_2) = \omega^l g_1^c g_2^d. \tag{37}$$

One can check that the map

$$\chi: \quad N \longrightarrow GL(2, \mathbb{Z}_3) \qquad n \longmapsto \begin{pmatrix} a & c \\ b & d \end{pmatrix} \tag{38}$$

is a homomorphism of groups, and define a short exact sequence

$$0 \longrightarrow D \cdot \Gamma \xrightarrow{i} N \xrightarrow{\chi} SL(2, \mathbb{Z}_3) \longrightarrow 0, \tag{39}$$

where i is the natural inclusion.
When mapped to $PGL(3, \mathbb{C})$ the sequence (39) gives a second exact sequence

$$0 \longrightarrow \tilde{\Gamma} \xrightarrow{\tilde{i}} \tilde{N} \xrightarrow{\tilde{\chi}} SL(2, \mathbb{Z}_3) \longrightarrow 0. \tag{40}$$

This proves that $\tilde{N}/\tilde{\Gamma} \simeq SL(2, \mathbb{Z}_3)$. Now, for any matrices in $GL(2, \mathbb{Z}_3)$ there are $9-1 = 8$ choices for the elements of the firs column and $9 - 3 = 6$ choices for the second column, that give 48 matrices with determinants equal to ± 1. Therefore, the group $SL(2, \mathbb{Z}_3)$ has cardinality $|SL(2, \mathbb{Z}_3)| = 24$, and the group $\tilde{N}$ is a finit group of order $9 \cdot 24 = 216$. □

The group $\tilde{N}$ has the same order of the Hessian group. It is generated by

$$\tilde{N} = < g_1, g_2, N_S, N_T > . \tag{41}$$

It remain to show that the two groups coincide. To this end, let us firs review the definitio of the Hessian group.
The Hesse pencil [Hes1, Hes2] is a 1-parameter family of plane cubic curves $E_\mu \subset \mathbb{P}^2$, $\mu \in \mathbb{P}^1$, passing through 9 points in particular position. We can choose the following base points for the pencil, arranging them in a square array:

$$\begin{array}{lll} p_0 = (0:1:-1), & p_1 = (1:0:-1), & p_2 = (1:-1:0), \\ p_3 = (0:1:-\omega), & p_4 = (1:0:-\omega^2), & p_5 = (1:-\omega:0), \\ p_6 = (0:1:-\omega^2), & p_7 = (1:0:-\omega), & p_8 = (1:-\omega^2:0). \end{array} \tag{42}$$

The pencil is define by

$$E_\mu : \qquad x_0^3 + x_1^3 + x_2^3 - 3\mu x_0 x_1 x_2 = 0. \tag{43}$$

There are twelve lines containing the points in the horizontal, vertical and diagonal rows of the above array. Each group of three lines is the support of one of the 4 singular cubic curves of the pencil

$$\begin{array}{ll} \mu = \infty : & x_0 x_1 x_2 = 0, \\ \mu = 1 : & (x_0 + x_1 + x_2)(x_0 + \omega x_1 + \omega^2 x_2)(x_0 + \omega^2 x_1 + \omega x_2) = 0, \\ \mu = \omega : & (x_0 + \omega x_1 + x_2)(\omega x_0 + x_1 + x_2)(x_0 + x_1 + \omega x_2) = 0, \\ \mu = \omega^2 : & (x_0 + \omega^2 x_1 + x_2)(x_0 + x_1 + \omega^2 x_2)(\omega^2 x_0 + x_1 + x_2) = 0. \end{array} \tag{44}$$

The lines in each singular fibe intersect in three points that determine a degenerate orbit of $\tilde{\Gamma}$:

$$\begin{array}{l} \text{Sing}(E_0) = \text{Fix}(< g_1 >) = \{(1:0:0), (0:1:0), (0:0:1)\}, \\ \text{Sing}(E_1) = \text{Fix}(< g_2 >) = \{(1:1:1), (1:\omega^2:\omega), (1:\omega:\omega^2)\}, \\ \text{Sing}(E_\omega) = \text{Fix}(< g_1^2 g_2 >) = \{(1:\omega^2:1), (\omega^2:1:1), (1:1:\omega^2)\}, \\ \text{Sing}(E_{\omega^2}) = \text{Fix}(< g_1 g_2 >) = \{(1:\omega:1), (1:1:\omega), (\omega:1:1)\}. \end{array} \tag{45}$$

The Hessian group [Jo, Mas, SS] $G \subset PGL(3, \mathbb{C})$ keeps the set of the base points of the Hesse pencil invariant. It has order 216 and its action mixes the curves E_μ of the pencil. Therefore to show that $\tilde{N}$, the automorphism group of $\tilde{\Gamma}$, coincides with the Hessian group it suffice to show that $\tilde{N} \subset G$, namely that $\tilde{N}$ preserves the base points of the pencil. A trivial computation shows that the generators 41 of $\tilde{N}$ satisfy such requirement and this completes the proof of the theorem 2.

3.3. Resolution of $\mathbb{P}^2/\tilde{\Gamma}$

The group $\tilde{\Gamma} \subset \tilde{N}$ preserves the Hesse pencil structure. The action of $\tilde{\Gamma}$ on any curve E_μ of the pencil is the translation by the base points of the pencil. Thus $E_\mu/\tilde{\Gamma} \simeq E_\mu$ for any curve suggesting a possible resolution of $\mathbb{P}^2/\tilde{\Gamma}$.
More explicitly, let us shortly review the definitio of the group structure on a cubic curve [Har]. Let X be a non singular cubic curve in $\mathbb{P}^2$ and let $\text{Pic}^0(X) \subset \text{Pic}(X)$ be the group of degree zero divisors on X. A point P_0 on X is said an inflectio point if the intersection multiplicity of the tangent line to X in P_0 is equal to 3. A plane cubic curve has exactly 9 inflectio points and any line intersecting any two of them intersects the curve in a third one. The map that associates the divisor $P - P_0 \in \text{Pic}^0(X)$ to any closed point $P \in X$ is bijective and gives a group structure to the set of closed points of X. This is the group structure of $\text{Pic}^0(X)$ with P_0 as the identity. Similar varieties are known as group varieties. The group law has a nice geometric interpretation. In $\mathbb{P}^2$ each couple of line L, L' are equivalent in the Picard group $\text{Pic}(\mathbb{P}^2)$, therefore, if $L \cap X = P, Q, R$ and $L' \cap X = P', Q', R'$, we have $P + Q + R = P' + Q' + R'$ in $\text{Pic}(X)$. In particular, since P_0 is an inflectio point, it follows that $P + Q + R = 3P_0$ in $\text{Pic}(X)$. This implies that $P + Q + R = 0$ in the group variety X. The sum of P and Q is equal to the point T such that $P + Q - T = 0$, or, equivalently, $-T = R \in X \cap L$. On the other hand $-T \in X \cap L''$ where L'' is the line

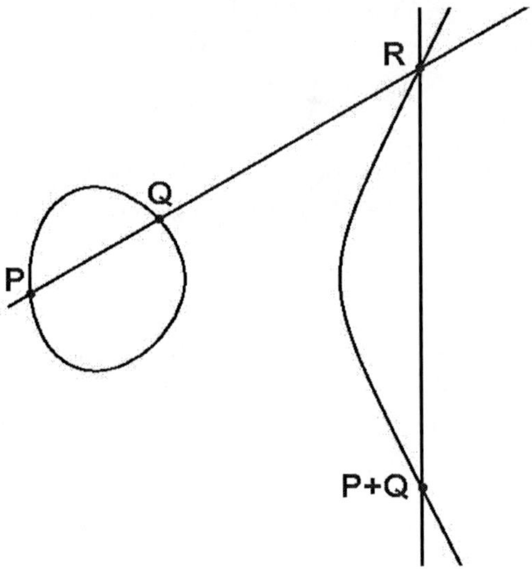

Figure 1. The group law on the cubic curve in $\mathbb{P}^2$ define by $Y^2Z = X^3 - XZ^2$. The inflectio point P_0 is the point at infinit .

passing for T and P_0. Hence, the sum of P and Q is given by the third intersection point T between the curve and the line L" passing for R and the fi ed inflectio point P_0.

On any smooth cubic curve of the Hesse pencil the 9 inflection points coincide with the base points of the pencil. If we fi a point, for example p_0, then on any E_μ the sum $p + p_i$ of any point $p \in E_\mu$ and the base point p_i is equal to the action of an element of $\tilde{\Gamma}$ on $p \in \mathbb{P}^2$ (it is sufficien to prove it for p a base points of the pencil). For any choice of the fi ed point p_i we have a group isomorphism H_i between the group of inflectio points (the base points of the pencil with group law the one from the group varieties E_μ) and $\tilde{\Gamma}$. For example, if we fi p_0 then:

$$H_0 : (p_0, p_1, p_2, p_3, p_4, p_5, p_6, p_7, p_8) \longmapsto (I, g_2, g_2^2, g_1^2 g_2, g_1 g_2, g_1 g_2^2, g_1^2 g_2^2, g_1, g_1^2). \tag{46}$$

Thus, the action of $\tilde{\Gamma}$ on $\mathbb{P}^2$ is the translation by the base points of the pencil, the points of order three. On any curve E_μ, such translation group is called $E_\mu[3]$. Thus the image $E_\mu/\tilde{\Gamma} \subset \mathbb{P}^2/\tilde{\Gamma}$ is just $E_\mu/E_\mu[3]$ which is well understood, being isomorphic to E_μ under the map

$$\begin{array}{ccc} E_\mu & \xrightarrow{\cdot 3} & E_\mu/E_\mu[3] \\ p & \longmapsto & 3p \end{array} \tag{47}$$

If we exclude the base points of the Hesse pencil, we can see the projective plane as a bundle of elliptic curves E_μ on $\mathbb{P}^1$. Indeed, we have just proved that the quotient map sends any fibe E_μ of the bundle to an elliptic curve in $\mathbb{P}^2/\tilde{\Gamma}$. Thus, also $\mathbb{P}^2/\tilde{\Gamma}$ contains a natural elliptic pencil, with any fibe isomorphic to one in $\mathbb{P}^2$. However, the projective plane $\mathbb{P}^2$ is not isomorphic to the singular $\mathbb{P}^2/\tilde{\Gamma}$: the pencil in $\mathbb{P}^2/\tilde{\Gamma}$ has only one base point, the image of the 9 base points of the Hesse pencil in $\mathbb{P}^2$ under the quotient. Therefore, the isomorphisms on the fiber are not compatible in the base points of the Hesse pencil and $\mathbb{P}^2$ is only birationally equivalent to $\mathbb{P}^2/\tilde{\Gamma}$. This suggest that by blowing up 8 of the 9 base points of the pencil one should obtain a regular morphism Ψ from a smooth surface S (actually a limit of del Pezzo surfaces) to $\mathbb{P}^2/\tilde{\Gamma}$ and hence a resolution of this variety. Now, we can state the main result

Theorem 4. *([CaCo-1], Theorem 3.9) Let* $S := Bl_{p1,\ldots,p8}(\mathbb{P}^2)$ *be the quasi del Pezzo surface obtained as the blow up* $\mathbb{P}^2$ *at eight base points of the Hesse pencil. Then we have the following commutative diagram:*

$$\begin{array}{ccccc} E_\mu & \hookrightarrow & \mathbb{P}^2 & \xleftarrow{\pi} & S \\ \downarrow \cdot 3 & & \downarrow & \swarrow \Psi & \\ E_\mu/\tilde{\Gamma} & \hookrightarrow & \mathbb{P}^2/\tilde{\Gamma} & & \end{array}$$

where Ψ *is a desingularization map given by the anticanonical sections on* S*. The total space of the canonical bundle* K_S *and the map* Ψ *define a crepant resolution of the orbifold* $\mathbb{C}^3/\Gamma$.

To prove this theorem we will construct explicitly the map Ψ. To this end, we will firs recall some generalities about del Pezzo surfaces. We refer to [Har, CoDo] and to the references therein. A del Pezzo surface is define to be a complex surface X whose anticanonical divisor class K_X is ample. This means that it exists a positive integer n such that nK_X is very ample, i.e. the global sections of nK_X give an embedding of X in a projective space. The Nakai-Moishezon criterion says that a divisor D on a surface X is ample if and only if $D^2 > 0$ and $D \cdot C > 0$ for all irreducible curves. Let S be the surface obtained as the blow up of n points $p_1, \ldots, p_n$ in $\mathbb{P}^2$. The divisor group of S is generated by H (the strict transform in S of the form definin the line in $\mathbb{P}^2$), and the exceptional divisors E_i. The intersection pairing is

$$H^2 = 1, \quad H \cdot E_i = 0 \quad \text{and} \quad E_i \cdot E_j = -\delta_{ij}. \tag{48}$$

The canonical divisor of the surface is $-K_S = 3H - E_1 - \ldots - E_n$. Thus, the Nakai-Moishezon criterion implies that S is a del Pezzo surface if $n \leq 8$ and the p_i are in general position, which means that no 3 of them are collinear and no 6 of them lie on a conic. A classical result states that every del Pezzo surfaces is either isomorphic to the blow up of $n \leq 8$ points in $\mathbb{P}^2$ or isomorphic to $\mathbb{P}^1 \times \mathbb{P}^1$.
From now on we specialize to the case $n = 8$. It is a well known fact that

$$\dim H^0(S, -nK_S) = 1 + \frac{n(n+1)}{2} \qquad (n \geq 1). \tag{49}$$

We report the essential properties of the anticanonical maps ϕ_n, the rational maps define by the global sections of $-nK_S$, for $n = 1, 2, 3$. We assume that the 8 points are all distinct and in general position. Thus the points defin a pencil of smooth cubic curves in $\mathbb{P}^2$ passing for them, all intersecting in a ninth points p_0. Any curve E in the pencil has a strict transform in S, again denoted by E, linearly equivalent to $-K_S$. Hence for any other E' in the pencil, $E \cdot E' = 1$ in S and p_0 is the unique points in S where they meet.

- The anticanonical bundle has $\dim H^0(-K_S) = 2$, therefore the cubic forms f_0, f_1

definin E and E' give a basis of $H^0(-K_S)$. The map ϕ_1 is define by

$$\begin{aligned} \phi_1 : \quad & S \dashrightarrow \mathbb{P}^1 \\ & q \mapsto (f_0(q) : f_1(q)) \end{aligned} \tag{50}$$

It is not define in p_0 where $f_0(p_0) = f_1(p_0) = 0$. Blowing up p_0 the map ϕ_1 define an elliptic fibratio over $\mathbb{P}^1$, where each fibe is an elliptic curve define by $\lambda f_0 + \mu f_1 = 0$ for some $(-\mu : \lambda) \in \mathbb{P}^1$. Note that the exceptional divisors E_i map onto the $\mathbb{P}^1$.

- Next we have dim $H^0(-2K_S) = 4$. The polynomial $f_0^2, f_0f_1, f_1^2 \in H^0(-2K_S)$ therefore it exists a homogeneous polynomial g of degree 6 such that $H^0(-2K_S) =< f_0^2, f_0f_1, f_1^2, g >$. The map ϕ_2 is define by

$$\begin{aligned} \phi_2 : \quad & S \dashrightarrow \mathbb{P}^3 \\ & q \mapsto (X_0 : X_1 : X_2 : X_3) = (f_0(q)^2 : f_0(q)f_1(q) : f_1(q)^2 : g(q)) \end{aligned} \tag{51}$$

Recall that $-2K_S = 6H - 2(E_1 + \ldots + E_8)$, thus the sextic curve $g = 0$ on $\mathbb{P}^2$ passes through $p_1, \ldots, p_8$ and it is singular there. Hence the sextic and the cubics $\lambda f_0 + \mu f_1 = 0$ intersect at $p_1, \ldots, p_8$ with multiplicity 2. There are two remaining intersection points, but it is not difficul to prove that p_0 is not one of them. This means that $g(p_0) \neq 0$ and ϕ_2 is a morphism. Any fibe of ϕ_1 is mapped $2:1$ to a $\mathbb{P}^1$:

$$\phi_2 : \lambda f_0 + \mu f_1 = 0 \longmapsto (f_0^2 : -\frac{\lambda}{\mu} f_0^2 : \frac{\lambda^2}{\mu^2} f_0^2 : g) = (1 : -\frac{\lambda}{\mu} : \frac{\lambda^2}{\mu^2} : \frac{g}{f_0^2}). \tag{52}$$

Hence ϕ_2 has degree two onto its image, which is the surface Q in $\mathbb{P}^3$ of equation $X_0X_2 = X_1^2$, a quadric with unique singular point $(0:0:0:1) = \phi_2(p_0)$. The fiber of ϕ_1 map to the lines passing through the vertex of the cone.

- Finally dim $H^0(-3K_S) = 7$, thus there is a homogeneous polynomial h of degree 9 such that $H^0(-3K_S) =< f_0^3, f_0^2f_1, f_0f_1^2, f_1^3, gf_0, gf_1, h >$. The curve $h = 0$ in $\mathbb{P}^2$ has triple points in $p_1, \ldots, p_8$. The map ϕ_3 is define by

$$\begin{aligned} \phi_3 : \quad & S \dashrightarrow \mathbb{P}^6 \\ & q \mapsto (f_0^3(q) : f_0^2(q)f_1(q) : f_0(q)f_1^2(q) : f_1^3(q) : g(q)f_0(q) : g(q)f_1(q) : h(q)) \end{aligned} \tag{53}$$

It is an embedding which sends each fibe of ϕ_1 to a smooth cubic in a $\mathbb{P}^2 \subset \mathbb{P}^6$.

Note that dim $H^0(-6K_S) = 22$, but also that $H^0(-6K_S)$ contains the 23 functions $f_0^6, f_0^5f_1, \ldots, f_1^6$, $gf_0^4, \ldots, gf_1^4$, $g^2f_0^2, \ldots, g^2f_1^2$, g^3, $hf_0^3, \ldots, hf_1^3$, h^2, f_0gh, f_1gh. Then, there must be a linear relation among these functions, which is a degree two polynomial in h and its coefficient are polynomials in f_0, f_1, g, reflectin the fact that the map define by f_0, f_1, g is $2:1$. It can be shown that this relation is unique. Thus, the generators f_0, f_1 of $H^0(-K_S)$, g of $H^0(-2K_S)$ and h of $H^0(-3K_S)$ defin an embedding

$$\begin{aligned} \Psi : \quad & S \longrightarrow \mathbb{P}_{1,1,2,3} \\ & q \longmapsto (f_0(q) : f_1(q) : g(q) : h(q)) \end{aligned} \tag{54}$$

that maps S into a hypersurface of degree six of the weighted projective space.

	L_{147}	L_{258}	L_{345}	L_{678}	L_{138}	L_{246}	L_{156}	L_{237}
L_{147}	-2	1	0	0	0	0	0	0
L_{258}	1	-2	0	0	0	0	0	0
L_{345}	0	0	-2	1	0	0	0	0
L_{678}	0	0	1	-2	0	0	0	0
L_{138}	0	0	0	0	-2	1	0	0
L_{246}	0	0	0	0	1	-2	0	0
L_{156}	0	0	0	0	0	0	-2	1
L_{237}	0	0	0	0	0	0	1	-2

Figure 2.

Let us now concentrate more on our specifi case. The surface S of theorem 4 is define as the blow up of $\mathbb{P}^2$ at eight of the base points of the Hesse pencil $p_1, \ldots, p_8$:

$$\pi : \quad S := Bl_{p1,\ldots,p8}(\mathbb{P}^2) \longrightarrow \mathbb{P}^2. \tag{55}$$

Obviously, these are not in general position as many of them are collinear. Let us consider the strict transform in S of the 8 lines in the singular fiber of the pencil which do not contain p_0:

$$\begin{aligned}
x_1 = 0 : &\qquad L_{147} = H - E_1 - E_4 - E_7,\\
x_2 = 0 : &\qquad L_{258} = H - E_2 - E_5 - E_8,\\
x_0 + \omega^2 x_1 + \omega x_2 = 0 : &\qquad L_{345} = H - E_3 - E_4 - E_5,\\
x_0 + \omega x_1 + \omega^2 x_2 = 0 : &\qquad L_{678} = H - E_6 - E_7 - E_8,\\
x_0 + \omega x_1 + x_2 = 0 : &\qquad L_{138} = H - E_1 - E_3 - E_8,\\
x_0 + x_1 + \omega x_2 = 0 : &\qquad L_{246} = H - E_2 - E_4 - E_6,\\
x_0 + \omega^2 x_1 + x_2 = 0 : &\qquad L_{156} = H - E_1 - E_5 - E_6,\\
x_0 + x_1 + \omega^2 x_2 = 0 : &\qquad L_{237} = H - E_2 - E_3 - E_7.
\end{aligned} \tag{56}$$

The intersection matrix between these curves is given in table 3.3.. It shows that the intersection graph for each pair of curve in the singular fiber is of type A_2. Moreover, the curves have zero intersection with the canonical divisor on S, $-K_S$ does not satisfy the second requirement of the Nakai-Moishezon criterion and it is not ample. The surface S is then a smooth variety that can be seen as a degenerate limit of del Pezzo surfaces. In particular, the anticanonical global sections are constant on the above L_{ijk}, therefore such curves get blown down by the anticanonical system discussed above. In this case the map (54) is a morphism from S to a singular surface of degree six in $\mathbb{P}_{1,1,2,3}$ which sends the curve L_{ijk} to 4 singular points of type A_2.
We can now conclude our proof. The forms

$$\begin{aligned}
f_0 &= XYZ\\
f_1 &= X^3 + Y^3 + Z^3\\
g &= X^6 + (Y^2 - YZ + Z^2)^3 + X^3(2Y^3 - 3Y^2Z - 3YZ^2 + 2Z^3)\\
h &= (X^3 + Y^3 + Z^3 + 3\omega Y^2 Z + 3\omega^2 YZ^2).
\end{aligned} \tag{57}$$

$$(X^6 + (Y^2 - YZ + Z^2)^3 + X^3(2Y^3 - 3Y^2Z - 3YZ^2 + 2Z^3))$$

defin four curves on $\mathbb{P}^2$. The cubics $f_0 = 0$ and $f_1 = 0$ defin the Hesse pencil, and intersect at $p_0, \dots, p_8$. Their strict transforms in S are divisors linearly equivalent to $-K_S$. The sextic $g = 0$ has double points in $p_1, \dots, p_8$ and does not contain p_0, then definin a divisor linearly equivalent to $-2K_S$. Finally, the curve $h = 0$ has triple points at $p_1, \dots, p_8$ and does not contain p_0, thus definin a divisor linearly equivalent to $-3K_S$. In this way, we see that the rational map

$$\begin{aligned} \overline{\Psi}: \quad \mathbb{P}^2 &\dashrightarrow \mathbb{P}_{1,1,2,3} \\ q &\mapsto (f_0(q) : f_1(q) : g(q) : h(q)) \end{aligned} \tag{58}$$

is not define in $p_1, \dots, p_8$, but it gives a morphism

$$\Psi: \quad S \longrightarrow \mathbb{P}_{1,1,2,3} \tag{59}$$

whose image is the surface of section 3.1.:

$$w^2 + y^3 - 27wx^3 - 3wyz + wz^3 = 0 \subset \mathbb{P}_{1,1,2,3}. \tag{60}$$

In particular, as expected, the (-2)-curves L_{ijk} on S are mapped to the singular points in 60. As the quotient $\mathbb{P}^2/\tilde{\Gamma}$ is isomorphic to (60), we have just proved that such quotient has a desingularization given by a quasi del Pezzo surface S. The orbifold $\mathbb{C}^3/\Gamma$ is isomorphic to the tautological cone over (60) in $\mathbb{C}^4$. Then, it has a desingularization given the variety X obtained as the total space of the canonical line bundle K_S over S, with desingularization map the one from Ψ. This complete the proof of theorem 4.

4. The Toric Analogue

To complete the geometrical analysis of the resolved orbifold, in order to be able to include D-branes geometry to construct the string version of the Standard Model, we should study the derived category or coherent sheaves over X. However, this would bring us to a technically more complicated field requiring a big amount of advanced topics in algebraic geometry, which we have no space to discuss here. For this reason, we prefer here to continue our analysis on a simpler example, a toric version of X. This consists in replacing the non abelian Heisenberg group Δ_{27} with its abelianization $\mathbb{Z}_3 \times \mathbb{Z}_3$. We will thus briefl study the geometry of the three dimensional orbifold $M = \mathbb{C}^3/(\mathbb{Z}_3 \times \mathbb{Z}_3)$. The analysis of the classical geometry will resemble that one of the previous section for $\mathbb{C}^3/\Delta_{27}$, but now takeing advantage of the powerful tools of toric geometry. Here we will not review the basics of toric geometry, referring to [Ful, Oda] for all the details of the computation.
The singular variety M admits several possible resolutions. In analogy with the previous section, we will show that M is a cone over a $\mathbb{P}^2/\mathbb{Z}_3$ orbifold. Next, we will firs consider that resolution $\tilde{M}$ of M which again result to be a cone over a resolution of $\mathbb{P}^2/\mathbb{Z}_3$. To determine the K-theory basis however, we have to start from the G-Hilbert resolution and then going back to the previous resolution by a series of flops We will compute the K-theory basis, which will provide also a basis for the derived category of coherent sheaves, for al that resolutions obtained by a flo in going from the G-Hilbert scheme to $\tilde{M}$. We will not explain the details of the computations. Explanations can be found in [CaCo-2] and references therein.

4.1. The Orbifold $M = \mathbb{C}^3/(\mathbb{Z}_3 \times \mathbb{Z}_3)$ and Its Tautological Resolution

The toric variety M is define by the action on $\mathbb{C}^3$ of the abelian group $\mathbb{Z}_3 \times \mathbb{Z}_3$, whose generators are

$$g_1 = \begin{pmatrix} 1 & 0 & 0 \\ 0 & \omega & 0 \\ 0 & 0 & \omega^2 \end{pmatrix}, \qquad g_2 = \begin{pmatrix} \omega & 0 & 0 \\ 0 & \omega & 0 \\ 0 & 0 & \omega \end{pmatrix}, \tag{61}$$

with $\omega = e^{\frac{2\pi i}{3}}$. Its fan Δ is generated by the vectors

$$v_1 = \begin{pmatrix} -1 \\ 1 \\ 1 \end{pmatrix}, \qquad v_2 = \begin{pmatrix} -1 \\ -2 \\ 1 \end{pmatrix}, \qquad v_3 = \begin{pmatrix} 2 \\ 1 \\ 1 \end{pmatrix}, \tag{62}$$

in $N \simeq \mathbb{Z}^3$ and in figur 3 we pict the two dimensional intersection of Δ with the plane $z = 1$. We note that there are four monomials of degree three that are invariant under the

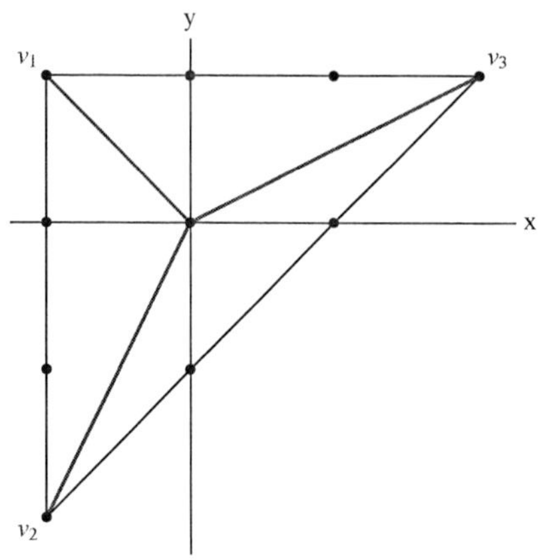

Figure 3. The fan of the orbifold $\mathbb{C}^3/\mathbb{Z}_3 \times \mathbb{Z}_3$. As a two dimensional fan it is the one of $\mathbb{P}^2/\mathbb{Z}_3$.

action of $\mathbb{Z}_3 \times \mathbb{Z}_3$:

$$\begin{aligned} x &= XYZ, \\ y &= X^3, \\ z &= Y^3, \\ w &= Z^3. \end{aligned} \tag{63}$$

They defin a map from the orbifold $\mathbb{C}^3/(\mathbb{Z}_3 \times \mathbb{Z}_3)$ to the tautological bundle over the projective space $\mathbb{P}^3$, and the homogeneous relation

$$x^3 - yzw = 0. \tag{64}$$

In particular, (63) defin an isomorphism between the orbifold M and the cone over the singular cubic surface define by equation 64. Any smooth cubic surface in $\mathbb{P}^3$ is a dP_6, a

del Pezzo surface of degree three isomorphic to the blown up of $\mathbb{P}^2$ in six points in general position. Therefore, we expect that the singular surface (64) should be related to some sort of limit in the moduli space of dP_6 surfaces.
Now, M is a non compact Calabi-Yau threefold with non isolated singularities. There are several crepant resolutions of such an orbifold, all related by flo transitions (see [CaCo-2] for a short guide to toric desingularization). Here we will consider the one corresponding to the fan in figur 4. This is the total space of the canonical bundle over the toric surface Σ

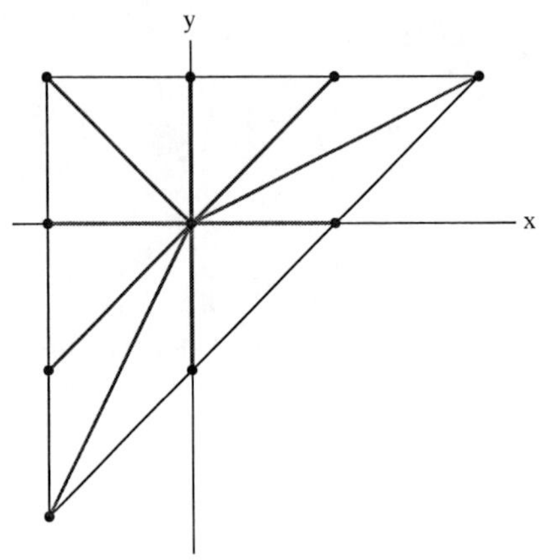

Figure 4. The fan of a crepant resolution of the orbifold $\mathbb{C}^3/\mathbb{Z}_3 \times \mathbb{Z}_3$. As a two dimensional fan it gives the minimal resolution of $\mathbb{P}^2/\mathbb{Z}_3$.

define by the same fan in figur 4 (thought as a two dimensional fan), with the zero section shrunk down to a point. In other words, it is the tautological cone over Σ, so that it will dub it the tautological resolution. Let us look at the relation between such a resolution and del Pezzo surfaces.
The action of the generator g of $\mathbb{Z}_3$ on the homogeneous coordinates of $\mathbb{P}^2$ is

$$g \cdot (z_1 : z_2 : z_3) = (z_1 : \omega z_2 : \omega^2 z_3), \qquad \omega^3 = 1. \tag{65}$$

It has three fi ed points

$$p_1 \equiv (1:0:0), \qquad p_2 \equiv (0:1:0), \qquad p_3 \equiv (0:0:1), \tag{66}$$

the toric invariant points. Using local charts near the p_i, one sees that the action of g is thus given by $\mathrm{diag}(\omega, \omega^2)$ on $\mathbb{P}^2$. Therefore, the quotient surface $\mathbb{P}^2/\mathbb{Z}_3$ has three singular points of type A_2. Since the orbifold M is locally the product $\mathbb{P}^2_3 \times A^1_{\mathbb{C}}$, in correspondence of the images of the p_i we have three curves of A_2 singular points in M. We can obtain a resolution of these singularities by performing a double blow-up of the singular points, then obtain the smooth toric surface S described by the fan depicted in figur 4. The total space of the canonical bundle K_S provides a crepant resolution of the initial orbifold M, the tautological resolution. This operation is represented in figur 6, where the blow up map π shows the birationality of the surface S with $\mathbb{P}^2$: starting from $\mathbb{P}^2$ we blow up the three toric invariant points $q_1 = (1:0:0)$, $q_2 = (0:1:0)$, $q_3 = (0:0:1)$, obtaining the toric del Pezzo surface dP_3. The three exceptional divisors E_i on dP_3 separate the three toric invariant lines D_i on $\mathbb{P}^2$ define by $x_i = 0$. Blowing up three intersections of the E_i with the strict transforms of the D_i in dP_3 we obtain the surface S. This is a degenerate limit in the moduli space of del Pezzo surfaces of degree three, and can be obtained by blowing up $\mathbb{P}^2$ in three couples of infinitel nearby points. Note that, for simplicity, we use the same name for the strict transforms of the relevant curves in the different varieties. In S we have the following intersection table:

	L_1	L_2	L_3	E_1	D_2	E_2	D_3	E_3	D_1
L_1	-1	0	0	0	1	0	0	1	0
L_2	0	-1	0	1	0	0	1	0	0
L_3	0	0	-1	0	0	1	0	0	1
E_1	0	1	0	-2	1	0	0	0	0
D_2	1	0	0	1	-2	0	0	0	0
E_2	0	0	1	0	0	-2	1	0	0
D_3	0	1	0	0	0	1	-2	0	0
E_3	1	0	0	0	0	0	0	-2	1
D_1	0	0	1	0	0	0	0	1	-2

Figure 5. Intersection table of S.

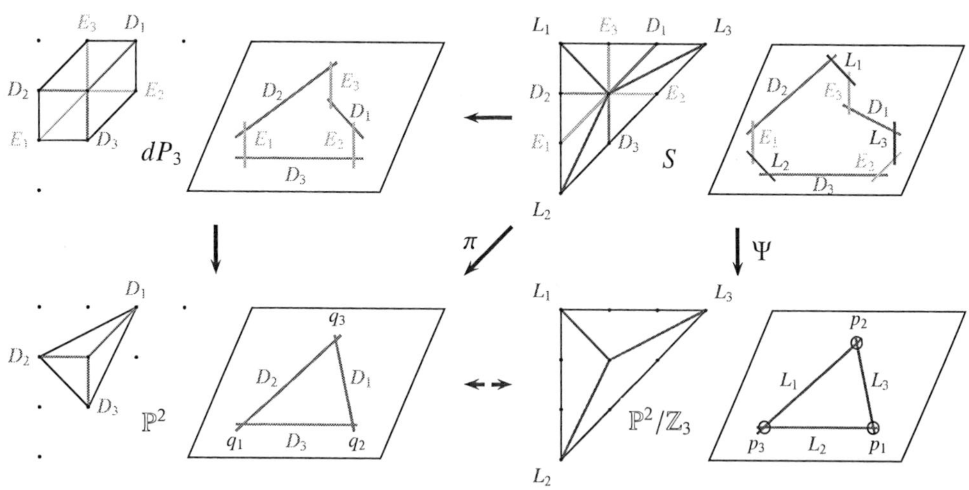

Figure 6. The minimal resolution of $\mathbb{P}^2/\mathbb{Z}_3$.

Again, as for the non abelian model, the intersection graph for each pair of curves (E_i, D_{i+1}) is A_2, and the resolution map from S to $\mathbb{P}^2_3$ sends such curves to the three singular points of type A_2.

The map from S to the singular cubic surface 64 is given by the anticanonical sections. The canonical divisor of the surface S is $K_S = -\sum D_i + E_i + L_i$, which has zero intersection with the curves (E_i, D_i). Hence, $-K_S$ does not satisfy the second requirement of the Nakai-Moishezon criterion and it is not ample. The anticanonical global sections

are constant on the above curves, and are then blown down by the anticanonical map. The resolution S is then a degenerate limit of a del Pezzo surface dP_6.
By means of toric methods we can obtain the explicit anticanonical map. In figur 7 we depict the polytope in the dual lattice M related to the anticanonical divisor $-K_S$. We then

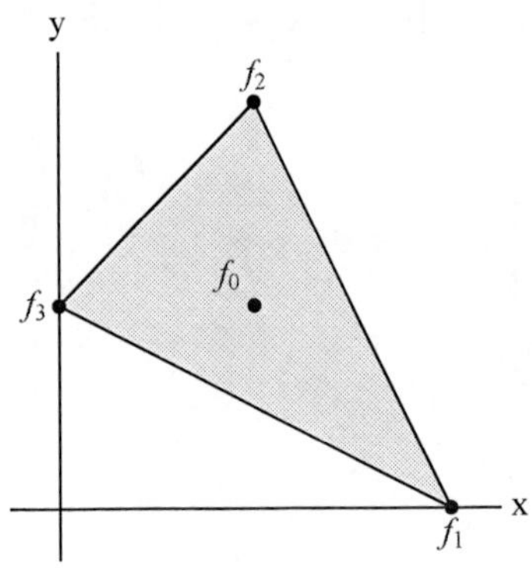

Figure 7. The polytope of $-K_S$.

see that a basis for $H^0(S, -K_S)$ is given by the four monomials

$$\begin{aligned} f_0 &= XY, \\ f_1 &= X^2, \\ f_2 &= XY^2, \\ f_3 &= Y, \end{aligned} \tag{67}$$

which defin the equivariant map

$$\begin{aligned} \Psi : \quad & S \to \mathbb{P}^3 \\ & q \mapsto (f_0(q) : f_1(q) : f_2(q) : f_3(q)) \end{aligned} \tag{68}$$

The image of Ψ is the singular cubic surface (64)

$$x^3 - yzw = 0. \tag{69}$$

This is a strictly semistable cubic surface with three double rational points of type A_2. Such surface arises naturally in the context of the complex ball uniformization of the moduli space of cubic surfaces, when boundary points are included (see [DGK], section 9 and [ACT]).
Then, we have the following commutative diagram:

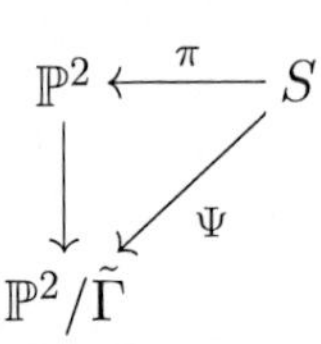

The total space of the canonical bundle K_S and the map Ψ give a crepant resolution of the starting orbifold M.

This concludes our analysis of the classical geometry of the toric model. However, to look at the brane geometry and compute the prepotential it is better to start with another resolution, the G-Hilbert scheme. We do it in the next subsections.

4.2. Mirror Symmetry, Cohomological Valued Hypergeometric Function, and Superpotentials

In order to give meaning to the computations listed in the next sections, here we review shortly how the superpotentials can be computed by using homological mirror symmetry. This section is essentially a short summary of part of the book [H& al].

Let us consider a string theory having a toric Calabi-Yau variety X as target space, no matter if it is compact or not. In this case one can interpret mirror symmetry as a T-duality transformation. Indeed, string theory on X can be described as a gauged linear sigma model, a two dimensional $U(1)^m$ supersymmetric gauge theory. It contains a certain number $n > m$ of complex scalar field $Z = \{Z_\alpha\}_{\alpha=1}^n$ having charges $Q_{\alpha,r}$ $r = 1, \ldots, m$, with respect to the gauge group $U(1)^m$, and having potential energy

$$U(Z) = \frac{1}{2}\sum_{r=1}^{m} g_r^2 \left(\sum_{\alpha=1}^{n} Q_{\alpha,r} Z_\alpha \bar{Z}_\alpha - \mathrm{r}_r\right)^2, \tag{70}$$

where g_r and r_r are the gauge couplings and the Fayet-Iliopoulos terms respectively. To supersymmetric ground states it correspond a vanishing potential energy:

$$\sum_{\alpha=1}^{n} Q_{\alpha,r} Z_\alpha \bar{Z}_\alpha = \mathrm{r}_r. \tag{71}$$

For a fi ed choice of the Fayet-Iliopoulos parameters, these equations defin a toric variety X associated to a fan, in an $(n-m)$–dimensional lattice N, generated by a suitable set $\Sigma(1) = v_1, \ldots, v_n$ of vectors in N. Thus, the supersymmetric vacua are identifie with the points of a toric variety X. Each vector v_α determines an invariant divisor D_{v_α}. Correspondingly, one can chose a basis $\{C_r\}_{r=1}^m$ of irreducible curves in $H_2(X, \mathbb{Z})$ such that the charges are given by the intersection numbers $Q_{\alpha,r} = D_{v_\alpha} \cdot C_r$. As the F–I parameters rescale as $|Z|^2$, if chosen to be positive, they will parameterize the points of the Kähler cone. This means that the supersymmetric configuration are completely characterized in geometrical terms.

Mirror symmetry can thus be realized as a T–duality transformation. Roughly speaking, T–duality on a circle transforms a type A string theory on a circle of radius R in a type B string theory on a circle of radius α'/R and viceversa. If Z_α take value on a complex variety, we can T–dualize their phases, which defin circles in the target manifold. The result is a Landau-Ginzburg theory for a set of chiral superfield related by the set of constraints

$$\sum_{\alpha=1}^{n} Q_{\alpha,r} Y_\alpha = t_r \tag{72}$$

where t_r are the complexifie Kähler parameters ($\mathrm{Re}(t_r) = \mathrm{r}_r$), and with superpotential

$$W(Y,t) = \sum_{\alpha=1}^{n} e^{-Y_\alpha}. \tag{73}$$

Thus, the mirror map maps the twodimensional sigma model in a Landau–Ginzburg model with superpotential $W(Y,t)$. We can take contact with the Batyrev's geometric construction of mirror manifolds, for the cases when the starting linear sigma model describes strings on a crepant resolution of some abelian quotient $\mathbb{C}_3/G$. It is described by a set of vectors $v_1, \ldots, v_n$ in a three-dimensional lattice such that for some isomorphism $\phi : N \longrightarrow \mathbb{Z}^3$ one has $\phi(v_\alpha) = (n_{\alpha,1}, n_{\alpha,2}, 1)$. The solutions of the constraints can thus be written in terms of three independent field y_0, y_1, y_2 as $Y_\alpha = y_0 + n_{\alpha,1}y_1 + n_{\alpha,2}y_2 + c_\alpha$, c_α being suitable constants satisfying $\sum_{\alpha=1}^{n} Q_{\alpha,r}Y_\alpha = t_r$. These linear redefinition do not affect the functional measure, and setting $w_a = \exp(-y_a)$, $a = 0, 1, 2$ and $a_\alpha = \exp(-c_\alpha)$ one gets for the superpotential

$$W(w,a) = w_0 \sum_{\alpha=1}^{n} a_\alpha w_1^{n_{\alpha,1}} w_2^{n_{\alpha,2}}, \qquad w_a \in \mathbb{C}_*. \tag{74}$$

This LG model is equivalent to another one, where $w_0 \in \mathbb{C}$, with two extra chiral field $U, V \in \mathbb{C}$, whose superpotential is

$$\tilde{W}(U,V;w;a) = W(w,a) - w_0 UV. \tag{75}$$

Integrating the fiel w_0 thus gives a delta function $\delta(\sum_{\alpha=1}^{n} a_\alpha w_1^{n_{\alpha,1}} w_2^{n_{\alpha,2}} - UV)$ that means that the mirror LG model is equivalent to a geometrical theory on a Calabi-Yau manifold

$$Y_a = \{(\vec{u}, \vec{w}) \in \mathbb{C}^2 \times \mathbb{C}_*^2 | F_a(\vec{u}, \vec{w}) = 0\}, \tag{76}$$

where

$$F_a(\vec{u}, \vec{w}) = u_1^2 + u_2^2 + f_a(\vec{u}, \vec{w}) = u_1^2 + u_2^2 + \sum_{\alpha=1}^{n} a_\alpha w_1^{n_{\alpha,1}} w_2^{n_{\alpha,2}}. \tag{77}$$

The Kähler parameters t now parameterize the complex moduli of Y. Up to now, however, we have not taken into account the presence of brane configurations Because we are looking for supersymmetric vacua, we need to know what kind of brane configuration are admitted on a Calabi-Yau manifold X. In other words, one must search for boundary condition compatible with supersymmetry. The answer depends on the type of string theory one is considering. For type A strings, supersymmetric branes are classically represented by halfdimensional subvarieties S, $\iota : S \hookrightarrow X$, over which the Kähler form ω of the C-Y manifold vanishes, $\iota^*\omega = 0$, and supporting fla vector bundles. Thus A–branes are Lagrangian submanifolds with respect to the symplectic structure ω. Instead, for type B strings one find that supersymmetric branes must wrap holomorphic cycles of X supporting holomorphic vector bundles. In our case it means that type B brane configuration will be described classically by compact divisors, curves of the Mori cone and points. Thus, mirror symmetry should map BPS states of a model into the BPS states of the mirror model, converting

A–branes to B–branes and viceversa. This gives rise to an odd asymmetry between A and B configurations indeed all A–branes have the same dimensions, whereas this does not happen for B–branes. The point is that in the LG model description branes configuration can change when moduli vary. In this picture, BPS states will correspond to critical points of the superpotential. Essentially they determine the points of Y_t around which the supersymmetric three–cycles are defined By varying t, the critical points move on the W–plane, and if some of these points moves around a branch point, a monodromy transformation can give rise to a new brane configuration In the geometric picture, the boundary states corresponding to the branes are described by the periods of the holomorphic three-form Ω of Y_t. The monodromy thus acts on a basis of cycles recasting them in some linear recombination or, equivalently, on the periods by the same linear recombination. On the mirror X it should correspond to a recombination of the holomorphic cycles, hard to understand in the naïve geometrical picture where they have different dimensions! A firs democracy is introduced by a K-theoretical description, where lower dimensional branes can be described in terms of the top dimensional branes and a tachyon field K-theory captures topological aspects of the problem, but it is quite limited from the geometrical point of view. A deeper geometrical understanding of stable brane configuratio in a topological limit of type B superstrings can be understood in terms of triangulated categories, in particular the derived category of coherent sheaves on the manifold. This provides a deep contact between physics and the "homological mirror symmetry," conjectured by Kontsevich, proposing an enhancement of the usual geometrical mirror symmetry to an homological level as an equivalence between triangulated categories: the derived category of coherent sheaves on a Calabi Yau manifold X with a fi ed complex structure on one side and the derived $\mathcal{A}^{\infty}$ Fukaya's category over the mirror manifold Y on the other side, essentially generated by the Lagrangian submanifolds of $\{Y, \omega\}$, where the symplectic structure ω is given by the fi ed Kähler form on Y, dual to the complex form on X

$$\text{Mir} : D^b Coh(X) \xrightarrow{\simeq} DFuk^o(Y, \omega). \tag{78}$$

The BPS states in the mirror type A string model are described by periods that are integrals of the holomorphic three form Ω on Y over the Lagrangian cycles. For noncompact abelian quotients, the holomorphic three form on the mirror Y is

$$\Omega = \frac{1}{4\pi^3} \mathrm{Res}_{F=0} \left[\frac{du^1 \wedge du^2 \wedge dw^1 \wedge dw^2}{w^1 w^2 F(\vec{u}; \vec{w}; a)} \right]. \tag{79}$$

Here the Kähler form is fi ed, but Ω depends explicitly on the complex moduli of Y, appearing in the explicit dependence on a of the polynomial F. Thus, the periods

$$\Pi_{C_i}(a) = \int_{C_i} \Omega, \tag{80}$$

of any set of Lagrangian cycles C_i, are locally holomorphic functions of the complex moduli. Actually, they are forced to satisfy a set of hypergeometric differential equations known as the GKZ hypergeometric system. For compact varieties, the knowledge of a complete set of solutions for the GKZ system correspond to an exhaustive description of the set of BPS brane configuration on the A side. Furthermore, the special Kähler geometry of the complex structure moduli space of a Calabi Yau manifold can be described in terms of periods. If x parameterizes the structure complex moduli of Y then the Kähler potential of the moduli space is

$$K(x,\bar{x}) = -\log\left[i\sum_{I=0}^{h^{2,1}(Y)}\left(X^I\frac{\partial\bar{G}}{\partial\bar{X}^I} - \bar{X}^I\frac{\partial G}{\partial X^I}\right)\right], \tag{81}$$

where $X^I(x) = \int_A^I \Omega(x)$ are the periods with respect to a canonical symplectic basis $\{A^I, B_I\}$ of $H_3(Y,\mathbb{Z})$, and $G(x)$ is the prepotential

$$G(x) = \frac{1}{2}\sum_{I=0}^{h^{1,2}(Y)}\int_{A^I}\Omega\int_{B_I}\Omega. \tag{82}$$

We thus obtain a correspondence between Kähler moduli t_i of X and complex moduli of Y, such that

$$t_i = \frac{X^i}{X^0}, \qquad i = 1,\ldots,h^{1,2}(Y) = h^{1,1}(X). \tag{83}$$

The Kähler moduli space of X is a special Kähler manifold itself, which can be described in terms of a prepotential function $F(t)$. At classical level such geometry is described by the prepotential

$$F^c(t) = \frac{1}{6}d_{ijk}t^it^jt^k \tag{84}$$

where $d_{ijk} = J_i\cdot J_j\cdot J_k$ are the intersection numbers associated to the Kähler cone generators. They determine the Yukawa couplings of the chiral fields These couplings receive non perturbative quantum corrections from worldsheet instantons. At lowest order they correspond to the wrapping of the worldsheet on rational curves in X. Their energy is given by the volume of the wrapped cycle as measured by the given Kähler metric. Any given class of rational curve of degree $\vec{d}$ will contribute to the prepotential with a term

$$n_{\vec{d}}\,Li_3(e^{2\pi i\vec{d}\cdot t}), \tag{85}$$

where $n_{\vec{d}}$ is the number of classes of curves with the given degree. Then, the full prepotential takes the form

$$F(t) = \frac{1}{6}d_{ijk}t^it^jt^k - \frac{1}{24}c_2(X)\cdot J_it^i - i\frac{\zeta(3)}{16\pi^3}c_3(X) + \sum_{d\in\mathbb{Z}_>^{h^{1,1}}} n_{\vec{d}}\,Li_3(e^{2\pi i\vec{d}\cdot t}). \tag{86}$$

The $n_{\vec{d}}$ are the Gromov-Witten invariants in the Gopakumar-Vafa interpretation. The identificatio

$$\left\{\int_{A^I}\Omega;\int_{B_I}\Omega\right\}_{I=0}^{h^{2,1}(Y)} = \left\{1,\ t^i;\ \partial_{t^i}F,\ 2F - \sum_{j=1}^{h^{1,1}(X)} t^j\partial_{t^i}F\right\}_{i=1}^{h^{1,1}(X)} \tag{87}$$

together with mirror symmetry gives a simple way to compute the GW–invariants of X. Remarkably, this procedure can be elegantly codifie in a suitable hypergeometric function having value in the cohomology ring. One can introduce a cohomological valued power series whose expansion in the Chow ring

$$A^*(X) \otimes \mathbb{C}[[x]][\log x]$$

gives a basis for the period integrals of the mirror manifold Y in the large complex structure limit (LCSL). This cohomological series encodes many geometrical information on both the manifolds X and Y so summarizing several fundamental aspects of mirror symmetry. Hosono [Hos1] has proposed an extension of this picture to local mirror symmetry for noncompact C-Y manifolds, in particular for resolutions of abelian quotients $\mathbb{C}^k/G$, with $k = 2, 3$. As the Hosono's conjecture encodes all the tools we will use to compute the superpotentials, we end this section by reporting it here, as stated in [CaCo-2].

The starting point is that the periods for the mirror manifold are solutions of a set of Picard-Fuchs equations, whose general solution can be expressed in terms of an hypergeometric function with value in the cohomology of X

$$w = w\left(x_1, \ldots, x_4; \frac{J_1}{2\pi i}, \ldots, \frac{J_4}{2\pi i}\right)$$

Let B_i be a basis of the compactly supported K-theory $K^c(X)$, and Φ_i the corresponding dual basis in $K(X)$ w.r.t. the natural pairing between K and K^c. Defin the basis for $H^*(X, \mathbb{Q})$

$$Q_i := \mathrm{ch}(\Phi_i)\,, \quad i = 0, \ldots, 5$$

and expand the cohomology-valued hypergeometric series w with respect to this basis

$$w\left(x_1, \ldots, x_4; \frac{J_1}{2\pi i}, \ldots, \frac{J_4}{2\pi i}\right) = \sum_{i=0}^{5} w_i(x_1, \ldots, x_4) Q_i\,.$$

The conjecture states

1. the coefficien hypergeometric series $w_i(x_1, \ldots, x_4)$ may be identifie with the period integrals over the cycles $mir(B_i)$,

$$w_i(x_1, \ldots, x_4) = \int_{mir(B_i)} \Omega(Y_x)\,;$$

2. the monodromy of the hypergeometric series is integral and symplectic with respect to the symplectic form define in $K^c(X)$

$$\chi(B_i, B_j) = \int_X \mathrm{ch}(B_i^\vee)\mathrm{ch}(B_j)\mathrm{td}(X)\,;$$

3. the central charge of an element $F \in K^c(X)$ is expressed in terms of the cohomology valued hypergeometric w as

$$Z(F) = \int_X \mathrm{ch}(F) w\left(x_1, \ldots, x_4; \frac{J_1}{2\pi i}, \ldots, \frac{J_4}{2\pi i}\right) \mathrm{td}(X)\,.$$

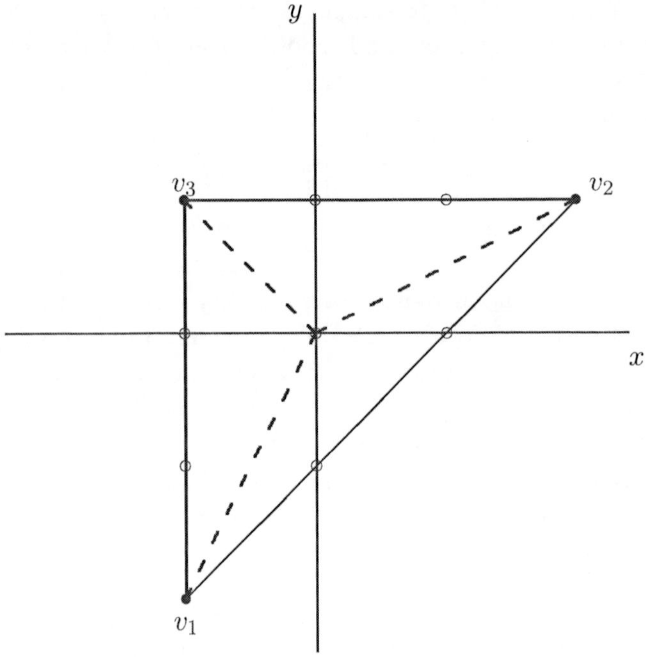

Figure 8. Fan for $\mathbb{C}^3_{3\times 3}$.

4.3. The $\mathbb{Z}_3 \times \mathbb{Z}_3$-Hilbert Resolution

The starting singular toric variety is define by the fan $\{0, \rho_1, \rho_2, \rho_3, A_{12}, A_{23}, A_{13}, V\}$, where ρ_i are the one dimensional cones generated by the vectors v_i,

$$v_1 = \begin{pmatrix} -1 \\ -2 \\ 1 \end{pmatrix}, \qquad v_2 = \begin{pmatrix} 2 \\ 1 \\ 1 \end{pmatrix}, \qquad v_3 = \begin{pmatrix} -1 \\ 1 \\ 1 \end{pmatrix} \tag{88}$$

in the lattice $N = \mathbb{Z}^3$, A_{ij} with $i < j$ are the two dimensional cones generated by v_i, v_j and V is the 3-D convex cone having A_{ij} as faces. In the plane $z = 1$ of $N_{\mathbb{R}} = \mathbb{R}^3 = \mathbb{Z}^3 \otimes \mathbb{R}$ it is represented as
Indeed, singular variety associated to this fan is $\mathbb{C}^3_{3\times 3} = \mathbb{C}^3/\mathbb{Z}_3 \times \mathbb{Z}_3$, where $\mathbb{Z}_3 \times \mathbb{Z}_3$ acts on the coordinates as

$$(z_1, z_2, z_3) \longmapsto (\epsilon_1 z_1, \epsilon_2 z_2, \epsilon_3 z_3), \qquad \epsilon_i^3 = 1, \ \epsilon_1\epsilon_2\epsilon_3 = 1. \tag{89}$$

The G-Hilbert resolution can be obtained by toric standard methods. We refer to [CaCo-2] and references therein for its definitio and the construction methods. The resulting variety is a toric one, define by the following fan Σ, where we added exceptional divisors (black dots) and curves (the lines), as in figur 9.
It consists of the origin, 10 one dimensional cones (generated by v_i), 18 two dimensional cones (the curves C_{ij}) and 9 three dimensional cones (the points). For future reference it is convenient to decorate the fan with the transition functions. Each 3-cone selects an affin space with the given ring of local monomials (note that they can have negative exponentials

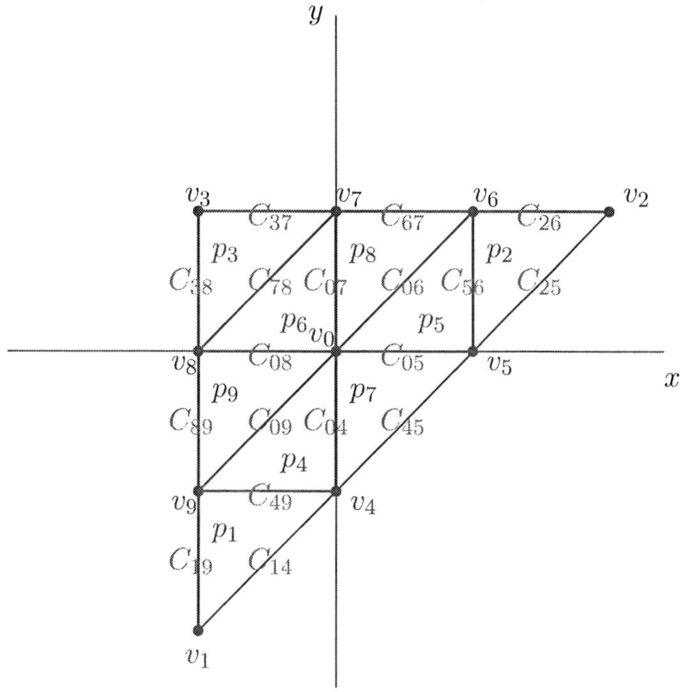

Figure 9. Fan for $\mathbb{Z}_3 \times \mathbb{Z}_3 - \mathbb{C}^3_{3\times 3}$.

and that in each 3-cone their product is xyz). Each monomial $m(x, y, z) = x^\alpha y^\beta z^\gamma$ is associated to a boundary facet by $x^\alpha y^\beta z^\gamma = 0$. Passing from an affin open set to another one intersecting it in the given boundary, the monomial transform as $m(x, y, z) \mapsto m(x, y, z)^{-1}$. For this reason in figur 10 the monomial are written across the lines with the numerator inside to the given cone and the denominator on the other side.

The Chow Ring $A^*(X_\Sigma)$

Let us call X_Σ the toric variety associated to Σ. It is then obvious that $A^3(X_\Sigma) = \mathbb{Z}X_\Sigma$. $A^2(X_\Sigma)$ contains ten generators, the divisors $D_i = V(\rho_i)$, which are the subvarieties of X_Σ invariant under the action of the cones $\rho_i =< v_i >, i = 0, 1, \ldots, 9$.
However, the divisors are related by three linear relations dictated by the vectors of the fan, which are

$$-D_1 + 2D_2 - D_3 + D_5 + D_6 - D_8 - D_9 = 0, \quad (90)$$

$$-2D_1 + D_2 + D_3 - D_4 + D_6 + D_7 - D_9 = 0, \quad (91)$$

$$D_0 + D_1 + D_2 + D_3 + D_4 + D_5 + D_6 + D_7 + D_8 + D_9 = 0. \quad (92)$$

Note that the last one is the vanishing condition for the firs Chern class.

Thus there are 7 free generators, let us say D_i, $i = 1, \ldots, 7$, so that

$$D_9 = -2D_1 + D_2 + D_3 - D_4 + D_6 + D_7, \tag{93}$$

$$D_8 = D_1 + D_2 - 2D_3 + D_4 + D_5 - D_7, \tag{94}$$

$$D_0 = -3D_2 - D_4 - 2D_5 - 2D_6 - D_7, \tag{95}$$

and $A^2(X_\Sigma) = \bigoplus_{i=1}^{7} \mathbb{Z}D_i \simeq \mathbb{Z}^7$. For $A^1(X_\Sigma)$ we have 18 generators given by the curves C_{ij}. However, they are related by the equivalences induced by the divisors they are contained in. Takeing account of all such relations, encoded in figure11 we fin only one free generator, let it be $C \equiv C_{45}$. Finally, $A^0(X_\Sigma) = 0$ because all points are equivalent, lying in pairs on some $\mathbb{P}^1$ curve, and are equivalent to 0, because some of them lies on an affin space $\mathbb{A}^1$. In conclusion, the Chow ring is

$$A^*(X_\Sigma) = \mathbb{Z}(X_\Sigma, D_1, D_2, D_3, D_4, D_5, D_6, D_7, C)/\mathcal{R}, \tag{96}$$

where $\mathcal{R}$ are the relations induced by the intersection products, given in table 1. For completeness we report the products between all toric invariant divisors.

The Compact Chow Modulus $A^c_*(X_\sigma)$

It is generated by the compact invariant subvarieties. These are the compact divisor D_0, the 9 curves $C_{94}, C_{56}, C_{78}, C_{04}, C_{05}, C_{06}, C_{07}, C_{08}, C_{09}$ and the points p_i, $i = 1, \ldots, 9$.

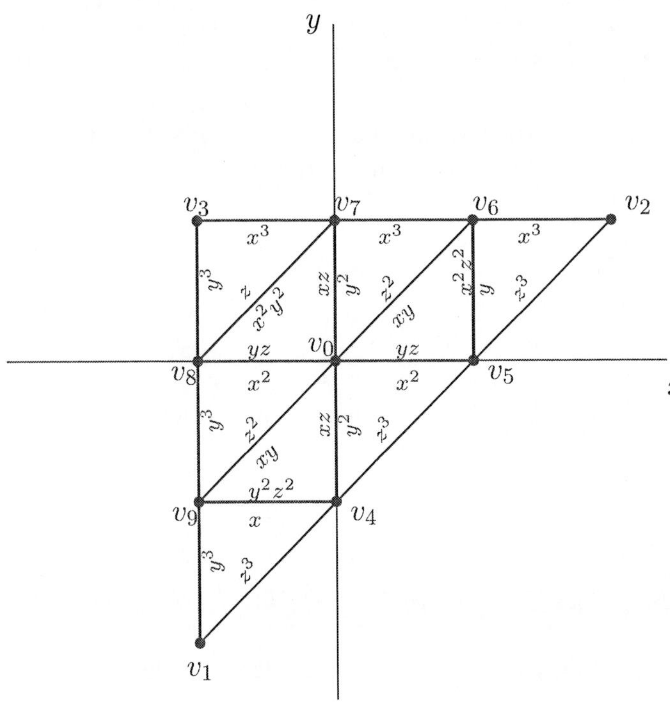

Figure 10. Decoration of the fan.

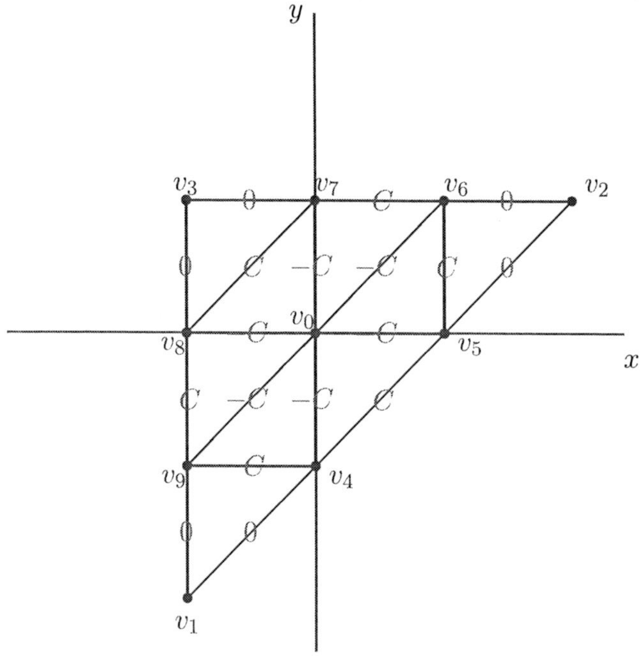

Figure 11. Equivalences in $A^1(X_\Sigma)$.

For each of them we have to consider the corresponding Chow ring. We will use an apex c referring to compactness, meaning that the definin relations are relative to the compact sets containing the given object. For example, for the points we get $A^*(p_i) = p_i^c$. For the curves $A^*(C_{ij}) = \mathbb{Z}C_{ij}^c \oplus \mathbb{Z}p_{ij}^c$; in this case there are in fact two points on each compact curve. However, such curves are $\mathbb{P}^1 s$ so the points are equivalent definin the unique element p_{ij}^c. For the compact divisor, note that it is a toric variety whose fan is obtained from the one of Σ, quotienting the lattice w.r.t. v_0. We obtain the fan of figur 12.
It contains 6 compact curves with two relations

$$C_{05} + C_{06} - C_{08} - C_{09} = 0 \tag{98}$$
$$C_{06} + C_{07} - C_{04} - C_{09} = 0, \tag{99}$$

so that

$$C_{08} = C_{04} + C_{05} - C_{07} \tag{100}$$
$$C_{09} = -C_{04} + C_{06} + C_{07}. \tag{101}$$

All points are equivalent to a unique point p_0^c, so that

$$A^*(D_0) = \mathbb{Z}D_0^c \oplus \mathbb{Z}C_{06}^c \oplus \mathbb{Z}C_{07}^c \oplus \mathbb{Z}C_{08}^c \oplus \mathbb{Z}C_{09}^c \oplus \mathbb{Z}p_0^c.$$

The compact Chow modulus is

$$A_*^c(X_\Sigma) = A^*(D_0) \oplus \bigoplus_{\substack{compact\\curves}} A^*(C_{ij}) \oplus \bigoplus_{i=1}^{9} A^*(p_i)/\sim$$

Table 1. Intersection product in $A^*(X_\Sigma)$.

	X_Σ	D_1	D_2	D_3	D_4	D_5	D_6	D_7	D_8	D_9	D_0	C
X_Σ	X_Σ	D_1	D_2	D_3	D_4	D_5	D_6	D_7	D_8	D_9	D_0	C
D_1	D_1	0	0	0	0	0	0	0	0	0	0	0
D_2	D_2	0	0	0	0	0	0	0	0	0	0	0
D_3	D_3	0	0	0	0	0	0	0	0	0	0	0
D_4	D_4	0	0	0	$-C$	C	0	0	0	C	$-C$	0
D_5	D_5	0	0	0	C	$-C$	C	0	0	0	$-C$	0
D_6	D_6	0	0	0	0	C	$-C$	C	0	0	$-C$	0
D_7	D_7	0	0	0	0	0	C	$-C$	C	0	$-C$	0
D_8	D_8	0	0	0	0	0	0	C	$-C$	C	$-C$	0
D_9	D_9	0	0	0	C	0	0	0	C	$-C$	$-C$	0
D_0	D_0	0	0	0	$-C$	$-C$	$-C$	$-C$	$-C$	$-C$	$6C$	0
C	C	0	0	0	0	0	0	0	0	0	0	0

(97)

where $\sim$ is the rational equivalence between elements contained in a common compact subvariety. At the end we get

$$A^c_*(X_\Sigma) = \mathbb{Z}D^c_0 \oplus \mathbb{Z}C^c_{94} \oplus \mathbb{Z}C^c_{56} \oplus \mathbb{Z}C^c_{78} \oplus \mathbb{Z}C^c_{04} \oplus \mathbb{Z}C^c_{05} \oplus \mathbb{Z}C^c_{06} \oplus \mathbb{Z}C^c_{07} \oplus p^c. \tag{102}$$

This is a modulus over the ring $A^*(\Sigma)$ by means of the map

$$A^*(X_\Sigma) \times A^c_*(X_\Sigma) \longrightarrow A^c_*(X_\Sigma)$$

define by the intersection product quotiented w.r.t. the rational equivalence in $A^c_*(X_\Sigma)$. It is represented by the following table, where we report all the intersections between invariant toric divisors and invariant compact curves. We defin $C^* = -C^c_{04} - 2C^c_{05} - 2C^c_{06} - C^c_{07}$. For toric varieties it holds the proposition

$$A^*(X_\Sigma) \simeq H^*(X_\Sigma, \mathbb{Z}), \qquad A^c_*(X_\Sigma) \simeq H_*(X_\Sigma, \mathbb{Z}). \tag{104}$$

Then $H_*(X_\Sigma, \mathbb{Z})$ is a modulus over $H^*(X_\Sigma, \mathbb{Z})$ determined by the intersection product as before.

The Dual Basis

We defin the compact curves

$$C_1 \equiv C^c_{94}, \quad C_2 \equiv C^c_{56}, \quad C_3 \equiv C^c_{78}, \quad C_4 \equiv C^c_{04}, \quad C_5 \equiv C^c_{05},$$
$$C_6 \equiv C^c_{06}, \quad C_7 \equiv C^c_{07}. \tag{105}$$

The dual divisors T_i in $A^*(X_\Sigma)$ represent elements such that $T_i \cdot C_i = \delta_{ij}$. These are given by

$$T_1 \equiv D_1, \quad T_2 \equiv D_2, \quad T_3 \equiv D_3,$$
$$T_4 \equiv -D_1 + D_2 + D_3 - D_4 + D_6 + D_7, \quad T_5 \equiv D_2 + D_3 + D_6 + D_7,$$
$$T_6 \equiv D_1 + D_2 + D_4 + D_5, \quad T_7 \equiv D_1 + D_2 - D_3 + D_4 + D_5 - D_7. \tag{106}$$

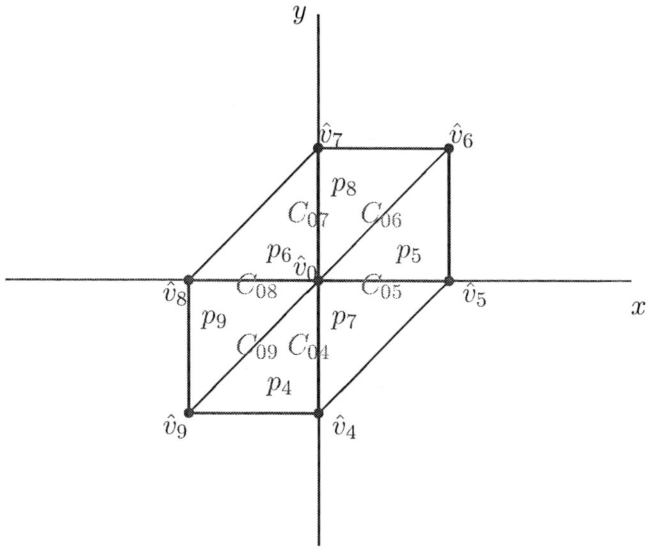

Figure 12. Fan for the compact divisor.

K-theory Generators

The generators of the group $K(X_\Sigma)$ are given by a result of Ito and Nakajima: they are represented by the tautological bundles associated to the irreducible representations of the group. Each affin open set is coordinatized by three invariant monomials as given in figur 10. Let us numerate the affin sets O_i as the points in figur 9, and let us call m_i^1, m_i^2, m_i^3 the associated monomials. Then the space $V_i = \mathbb{C}[x, y, z]/\{m_i^1, m_i^2, m_i^3\}$ for each i is a 9−dimensional vector space generated by the monomials which transform along the nine representations of $\mathbb{Z}_3 \times \mathbb{Z}_3$, which we will call the representations space. The representations are $(i, j) = \epsilon_1^i \otimes \epsilon_2^j$, $(i, j) \in \mathbb{Z}_3 \times \mathbb{Z}_3$, where

$$\epsilon_1^i \otimes \epsilon_2^j : (x, y, z) \longmapsto (\omega_1^i x, \omega_2^j y, \omega_1^{2i} \omega_2^{2j} z), \qquad \omega_1^3 = \omega_2^3 = 1. \tag{107}$$

The representation spaces are

$$V_1 = \langle 1, y, y^2, z, z^2, yz, y^2z, yz^2, y^2z^2 \rangle_{\mathbb{C}}, \tag{108}$$
$$V_2 = \langle 1, x, x^2, z, z^2, xz, x^2z, xz^2, x^2z^2 \rangle_{\mathbb{C}}, \tag{109}$$
$$V_3 = \langle 1, x, x^2, y, y^2, xy, x^2y, xy^2, x^2y^2 \rangle_{\mathbb{C}}, \tag{110}$$
$$V_4 = \langle 1, x, y, z, yz, y^2, z^2, yz^2, zy^2 \rangle_{\mathbb{C}}, \tag{111}$$
$$V_5 = \langle 1, x, y, z, xz, x^2, z^2, xz^2, zx^2 \rangle_{\mathbb{C}}, \tag{112}$$
$$V_6 = \langle 1, x, y, z, xy, y^2, x^2, yx^2, xy^2 \rangle_{\mathbb{C}}, \tag{113}$$
$$V_7 = \langle 1, x, y, z, z^2, xz, yz, xz^2, yz^2 \rangle_{\mathbb{C}}, \tag{114}$$
$$V_8 = \langle 1, x, y, z, x^2, xy, xz, zx^2, yx^2 \rangle_{\mathbb{C}}, \tag{115}$$
$$V_9 = \langle 1, x, y, z, y^2, yx, yz, xy^2, zy^2 \rangle_{\mathbb{C}}. \tag{116}$$

Looking at the transformations of the monomials under the representations we can then

Table 2. The map $A^*(X_\Sigma) \times A_*^c(X_\Sigma) \longrightarrow A_*^c(X_\Sigma)$.

	D_0^c	C_{94}^c	C_{56}^c	C_{78}^c	C_{04}^c	C_{05}^c	C_{06}^c	C_{07}^c	C_{08}^c	C_{09}^c	p^c
X_Σ	D_0^c	C_{94}^c	C_{56}^c	C_{78}^c	C_{04}^c	C_{05}^c	C_{06}^c	C_{07}^c	C_{08}^c	C_{09}^c	p^c
D_1	0	p^c	0	0	0	0	0	0	0	0	0
D_2	0	0	p^c	0	0	0	0	0	0	0	0
D_3	0	0	0	p^c	0	0	0	0	0	0	0
D_4	C_{04}^c	$-p^c$	0	0	$-p^c$	p^c	0	0	0	p^c	0
D_5	C_{05}^c	0	$-p^c$	0	p^c	$-p^c$	p^c	0	0	0	0
D_6	C_{06}^c	0	$-p^c$	0	0	p^c	$-p^c$	p^c	0	0	0
D_7	C_{07}^c	0	0	$-p^c$	0	0	p^c	$-p^c$	p^c	0	0
D_8	C_{08}^c	0	0	$-p^c$	0	0	0	p^c	$-p^c$	p^c	0
D_9	C_{09}^c	$-p^c$	0	0	p^c	0	0	0	p^c	$-p^c$	0
D_0	C^*	p^c	p^c	p^c	$-p^c$	$-p^c$	$-p^c$	$-p^c$	$-p^c$	$-p^c$	0
C	p^c	0	0	0	0	0	0	0	0	0	0

(103)

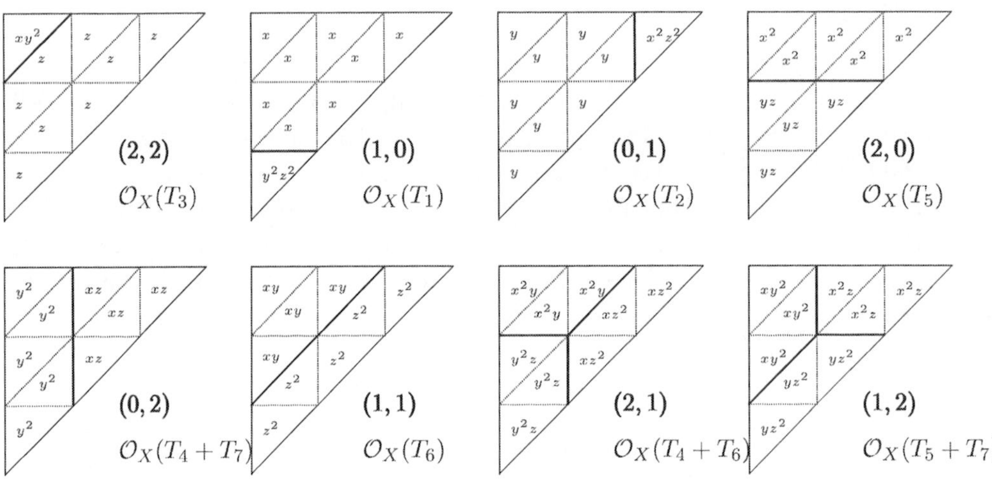

Figure 13. Tautological bundles.

construct the sheaf of monomials associated to each representation. For example the trivial representation $(0, 0)$ has all monomials equal to 1, representing the sheaf $\mathcal{O}_X$. The non trivial representation are collected in figur 13. They are labeled with the corresponding representation and the realization as divisorial sheaves. The last ones are determined as follows. The thick lines represent the boundaries where the monomials change in passing from an affin open set to the next one. Compared to the corresponding coordinate transformation one determines the degree of the sheaf on the given curve (it results to be 1 over all thick lines and 0 elsewhere). The divisorial sheaf is the one associated to the divisor having the same intersections with the curves as given by the degrees. Thus a basis for $K(X_\Sigma)$ is given by

$$\begin{array}{lll} \mathcal{R}_{00} = \mathcal{O}_X, & \mathcal{R}_{01} = \mathcal{O}_X(T_2), & \mathcal{R}_{02} = \mathcal{O}_X(T_4 + T_7), \\ \mathcal{R}_{10} = \mathcal{O}_X(T_1), & \mathcal{R}_{11} = \mathcal{O}_X(T_6), & \mathcal{R}_{12} = \mathcal{O}_X(T_5 + T_7), \\ \mathcal{R}_{20} = \mathcal{O}_X(T_5), & \mathcal{R}_{21} = \mathcal{O}_X(T_4 + T_6), & \mathcal{R}_{22} = \mathcal{O}_X(T_3). \end{array} \tag{117}$$

The only moltiplicative relation between the elements of such basis is

$$\mathcal{R}_{02} \otimes \mathcal{R}_{11} \otimes \mathcal{R}_{20} = \mathcal{R}_{12} \otimes \mathcal{R}_{21}\,. \tag{118}$$

Now we study the GIT chamber of this resolution. We write parameters θ as $(\theta_{00}, \theta_{01}, \ldots, \theta_{22})$, where $\theta_{ij} := (\theta(\rho_{ij}))$. The inequalities definin the G-Hilb chamber are

$\theta_{10} > 0$	wall of type I related to the flo of the curve C_{94};	(119)
$\theta_{01} > 0$	wall of type I related to the flo of the curve C_{56};	(120)
$\theta_{22} > 0$	wall of type I related to the flo of the curve C_{78};	(121)
$\theta_{02} + \theta_{21} > 0$	wall of type I related to the flo of the curve C_{04};	(122)
$\theta_{12} + \theta_{20} > 0$	wall of type I related to the flo of the curve C_{05};	(123)
$\theta_{11} + \theta_{21} > 0$	wall of type I related to the flo of the curve C_{06};	(124)
$\theta_{02} + \theta_{12} > 0$	wall of type I related to the flo of the curve C_{07};	(125)
$\theta_{20} + \theta_{21} > 0$	wall of type I related to the flo of the curve C_{08};	(126)
$\theta_{11} + \theta_{12} > 0$	wall of type I related to the flo of the curve C_{09};	(127)
$\theta_{12} > 0$	wall of type 0 define by $\theta(\varphi_C(\mathcal{R}_{12}^{-1}\vert_{D_0})) > 0$;	(128)
$\theta_{21} > 0$	wall of type 0 define by $\theta(\varphi_C(\mathcal{R}_{21}^{-1}\vert_{D_0})) > 0$;	(129)
$\theta_{02} + \theta_{11} + \theta_{12}$		(130)
$+ \theta_{20} + \theta_{21} > 0$	wall of type 0 define by $\theta(\varphi_C(\mathcal{R}_{00}^{-1} \otimes \omega_{D_0})) < 0$.	(131)

Any other inequality is redundant.

The Mori and Kähler Cones

The Mori cone is the polyhedral cone in $A_2^c(X) \otimes \mathbb{Q}$ generated by effective toric invariant compact curves of X, which are the compact algebraic cycles $\sum a_{ij} C_{ij}^c$ where all the a_{ij} are non-negative. For $X = \mathrm{G}{-}\mathbb{C}^3_{3\times 3}$, in view of the relations (101), the Mori cone is generated by all the nine compact curves C_{ij}^c:

$$\begin{array}{lllll} C_1 \equiv C_{94}^c, & C_2 \equiv C_{56}^c, & C_3 \equiv C_{78}^c, & C_4 \equiv C_{04}^c, & C_5 \equiv C_{05}^c, \\ C_6 \equiv C_{06}^c, & C_7 \equiv C_{07}^c, & C_8 \equiv C_{08}^c, & C_9 \equiv C_{09}^c. \end{array} \tag{132}$$

Since $A_2^c(X) \otimes \mathbb{Q}$ has dimension seven, the Mori cone is not simplicial.

The Kähler cone of X is the set of all forms J in $H^2(X, \mathbb{Q})$ such that

$$\int_C J \geq 0$$

for all effective cycles in $H_2^c(X,\mathbb{Q})$. Using the isomorphisms (104) we can think at the Kähler cone of X as the dual polyhedral cone in $A^2(X)\otimes\mathbb{Q}$ of the Mori cone with respect to the intersection pairing. However also the Kähler cone is not simplicial and it has the following generators $G_a,\ a=\{1,\ldots,8\}$, that satisfy $G_a\cdot C_b\geq 0$ for all the C_b:

$$\begin{aligned}&G_1=D_3,\quad G_2=D_2,\quad G_3=D_1,\quad G_4=2D_2+D_5+D_6,\\&G_5=D_2+D_3+D_6+D_7,\quad G_6=D_1+D_2+D_4+D_5,\\&G_7=2D_2+D_3+D_5+D_6+D_7,\quad G_8=D_1+2D_2+D_4+D_5+D_6.\end{aligned}\tag{133}$$

Mirror Symmetry and the Prepotential

In order to calculate the prepotential using mirror symmetry we have to choose a large complex structure limit. We select the following basis between the generators of the Kähler cone:

$$\begin{aligned}&G_1=D_3,\quad G_2=D_2,\quad G_3=D_1,\quad G_4=2D_2+D_5+D_6,\\&G_5=D_2+D_3+D_6+D_7,\quad G_6=D_1+D_2+D_4+D_5,\\&G_7=2D_2+D_3+D_5+D_6+D_7,\end{aligned}\tag{134}$$

and we name $\{J_1,\ldots,J_7\}$ the corresponding forms in $H^2(X,\mathbb{Q})$. The dual basis in $A_2^c(X)\otimes\mathbb{Q}$ is:

$$\begin{aligned}&\tilde{C}_1=C_3,\quad \tilde{C}_2=C_2,\quad \tilde{C}_3=C_1,\quad \tilde{C}_4=C_7,\\&\tilde{C}_5=C_5,\quad \tilde{C}_6=-C_4+C_6+C_7,\quad \tilde{C}_7=C_4-C_7.\end{aligned}\tag{135}$$

Following Hosono, the vectors ℓ_a, $a=1,\ldots,7$ are given by the intersection numbers between the $\tilde{C}_i$ and the invariant divisors of X, so that we fin

$$\begin{array}{lcl}\tilde{C}_1 & : & \ell_1=(0,0,1,0,0,0,-1,-1,0,1)\,,\\ \tilde{C}_2 & : & \ell_2=(0,1,0,0,-1,-1,0,0,0,1)\,,\\ \tilde{C}_3 & : & \ell_3=(1,0,0,-1,0,0,0,0,-1,1)\,,\\ \tilde{C}_4 & : & \ell_4=(0,0,0,0,0,1,-1,1,0,-1)\,,\\ \tilde{C}_5 & : & \ell_5=(0,0,0,1,-1,1,0,0,0,-1)\,,\\ \tilde{C}_6 & : & \ell_6=(0,0,0,1,0,0,0,1,-1,-1)\,,\\ \tilde{C}_7 & : & \ell_7=(0,0,0,-1,1,-1,1,-1,1,0)\,.\end{array}\tag{136}$$

The cohomology-valued hypergeometric series in respect to the basis $\{1,\vec{J},C\}$ of $H^*(X,\mathbb{Q})$ is then

$$\begin{aligned}w\left(\vec{x},\frac{\vec{J}}{2\pi i}\right)=\;&1+\left(-\frac{ix_1x_4}{2\pi}-\frac{i\log x_1}{2\pi}\right)J_1+\left(-\frac{ix_2x_5}{2\pi}-\frac{i\log x_2}{2\pi}\right)J_2\\&+\left(-\frac{ix_3x_6}{2\pi}-\frac{i\log x_3}{2\pi}\right)J_3+\left(-\frac{ix_1x_4}{2\pi}-\frac{i\log x_4}{2\pi}\right)J_4\\&+\left(-\frac{ix_2x_5}{2\pi}-\frac{i\log x_5}{2\pi}\right)J_5+\left(-\frac{ix_3x_6}{2\pi}-\frac{i\log x_6}{2\pi}\right)J_6\\&+\left(\frac{ix_1x_4}{2\pi}+\frac{ix_2x_5}{2\pi}+\frac{ix_3x_6}{2\pi}-\frac{i\log x_7}{2\pi}\right)J_7\\&+\left(-\frac{x_1}{4\pi^2}-\frac{{x_1}^2}{16\pi^2}-\frac{x_2}{4\pi^2}-\frac{{x_2}^2}{16\pi^2}-\frac{x_3}{4\pi^2}-\frac{{x_3}^2}{16\pi^2}\right.\end{aligned}$$

$$+\frac{x_4}{4\pi^2}+\frac{x_4{}^2}{16\pi^2}+\frac{x_5}{4\pi^2}+\frac{x_5{}^2}{16\pi^2}+\frac{x_6}{4\pi^2}+\frac{x_6{}^2}{16\pi^2}+\frac{x_4x_7}{4\pi^2}$$
$$+\frac{x_5x_7}{4\pi^2}+\frac{x_6x_7}{4\pi^2}+\frac{x_1x_4\log x_4}{4\pi^2}+\frac{x_2x_5\log x_5}{4\pi^2}-\frac{\log x_4\log x_5}{4\pi^2}$$
$$+\frac{x_3x_6\log x_6}{4\pi^2}-\frac{\log x_4\log x_6}{4\pi^2}-\frac{\log x_5\log x_6}{4\pi^2}-\frac{\log x_4\log x_7}{4\pi^2}$$
$$-\frac{\log x_5\log x_7}{4\pi^2}-\frac{\log x_6\log x_7}{4\pi^2}-\frac{\log x_7{}^2}{8\pi^2}\bigg)C\,. \tag{137}$$

However we stopped our calculation at the second order in $\vec{x}$. Finally the prepotential is

$$\begin{aligned}F(\vec{t}) = \; & -\frac{iq_1}{8\pi^3}-\frac{iq_1{}^2}{64\pi^3}-\frac{iq_2}{8\pi^3}-\frac{iq_2{}^2}{64\pi^3}-\frac{iq_3}{8\pi^3}-\frac{iq_3{}^2}{64\pi^3}\\ & -\frac{iq_4}{8\pi^3}-\frac{iq_4{}^2}{64\pi^3}-\frac{iq_5}{8\pi^3}-\frac{iq_5{}^2}{64\pi^3}-\frac{iq_6}{8\pi^3}-\frac{iq_6{}^2}{64\pi^3}\\ & -\frac{iq_4q_7}{8\pi^3}-\frac{iq_5q_7}{8\pi^3}-\frac{iq_6q_7}{8\pi^3}-t_1-\frac{3t_1{}^2}{2}-t_1{}^3\\ & -t_1t_4-t_1{}^2t_4-t_1t_5-t_1{}^2t_5-t_1t_4t_5-t_1t_6\\ & -t_1{}^2t_6-t_1t_4t_6-t_1t_5t_6-\frac{3t_1t_7}{2}-\frac{3t_1{}^2t_7}{2}\\ & -t_1t_4t_7-t_1t_5t_7-t_1t_6t_7-\frac{t_1t_7{}^2}{2}\\ & +\text{Gclass}[-t_1+t_2,-t_1+t_3,t_1+t_4,t_1+t_5,t_1+t_6,t_7]\\ & +\text{Gistant}\left[\frac{q_2}{q_1},\frac{q_3}{q_1},q_1q_4,q_1q_5,q_1q_6,q_7\right]\,.\end{aligned} \tag{138}$$

The GW–invariants

For this resolution we have computed the prepotential for all the large complex structure limits, then we determined the GW–invariants. They are all integer numbers and we found the same results for those invariants calculated in many different LCLS points. As observed in [CaCo-2], the invariants associated to curves having zero intersection with the unique compact divisor D_0 are uncomputable. Moreover all the curves under investigation have negative intersection with some noncompact divisor, therefore all invariants except the ones of degree one vanish. We list the GW–invariants up to degree two, in term of the curves 132.

$$\begin{aligned}& GW_{[1,0,0,0,0,0,0]}=GW_{[0,1,0,0,0,0,0]}=GW_{[0,0,1,0,0,0,0]}=GW_{[0,0,0,1,0,0,0]}=\\ & GW_{[0,0,0,0,1,0,0]}=GW_{[0,0,0,0,0,1,0]}=GW_{[0,0,0,0,0,0,1]}=GW_{[0,0,0,1,1,0,-1]}=\\ & GW_{[0,0,0,-1,0,1,1]}=-1\\ \\ & GW_{[2,0,0,0,0,0,0]}=GW_{[0,2,0,0,0,0,0]}=GW_{[0,0,2,0,0,0,0]}=GW_{[0,0,0,2,0,0,0]}=\\ & GW_{[0,0,0,0,2,0,0]}=GW_{[0,0,0,0,0,2,0]}=GW_{[0,0,0,0,0,0,2]}=GW_{[0,0,0,2,2,0,-2]}=\\ & GW_{[0,0,0,-2,0,2,2]}=0\\ \\ & GW_{[1,1,0,0,0,0,0]}=GW_{[1,0,1,0,0,0,0]}=GW_{[0,1,1,0,0,0,0]}=GW_{[0,0,0,1,0,1,0]}=\\ & GW_{[0,0,0,1,0,0,1]}=GW_{[0,0,0,2,1,0,-1]}=GW_{[0,0,0,0,1,0,1]}=GW_{[0,0,0,1,2,0,-1]}=\\ & GW_{[0,0,0,-1,1,1,1]}=GW_{[0,0,0,1,1,1,-1]}=GW_{[0,0,0,-1,0,2,1]}=GW_{[0,0,0,-1,0,1,2]}=0.\end{aligned} \tag{139}$$

We consider these results a positive check for the procedure we used to defin the cohomology-valued hypergeometric series in the non simplicial Mori cone case.

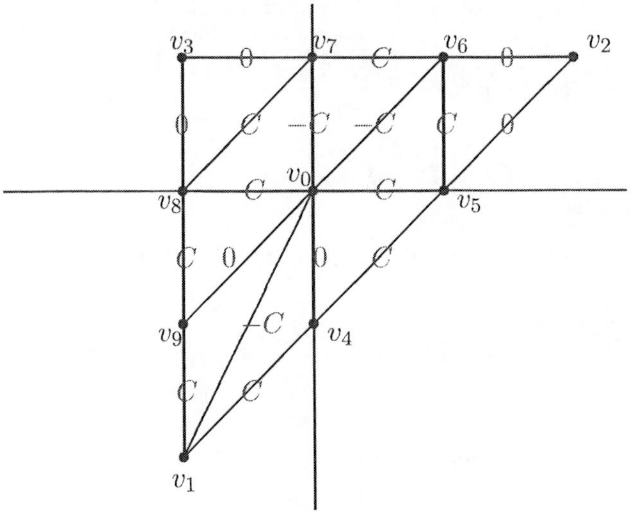

Figure 14. Equivalences in $A^1(X_\Sigma)$.

4.4. The Flopped Resolutions

Resolution II

This resolution differs from the G-Hilbert one by the flo

$$C_{94} \longrightarrow C_{01} \,. \tag{140}$$

We defin $C = [C_{45}]$ and we report the relations between any other toric curve and C in the decorated fan of figur 14. The Chow ring is

$$A^*(X_\Sigma) = \mathbb{Z}(X_\Sigma, D_1, D_2, D_3, D_4, D_5, D_6, D_7, C)/\mathcal{R}, \tag{141}$$

where $\mathcal{R}$ are the relations induced by the intersection products, given in table 3. For completeness we report the products between all toric invariant divisors. The group of compact subvarieties of X is

$$A_*^c(X_\Sigma) = \mathbb{Z}D_0^c \oplus \mathbb{Z}C_{01}^c \oplus \mathbb{Z}C_{56}^c \oplus \mathbb{Z}C_{78}^c \oplus \mathbb{Z}C_{04}^c \oplus \mathbb{Z}C_{05}^c \oplus \mathbb{Z}C_{06}^c \oplus \mathbb{Z}C_{07}^c \oplus p^c. \tag{143}$$

with relations to other compact curves

$$C_{08} = C_{01} + C_{04} + C_{05} - C_{07} \tag{144}$$

$$C_{09} = -2C_{01} - C_{04} + C_{06} + C_{07}. \tag{145}$$

In table 4 we summarize the intersection pairing. We defin $C^* = -C_{04}^c - 2C_{05}^c - 2C_{06}^c - C_{07}^c$.

Starting from the G-Hilb chamber we obtain this resolution by crossing the wall (119). The tautological bundles are again

$$\mathcal{R}_{00} = \mathcal{O}_X, \qquad \mathcal{R}_{01} = \mathcal{O}_X(T_2), \qquad \mathcal{R}_{02} = \mathcal{O}_X(T_4 + T_7),$$
$$\mathcal{R}_{10} = \mathcal{O}_X(T_1), \qquad \mathcal{R}_{11} = \mathcal{O}_X(T_6), \qquad \mathcal{R}_{12} = \mathcal{O}_X(T_5 + T_7),$$
$$\mathcal{R}_{20} = \mathcal{O}_X(T_5), \qquad \mathcal{R}_{21} = \mathcal{O}_X(T_4 + T_6), \qquad \mathcal{R}_{22} = \mathcal{O}_X(T_3). \tag{147}$$

Table 3. Intersection product in $A^*(X_\Sigma)$

	X_Σ	D_1	D_2	D_3	D_4	D_5	D_6	D_7	D_8	D_9	D_0	C
X_Σ	X_Σ	D_1	D_2	D_3	D_4	D_5	D_6	D_7	D_8	D_9	D_0	C
D_1	D_1	$-C$	0	0	C	0	0	0	0	C	$-C$	0
D_2	D_2	0	0	0	0	0	0	0	0	0	0	0
D_3	D_3	0	0	0	0	0	0	0	0	0	0	0
D_4	D_4	C	0	0	$-2C$	C	0	0	0	0	0	0
D_5	D_5	0	0	0	C	$-C$	C	0	0	0	$-C$	0
D_6	D_6	0	0	0	0	C	$-C$	C	0	0	$-C$	0
D_7	D_7	0	0	0	0	0	C	$-C$	C	0	$-C$	0
D_8	D_8	0	0	0	0	0	0	C	$-C$	C	$-C$	0
D_9	D_9	C	0	0	0	0	0	0	C	$-2C$	0	0
D_0	D_0	$-C$	0	0	0	$-C$	$-C$	$-C$	$-C$	0	$5C$	0
C	C	0	0	0	0	0	0	0	0	0	0	0

(142)

Table 4. The map $A^*(X_\Sigma) \times A_*^c(X_\Sigma) \longrightarrow A_*^c(X_\Sigma)$

	D_0^c	C_{01}^c	C_{56}^c	C_{78}^c	C_{04}^c	C_{05}^c	C_{06}^c	C_{07}^c	C_{08}^c	C_{09}^c	p^c
X_Σ	D_0^c	C_{01}^c	C_{56}^c	C_{78}^c	C_{04}^c	C_{05}^c	C_{06}^c	C_{07}^c	C_{08}^c	C_{09}^c	p^c
D_1	C_{01}^c	$-p^c$	0	0	p^c	0	0	0	0	p^c	0
D_2	0	0	p^c	0	0	0	0	0	0	0	0
D_3	0	0	0	p^c	0	0	0	0	0	0	0
D_4	C_{04}^c	p^c	0	0	$-2p^c$	p^c	0	0	0	0	0
D_5	C_{05}^c	0	$-p^c$	0	p^c	$-p^c$	p^c	0	0	0	0
D_6	C_{06}^c	0	$-p^c$	0	0	p^c	$-p^c$	p^c	0	0	0
D_7	C_{07}^c	0	0	$-p^c$	0	0	p^c	$-p^c$	p^c	0	0
D_8	C_{08}^c	0	0	$-p^c$	0	0	0	p^c	$-p^c$	p^c	0
D_9	C_{09}^c	p^c	0	0	0	0	0	0	p^c	$-2p^c$	0
D_0	C^*	$-p^c$	p^c	p^c	0	$-p^c$	$-p^c$	$-p^c$	$-p^c$	0	0
C	p^c	0	0	0	0	0	0	0	0	0	0

(146)

The inequalities definin the GIT chamber of this resolution are

$$
\begin{array}{lll}
\theta_{10} < 0 & \text{wall of type I related to the flo of the curve } C_{01}; & (148)\\
\theta_{01} > 0 & \text{wall of type I related to the flo of the curve } C_{56}; & (149)\\
\theta_{22} > 0 & \text{wall of type I related to the flo of the curve } C_{78}; & (150)\\
\theta_{02} + \theta_{21} > -\theta_{10} & \text{wall of type III related to the contraction of the divisor } D_4; & (151)\\
\theta_{12} + \theta_{20} > 0 & \text{wall of type I related to the flo of the curve } C_{05}; & (152)\\
\theta_{11} + \theta_{21} > 0 & \text{wall of type I related to the flo of the curve } C_{06}; & (153)\\
\theta_{02} + \theta_{12} > 0 & \text{wall of type I related to the flo of the curve } C_{07}; & (154)\\
 & & (155)
\end{array}
$$

$$
\begin{array}{lll}
\theta_{20} + \theta_{21} > 0 & \text{wall of type I related to the flo of the curve } C_{08}; & (156)\\
\theta_{11} + \theta_{12} > -\theta_{10} & \text{wall of type III related to the contraction of the divisor } D_9; & (157)\\
\theta_{12} > 0 & \text{wall of type 0 define by } \theta(\varphi_C(\mathcal{R}_{12}^{-1}|_{D_0})) > 0; & (158)\\
\theta_{21} > 0 & \text{wall of type 0 define by } \theta(\varphi_C(\mathcal{R}_{21}^{-1}|_{D_0})) > 0; & (159)\\
\theta_{02} + \theta_{11} + \theta_{12} & & (160)\\
\quad + \theta_{20} + \theta_{21} > -\theta_{10} & \text{wall of type 0 define by } \theta(\varphi_C(\mathcal{R}_{00}^{-1} \otimes \omega_{D_0})) < 0. & (161)
\end{array}
$$

Any other inequality is redundant.
In view of the relations (145), the Mori cone is generated by all the nine compact curves C^c_{ij}:

$$
\begin{array}{l}
C_1 \equiv C^c_{01}, \quad C_2 \equiv C^c_{56}, \quad C_3 \equiv C^c_{78}, \quad C_4 \equiv C^c_{04}, \quad C_5 \equiv C^c_{05},\\
C_6 \equiv C^c_{06}, \quad C_7 \equiv C^c_{07}, \quad C_8 \equiv C^c_{08}, \quad C_9 \equiv C^c_{09}.
\end{array} \tag{162}
$$

Since $A^c_2(X) \otimes \mathbb{Q}$ has dimension seven, the Mori cone is not simplicial. Also the Kähler cone is not simplicial, indeed it has 10 generators. In order to calculate the prepotential using mirror symmetry we have to choose a large complex structure limit. We select the following basis between the generators of the Kähler cone:

$$
\begin{array}{l}
G_1 = D_3, \quad G_2 = D_2, \quad G_3 = 2D_2 + D_5 + D_6,\\
G_4 = 4D_2 + D_4 + 2D_5 + 2D_6, \quad G_5 = D_2 + D_3 + D_6 + D_7,\\
G_6 = 3D_2 + D_4 + 2D_5 + D_6, \quad G_7 = D_1 + 2D_2 + D_4 + D_5 + D_6,
\end{array} \tag{163}
$$

and we name $\{J_1, \ldots, J_7\}$ the corresponding forms in $H^2(X, \mathbb{Q})$. The dual basis in $A^c_2(X) \otimes \mathbb{Q}$ is:

$$
\begin{array}{l}
\tilde{C}_1 = C_3, \quad \tilde{C}_2 = C_2, \quad \tilde{C}_3 = C_4, \quad \tilde{C}_4 = C_1 - C_6,\\
\tilde{C}_5 = C_1 + C_4 + C_5 - C_7, \quad \tilde{C}_6 = C_6, \quad \tilde{C}_7 = -2C_1 - C_4 + C_6 + C_7.
\end{array} \tag{164}
$$

The vectors ℓ_a, $a = 1, \ldots, 7$ are given by the intersection numbers between the $\tilde{C}_i$ and the invariant divisors of X:

$$
\begin{array}{lll}
\tilde{C}_1 & : & \ell_1 = (0, 0, 1, 0, 0, 0, -1, -1, 0, 1)\,,\\
\tilde{C}_2 & : & \ell_2 = (0, 1, 0, 0, -1, -1, 0, 0, 0, 1)\,,\\
\tilde{C}_3 & : & \ell_3 = (1, 0, 0, -2, 1, 0, 0, 0, 0, 0)\,,\\
\tilde{C}_4 & : & \ell_4 = (-1, 0, 0, 1, -1, 1, -1, 0, 1, 0)\,,\\
\tilde{C}_5 & : & \ell_5 = (0, 0, 0, 0, 0, 0, 1, -1, 1, -1)\,,
\end{array}
$$

$$\begin{array}{ccl} \tilde{C}_6 & : & \ell_6 = (0,0,0,0,1,-1,1,0,0,-1)\,, \\ \tilde{C}_7 & : & \ell_7 = (1,0,0,0,0,0,0,1,-2,0)\,. \end{array} \tag{165}$$

The cohomology-valued hypergeometric series in respect to the basis $\{1, \vec{J}, C\}$ of $H^*(X, \mathbb{Q})$ is then

$$\begin{aligned} w\left(\vec{x}, \frac{\vec{J}}{2\pi i}\right) = \; & 1 + \left(-\frac{ix_1x_5}{2\pi} - \frac{i\log x_1}{2\pi}\right) J_1 + \left(-\frac{ix_2x_6}{2\pi} - \frac{i\log x_2}{2\pi}\right) J_2 \\ & + \left(-\frac{ix_3}{\pi} - \frac{3ix_3{}^2}{2\pi} - \frac{i\log x_3}{2\pi}\right) J_3 \\ & + \left(\frac{ix_3}{2\pi} + \frac{3ix_3{}^2}{4\pi} + \frac{ix_2x_6}{2\pi} + \frac{ix_7}{2\pi} + \frac{3ix_7{}^2}{4\pi} - \frac{i\log x_4}{2\pi}\right) J_4 \\ & + \left(-\frac{ix_1x_5}{2\pi} + \frac{ix_7}{2\pi} + \frac{3ix_7{}^2}{4\pi} - \frac{i\log x_5}{2\pi}\right) J_5 \\ & + \left(-\frac{ix_2x_6}{2\pi} - \frac{i\log x_6}{2\pi}\right) J_6 + \left(\frac{ix_1x_5}{2\pi} - \frac{ix_7}{\pi} - \frac{3ix_7{}^2}{2\pi} - \frac{i\log x_7}{2\pi}\right) J_7 \\ & + \left(-\frac{1}{12} - \frac{x_1}{4\pi^2} - \frac{x_1{}^2}{16\pi^2} - \frac{x_2}{4\pi^2} - \frac{x_2{}^2}{16\pi^2} + \frac{x_3{}^2}{4\pi^2} + \frac{x_5}{4\pi^2} + \frac{x_5{}^2}{16\pi^2}\right. \\ & + \frac{x_6}{4\pi^2} + \frac{x_4x_6}{4\pi^2} + \frac{x_6{}^2}{16\pi^2} + \frac{x_7{}^2}{4\pi^2} \\ & + \frac{x_3\log x_3}{4\pi^2} + \frac{3x_3{}^2\log x_3}{8\pi^2} - \frac{\log x_3\log x_4}{4\pi^2} - \frac{\log x_4{}^2}{4\pi^2} \\ & + \frac{x_1x_5\log x_5}{4\pi^2} - \frac{\log x_3\log x_5}{4\pi^2} - \frac{\log x_4\log x_5}{2\pi^2} \\ & + \frac{x_2x_6\log x_6}{4\pi^2} - \frac{\log x_3\log x_6}{4\pi^2} - \frac{\log x_4\log x_6}{2\pi^2} \\ & - \frac{\log x_5\log x_6}{2\pi^2} - \frac{\log x_6{}^2}{8\pi^2} + \frac{x_7\log x_7}{4\pi^2} + \frac{3x_7{}^2\log x_7}{8\pi^2} \\ & - \frac{\log x_3\log x_7}{4\pi^2} - \frac{\log x_4\log x_7}{2\pi^2} - \frac{\log x_5\log x_7}{4\pi^2} \\ & \left. - \frac{\log x_6\log x_7}{2\pi^2} - \frac{\log x_7{}^2}{8\pi^2}\right) C\,. \end{aligned} \tag{166}$$

However we stopped our calculation at the second order in $\vec{x}$. Finally the prepotential is

$$\begin{aligned} F(\vec{t}) = \; & -\frac{iq_1}{8\pi^3} - \frac{iq_1{}^2}{64\pi^3} - \frac{iq_2}{8\pi^3} - \frac{iq_2{}^2}{64\pi^3} - \frac{iq_5}{8\pi^3} \\ & - \frac{iq_5{}^2}{64\pi^3} - \frac{iq_6}{8\pi^3} - \frac{iq_4q_6}{8\pi^3} - \frac{iq_6{}^2}{64\pi^3} - \frac{iq_5q_7}{8\pi^3} \\ & - \frac{11t_1}{12} - \frac{5t_1{}^2}{4} - \frac{5t_1{}^3}{6} - t_1t_3 - t_1{}^2t_3 - 2t_1t_4 - 2t_1{}^2t_4 \\ & - t_1t_3t_4 - t_1t_4{}^2 - t_1t_5 - t_1{}^2t_5 - t_1t_3t_5 \\ & - 2t_1t_4t_5 - \frac{3t_1t_6}{2} - \frac{3t_1{}^2t_6}{2} - t_1t_3t_6 - 2t_1t_4t_6 \\ & - 2t_1t_5t_6 - \frac{t_1t_6{}^2}{2} - \frac{3t_1t_7}{2} - \frac{3t_1{}^2t_7}{2} \\ & - t_1t_3t_7 - 2t_1t_4t_7 - t_1t_5t_7 - 2t_1t_6t_7 - \frac{t_1t_7{}^2}{2} \\ & + \text{Gclass}[-t_1 + t_2, t_3, t_4, t_1 + t_5, t_1 + t_6, t_7] \\ & + \text{Gistant}\left[\frac{q_2}{q_1}, q_3, q_4, q_1q_5, q_1q_6, q_7\right]\,. \end{aligned} \tag{167}$$

Resolution III

This resolution differs from the II by the flo

$$C_{56} \longrightarrow C_{02} \,. \tag{168}$$

We defin $C = [C_{45}]$ and we report the relations between any other toric curve and C in the decorated fan of figur 15.

The Chow ring is

$$A^*(X_\Sigma) = \mathbb{Z}(X_\Sigma, D_1, D_2, D_3, D_4, D_5, D_6, D_7, C)/\mathcal{R}, \tag{169}$$

where $\mathcal{R}$ are the relations induced by the intersection products, given in table 5. For completeness we report the products between all toric invariant divisors. The group of compact subvarieties of X is

Table 5. Intersection product in $A^*(X_\Sigma)$

	X_Σ	D_1	D_2	D_3	D_4	D_5	D_6	D_7	D_8	D_9	D_0	C
X_Σ	X_Σ	D_1	D_2	D_3	D_4	D_5	D_6	D_7	D_8	D_9	D_0	C
D_1	D_1	$-C$	0	0	C	0	0	0	0	C	$-C$	0
D_2	D_2	0	$-C$	0	0	C	C	0	0	0	$-C$	0
D_3	D_3	0	0	0	0	0	0	0	0	0	0	0
D_4	D_4	C	0	0	$-2C$	C	0	0	0	0	0	0
D_5	D_5	0	C	0	C	$-2C$	0	0	0	0	0	0
D_6	D_6	0	C	0	0	0	$-2C$	C	0	0	0	0
D_7	D_7	0	0	0	0	0	C	$-C$	C	0	$-C$	0
D_8	D_8	0	0	0	0	0	0	C	$-C$	C	$-C$	0
D_9	D_9	C	0	0	0	0	0	0	C	$-2C$	0	0
D_0	D_0	$-C$	$-C$	0	0	0	0	$-C$	$-C$	0	$4C$	0
C	C	0	0	0	0	0	0	0	0	0	0	0

(170)

$$A^c_*(X_\Sigma) = \mathbb{Z}D^c_0 \oplus \mathbb{Z}C^c_{01} \oplus \mathbb{Z}C^c_{02} \oplus \mathbb{Z}C^c_{78} \oplus \mathbb{Z}C^c_{04} \oplus \mathbb{Z}C^c_{05} \oplus \mathbb{Z}C^c_{06} \oplus \mathbb{Z}C^c_{07} \oplus p^c. \tag{171}$$

with relations to other compact curves

$$C_{08} = C_{01} + C_{02} + C_{04} + C_{05} - C_{07} \tag{172}$$

$$C_{09} = -2C_{01} + C_{02} - C_{04} + C_{06} + C_{07}. \tag{173}$$

In table 6 we summarize the intersection pairing. We defin $C^* = -3C^c_{02} - C^c_{04} - 2C^c_{05} - 2C^c_{06} - C^c_{07}$.

Starting from the resolution II chamber we obtain this resolution by crossing the wall (149). The tautological bundles are again

$$\begin{aligned} \mathcal{R}_{00} &= \mathcal{O}_X, & \mathcal{R}_{01} &= \mathcal{O}_X(T_2), & \mathcal{R}_{02} &= \mathcal{O}_X(T_4 + T_7), \\ \mathcal{R}_{10} &= \mathcal{O}_X(T_1), & \mathcal{R}_{11} &= \mathcal{O}_X(T_6), & \mathcal{R}_{12} &= \mathcal{O}_X(T_5 + T_7), \\ \mathcal{R}_{20} &= \mathcal{O}_X(T_5), & \mathcal{R}_{21} &= \mathcal{O}_X(T_4 + T_6), & \mathcal{R}_{22} &= \mathcal{O}_X(T_3). \end{aligned} \tag{175}$$

Table 6. The map $A^*(X_\Sigma) \times A^c_*(X_\Sigma) \longrightarrow A^c_*(X_\Sigma)$

	D^c_0	C^c_{01}	C^c_{02}	C^c_{78}	C^c_{04}	C^c_{05}	C^c_{06}	C^c_{07}	C^c_{08}	C^c_{09}	p^c
X_Σ	D^c_0	C^c_{01}	C^c_{02}	C^c_{78}	C^c_{04}	C^c_{05}	C^c_{06}	C^c_{07}	C^c_{08}	C^c_{09}	p^c
D_1	C^c_{01}	$-p^c$	0	0	p^c	0	0	0	0	p^c	0
D_2	C^c_{02}	0	$-p^c$	0	0	p^c	p^c	0	0	0	0
D_3	0	0	0	p^c	0	0	0	0	0	0	0
D_4	C^c_{04}	p^c	0	0	$-2p^c$	p^c	0	0	0	0	0
D_5	C^c_{05}	0	p^c	0	p^c	$-2p^c$	0	0	0	0	0
D_6	C^c_{06}	0	p^c	0	0	0	$-2p^c$	p^c	0	0	0
D_7	C^c_{07}	0	0	$-p^c$	0	0	p^c	$-p^c$	p^c	0	0
D_8	C^c_{08}	0	0	$-p^c$	0	0	0	p^c	$-p^c$	p^c	0
D_9	C^c_{09}	p^c	0	0	0	0	0	0	p^c	$-2p^c$	0
D_0	C^*	$-p^c$	$-p^c$	p^c	0	0	0	$-p^c$	$-p^c$	0	0
C	p^c	0	0	0	0	0	0	0	0	0	0

(174)

The inequalities definin the GIT chamber of this resolution are

$\theta_{10} < 0$	wall of type I related to the flo of the curve C_{01};	(176)
$\theta_{01} < 0$	wall of type I related to the flo of the curve C_{02};	(177)
$\theta_{22} > 0$	wall of type I related to the flo of the curve C_{78};	(178)
$\theta_{02} + \theta_{21} > -\theta_{10}$	wall of type III related to the contraction of the divisor D_4;	(179)
$\theta_{12} + \theta_{20} > -\theta_{01}$	wall of type III related to the contraction of the divisor D_5;	(180)
$\theta_{11} + \theta_{21} > -\theta_{01}$	wall of type III related to the contraction of the divisor D_6;	(181)

$\theta_{02} + \theta_{12} > 0$	wall of type I related to the flo of the curve C_{07};	(182)
$\theta_{20} + \theta_{21} > 0$	wall of type I related to the flo of the curve C_{08};	(183)
$\theta_{11} + \theta_{12} > -\theta_{10}$	wall of type III related to the contraction of the divisor D_9;	(184)
$\theta_{12} > 0$	wall of type 0 define by $\theta(\varphi_C(\mathcal{R}^{-1}_{12}\vert_{D_0})) > 0$;	(185)
$\theta_{21} > 0$	wall of type 0 define by $\theta(\varphi_C(\mathcal{R}^{-1}_{21}\vert_{D_0})) > 0$;	(186)

$\theta_{02} + \theta_{11} + \theta_{12} + \theta_{20}$ $+\theta_{21} > -(\theta_{10} + \theta_{01})$	wall of type 0 define by $\theta(\varphi_C(\mathcal{R}^{-1}_{00} \otimes \omega_{D_0})) < 0$.	(187)

Any other inequality is redundant.

In view of the relations (173), the Mori cone is generated by all the nine compact curves C^c_{ij}:

$$\begin{array}{lllll} C_1 \equiv C^c_{01}, & C_2 \equiv C^c_{02}, & C_3 \equiv C^c_{78}, & C_4 \equiv C^c_{04}, & C_5 \equiv C^c_{05}, \\ C_6 \equiv C^c_{06}, & C_7 \equiv C^c_{07}, & C_8 \equiv C^c_{08}, & C_9 \equiv C^c_{09}. & \end{array} \tag{188}$$

Since $A^c_2(X) \otimes \mathbb{Q}$ has dimension seven, the Mori cone is not simplicial. Also the Kähler cone is not simplicial, indeed it has 14 generators. In order to calculate the prepotential

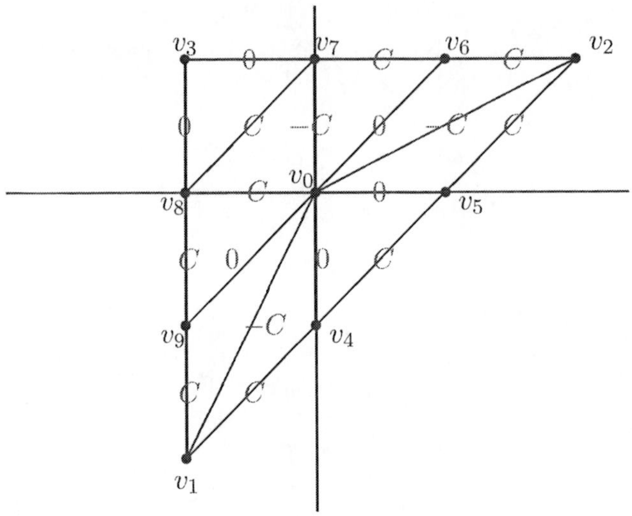

Figure 15. Equivalences in $A^1(X_\Sigma)$.

using mirror symmetry we have to choose a large complex structure limit. We select the following basis between the generators of the Kähler cone:

$$\begin{aligned} &G_1 = D_3, \quad G_2 = 6D_2 + 2D_4 + 4D_5 + 3D_6, \quad G_3 = 2D_2 + D_5 + D_6, \\ &G_4 = 4D_2 + D_4 + 2D_5 + 2D_6, \quad G_5 = D_2 + D_3 + D_6 + D_7, \\ &G_6 = D_1 + D_2 + D_4 + D_5, \quad G_7 = 2D_2 + D_3 + D_5 + D_6 + D_7, \end{aligned} \tag{189}$$

and we name $\{J_1, \ldots, J_7\}$ the corresponding forms in $H^2(X, \mathbb{Q})$. The dual basis in $A_2^c(X) \otimes \mathbb{Q}$ is:

$$\begin{aligned} &\tilde{C}_1 = C_3, \quad \tilde{C}_2 = C_2, \quad \tilde{C}_3 = -2C_1 + C_2 + C_7, \quad \tilde{C}_4 = C_1 - 2C_2, \\ &\tilde{C}_5 = -C_1 + 2C_2 + C_5, \quad \tilde{C}_6 = -2C_1 + C_2 - C_4 + C_6 + C_7, \\ &\tilde{C}_7 = 2C_1 - C_2 + C_4 - C_7. \end{aligned} \tag{190}$$

The vectors ℓ_a, $a = 1, \ldots, 7$ are given by the intersection numbers between the $\tilde{C}_i$ and the invariant divisors of X:

$$\begin{array}{lcl} \tilde{C}_1 & : & \ell_1 = (0,0,1,0,0,0,-1,-1,0,1)\,, \\ \tilde{C}_2 & : & \ell_2 = (0,-1,0,0,1,1,0,0,0,-1)\,, \\ \tilde{C}_3 & : & \ell_3 = (2,-1,0,-2,1,2,-1,1,-2,0)\,, \\ \tilde{C}_4 & : & \ell_4 = (-1,2,0,1,-2,-2,0,0,1,1)\,, \\ \tilde{C}_5 & : & \ell_5 = (1,-1,0,0,0,2,0,0,-1,-1)\,, \\ \tilde{C}_6 & : & \ell_6 = (1,0,0,0,0,0,0,1,-2,0)\,, \\ \tilde{C}_7 & : & \ell_7 = (-1,1,0,0,0,-2,1,-1,2,0)\,. \end{array} \tag{191}$$

The cohomology-valued hypergeometric series in respect to the basis $\{1, \vec{J}, C\}$ of $H^*(X, \mathbb{Q})$ is then

$$w\left(\vec{x}, \frac{\vec{J}}{2\pi i}\right) = \quad 1 + \left(-\frac{i\log x_1}{2\pi}\right) J_1 + \left(\frac{i x_4 x_5}{2\pi} + \frac{i x_6 x_7}{2\pi} - \frac{i\log x_2}{2\pi}\right) J_2$$

$$+\left(\frac{ix_4x_5}{2\pi}-\frac{ix_6}{\pi}-\frac{3ix_6{}^2}{2\pi}-\frac{ix_3x_7}{\pi}+\frac{ix_6x_7}{\pi}-\frac{i\log x_3}{2\pi}\right)J_3$$
$$+\left(-\frac{ix_4x_5}{\pi}+\frac{ix_6}{2\pi}+\frac{3ix_6{}^2}{4\pi}+\frac{ix_3x_7}{2\pi}-\frac{ix_6x_7}{\pi}-\frac{i\log x_4}{2\pi}\right)J_4$$
$$+\left(-\frac{ix_6}{2\pi}-\frac{3ix_6{}^2}{4\pi}+\frac{ix_6x_7}{\pi}-\frac{i\log x_5}{2\pi}\right)J_5$$
$$+\left(-\frac{ix_6}{\pi}-\frac{3ix_6{}^2}{2\pi}-\frac{i\log x_6}{2\pi}\right)J_6$$
$$+\left(\frac{ix_6}{\pi}+\frac{3ix_6{}^2}{2\pi}-\frac{ix_6x_7}{\pi}-\frac{i\log x_7}{2\pi}\right)J_7$$
$$+\left(-\frac{1}{6}-\frac{x_1}{4\pi^2}-\frac{x_1{}^2}{16\pi^2}+\frac{x_2}{4\pi^2}+\frac{x_2{}^2}{16\pi^2}+\frac{x_6{}^2}{4\pi^2}\right.$$
$$+\frac{x_5x_7}{4\pi^2}-\frac{3\log x_2{}^2}{4\pi^2}+\frac{x_3x_7\log x_3}{4\pi^2}-\frac{\log x_2\log x_3}{2\pi^2}$$
$$+\frac{x_4x_5\log x_4}{4\pi^2}-\frac{\log x_2\log x_4}{\pi^2}-\frac{\log x_3\log x_4}{4\pi^2}-\frac{\log x_4{}^2}{4\pi^2}$$
$$+\frac{x_4x_5\log x_5}{4\pi^2}-\frac{\log x_2\log x_5}{\pi^2}-\frac{\log x_3\log x_5}{4\pi^2}$$
$$-\frac{\log x_4\log x_5}{2\pi^2}+\frac{x_6\log x_6}{4\pi^2}+\frac{3x_6{}^2\log x_6}{8\pi^2}+\frac{x_6x_7\log x_6}{4\pi^2}$$
$$-\frac{3\log x_2\log x_6}{4\pi^2}-\frac{\log x_3\log x_6}{4\pi^2}$$
$$-\frac{\log x_4\log x_6}{2\pi^2}-\frac{\log x_5\log x_6}{4\pi^2}+\frac{x_3x_7\log x_7}{4\pi^2}$$
$$+\frac{x_6x_7\log x_7}{4\pi^2}-\frac{5\log x_2\log x_7}{4\pi^2}-\frac{\log x_3\log x_7}{4\pi^2}$$
$$\left.-\frac{3\log x_4\log x_7}{4\pi^2}-\frac{\log x_5\log x_7}{4\pi^2}-\frac{\log x_6\log x_7}{4\pi^2}-\frac{\log x_7{}^2}{8\pi^2}\right)C\,. \tag{192}$$

However we stopped our calculation at the second order in $\vec{x}$. Finally the prepotential is

$$\begin{aligned}F(\vec{t}) = \quad & -\frac{iq_1}{8\pi^3}-\frac{iq_1{}^2}{64\pi^3}-\frac{iq_2}{8\pi^3}-\frac{iq_2{}^2}{64\pi^3}-\frac{iq_5q_7}{8\pi^3}\\
&-\frac{5t_1}{6}-t_1{}^2-\frac{2t_1{}^3}{3}-3t_1t_2-3t_1{}^2t_2-3t_1t_2{}^2\\
&-t_1t_3-t_1{}^2t_3-2t_1t_2t_3-2t_1t_4-2t_1{}^2t_4\\
&-4t_1t_2t_4-t_1t_3t_4-t_1t_4{}^2-t_1t_5-t_1{}^2t_5\\
&-4t_1t_2t_5-t_1t_3t_5-2t_1t_4t_5-t_1t_6-t_1{}^2t_6-3t_1t_2t_6\\
&-t_1t_3t_6-2t_1t_4t_6-t_1t_5t_6-\frac{3t_1t_7}{2}-\frac{3t_1{}^2t_7}{2}\\
&-5t_1t_2t_7-t_1t_3t_7-3t_1t_4t_7-t_1t_5t_7-t_1t_6t_7-\frac{t_1t_7{}^2}{2}\\
&+\text{Gclass}[t_1+t_2,t_3,-t_1+t_4,t_1+t_5,t_6,t_7]\\
&+\text{Gistant}\left[q_1q_2,q_3,\frac{q_4}{q_1},q_1q_5,q_6,q_7\right]\,.\end{aligned} \tag{193}$$

Resolution IV

This resolution differs from the III by the flo

$$C_{78}\longrightarrow C_{03}\,. \tag{194}$$

This variety is the total space of the canonical bundle over the resolution of $\mathbb{P}^2/\mathbb{Z}_3$. We defin $C=[C_{45}]$ and we report the relations between any other toric curve and C in the decorated fan of figur 16.

The Chow ring is

$$A^*(X_\Sigma) = \mathbb{Z}(X_\Sigma, D_1, D_2, D_3, D_4, D_5, D_6, D_7, C)/\mathcal{R}, \tag{195}$$

where $\mathcal{R}$ are the relations induced by the intersection products, given in table 7. For completeness we report the products between all toric invariant divisors.

Table 7. Intersection product in $A^*(X_\Sigma)$

	X_Σ	D_1	D_2	D_3	D_4	D_5	D_6	D_7	D_8	D_9	D_0	C
X_Σ	X_Σ	D_1	D_2	D_3	D_4	D_5	D_6	D_7	D_8	D_9	D_0	C
D_1	D_1	$-C$	0	0	C	0	0	0	0	C	$-C$	0
D_2	D_2	0	$-C$	0	0	C	C	0	0	0	$-C$	0
D_3	D_3	0	0	$-C$	0	0	0	C	C	0	$-C$	0
D_4	D_4	C	0	0	$-2C$	C	0	0	0	0	0	0
D_5	D_5	0	C	0	C	$-2C$	0	0	0	0	0	0
D_6	D_6	0	C	0	0	0	$-2C$	C	0	0	0	0
D_7	D_7	0	0	C	0	0	C	$-2C$	0	0	0	0
D_8	D_8	0	0	C	0	0	0	0	$-2C$	C	0	0
D_9	D_9	C	0	0	0	0	0	0	C	$-2C$	0	0
D_0	D_0	$-C$	$-C$	$-C$	0	0	0	0	0	0	$3C$	0
C	C	0	0	0	0	0	0	0	0	0	0	0

(196)

The group of compact subvarieties of X is

$$A^c_*(X_\Sigma) = \mathbb{Z}D^c_0 \oplus \mathbb{Z}C^c_{01} \oplus \mathbb{Z}C^c_{02} \oplus \mathbb{Z}C^c_{03} \oplus \mathbb{Z}C^c_{04} \oplus \mathbb{Z}C^c_{05} \oplus \mathbb{Z}C^c_{06} \oplus \mathbb{Z}C^c_{07} \oplus p^c. \tag{197}$$

with relations to other compact curves

$$C_{08} = C_{01} + C_{02} - 2C_{03} + C_{04} + C_{05} - C_{07} \tag{198}$$

$$C_{09} = -2C_{01} + C_{02} + C_{03} - C_{04} + C_{06} + C_{07}. \tag{199}$$

In table 8 we summarize the intersection pairing. We defin $C^* = -3C^c_{02} - C^c_{04} - 2C^c_{05} - 2C^c_{06} - C^c_{07}$. Starting from the resolution III chamber we obtain this resolution by

Table 8. The map $A^*(X_\Sigma) \times A^c_*(X_\Sigma) \longrightarrow A^c_*(X_\Sigma)$

	D_0^c	C_{01}^c	C_{02}^c	C_{03}^c	C_{04}^c	C_{05}^c	C_{06}^c	C_{07}^c	C_{08}^c	C_{09}^c	p^c
X_Σ	D_0^c	C_{01}^c	C_{02}^c	C_{03}^c	C_{04}^c	C_{05}^c	C_{06}^c	C_{07}^c	C_{08}^c	C_{09}^c	p^c
D_1	C_{01}^c	$-p^c$	0	0	p^c	0	0	0	0	p^c	0
D_2	C_{02}^c	0	$-p^c$	0	0	p^c	p^c	0	0	0	0
D_3	C_{03}^c	0	0	$-p^c$	0	0	0	p^c	p^c	0	0
D_4	C_{04}^c	p^c	0	0	$-2p^c$	p^c	0	0	0	0	0
D_5	C_{05}^c	0	p^c	0	p^c	$-2p^c$	0	0	0	0	0
D_6	C_{06}^c	0	p^c	0	0	0	$-2p^c$	p^c	0	0	0
D_7	C_{07}^c	0	0	p^c	0	0	p^c	$-2p^c$	0	0	0
D_8	C_{08}^c	0	0	p^c	0	0	0	0	$-2p^c$	p^c	0
D_9	C_{09}^c	p^c	0	0	0	0	0	0	p^c	$-2p^c$	0
D_0	C^*	$-p^c$	$-p^c$	$-p^c$	0	0	0	0	0	0	0
C	p^c	0	0	0	0	0	0	0	0	0	0

(200)

crossing the wall (178). The tautological bundles are again

$$\begin{aligned}
&\mathcal{R}_{00} = \mathcal{O}_X, && \mathcal{R}_{01} = \mathcal{O}_X(T_2), && \mathcal{R}_{02} = \mathcal{O}_X(T_4 + T_7),\\
&\mathcal{R}_{10} = \mathcal{O}_X(T_1), && \mathcal{R}_{11} = \mathcal{O}_X(T_6), && \mathcal{R}_{12} = \mathcal{O}_X(T_5 + T_7),\\
&\mathcal{R}_{20} = \mathcal{O}_X(T_5), && \mathcal{R}_{21} = \mathcal{O}_X(T_4 + T_6), && \mathcal{R}_{22} = \mathcal{O}_X(T_3).
\end{aligned} \tag{201}$$

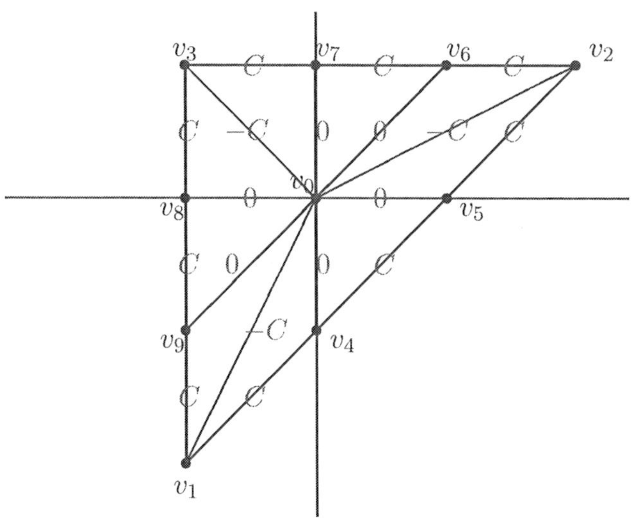

Figure 16. Equivalences in $A^1(X_\Sigma)$.

The inequalities definin the GIT chamber of this resolution are

$\theta_{10} < 0$	wall of type I related to the flo of the curve C_{01};	(202)
$\theta_{01} < 0$	wall of type I related to the flo of the curve C_{02};	(203)
$\theta_{22} < 0$	wall of type I related to the flo of the curve C_{78};	(204)
$\theta_{02} + \theta_{21} > -\theta_{10}$	wall of type III related to the contraction of the divisor D_4;	(205)
$\theta_{12} + \theta_{20} > -\theta_{01}$	wall of type III related to the contraction of the divisor D_5;	(206)
$\theta_{11} + \theta_{21} > -\theta_{01}$	wall of type III related to the contraction of the divisor D_6;	(207)
$\theta_{02} + \theta_{12} > -\theta_{22}$	wall of type III related to the contraction of the divisor D_7;	(208)
$\theta_{20} + \theta_{21} > -\theta_{22}$	wall of type III related to the contraction of the divisor D_8;	(209)
$\theta_{11} + \theta_{12} > -\theta_{10}$	wall of type III related to the contraction of the divisor D_9;	(210)

$\theta_{12} > 0$	wall of type 0 define by $\theta(\varphi_C(\mathcal{R}_{12}^{-1}\|_{D_o})) > 0$;	(211)
$\theta_{21} > 0$	wall of type 0 define by $\theta(\varphi_C(\mathcal{R}_{21}^{-1}\|_{D_o})) > 0$;	(212)

$\theta_{02} + \theta_{11} + \theta_{12} + \theta_{20} + \theta_{21} > -(\theta_{10} + \theta_{01} + \theta_{22})$	wall of type 0 define by $\theta(\varphi_C(\mathcal{R}_{00}^{-1} \otimes \omega_{D_o})) < 0$.	(213)

Any other inequality is redundant.
In view of the relations (199), the Mori cone is generated by all the nine compact curves C_{ij}^c:

$$\begin{aligned} &C_1 \equiv C_{01}^c, \quad C_2 \equiv C_{02}^c, \quad C_3 \equiv C_{03}^c, \quad C_4 \equiv C_{04}^c, \quad C_5 \equiv C_{05}^c, \\ &C_6 \equiv C_{06}^c, \quad C_7 \equiv C_{07}^c, \quad C_8 \equiv C_{08}^c, \quad C_9 \equiv C_{09}^c. \end{aligned} \tag{214}$$

Since $A_2^c(X) \otimes \mathbb{Q}$ has dimension seven, the Mori cone is not simplicial. Also the Kähler cone is not simplicial, indeed it has 21 generators. In order to calculate the prepotential using mirror symmetry we have to choose a large complex structure limit. We select the following basis between the generators of the Kähler cone:

$$\begin{aligned} &G_1 = 3D_2 + D_4 + 2D_5 + 2D_6 + D_7, \quad G_2 = 6D_2 + 2D_4 + 4D_5 + 3D_6, \\ &G_3 = 2D_2 + D_5 + D_6, \quad G_4 = 4D_2 + D_4 + 2D_5 + 2D_6, \\ &G_5 = D_2 + D_3 + D_6 + D_7, \quad G_6 = 3D_2 + D_4 + 2D_5 + D_6, \\ &G_7 = D_1 + D_2 + D_4 + D_5, \end{aligned} \tag{215}$$

and we name $\{J_1, \ldots, J_7\}$ the corresponding forms in $H^2(X, \mathbb{Q})$. The dual basis in $A_2^c(X) \otimes \mathbb{Q}$ is:

$$\begin{aligned} &\tilde{C}_1 = C_3, \quad \tilde{C}_2 = C_2 - C_3, \quad \tilde{C}_3 = C_4, \quad \tilde{C}_4 = -C_1 - C_2 + 2C_3 - C_4 + C_7, \\ &\tilde{C}_5 = C_1 + C_2 - 2C_3 + C_4 + C_5 - C_7, \quad \tilde{C}_6 = 2C_1 - C_2 - C_3 + C_4 - C_7, \\ &\tilde{C}_7 = -2C_1 + C_2 + C_3 - C_4 + C_6 + C_7. \end{aligned} \tag{216}$$

The vectors ℓ_a, $a = 1, \ldots, 7$ are given by the intersection numbers between the $\tilde{C}_i$ and the invariant divisors of X:

$$\tilde{C}_1 \quad : \quad \ell_1 = (0, 0, -1, 0, 0, 0, 1, 1, 0, -1)\,,$$

$$\begin{aligned}
\tilde{C_2} &\quad : \quad \ell_2 = (0,-1,1,0,1,1,-1,-1,0,0)\,,\\
\tilde{C_3} &\quad : \quad \ell_3 = (1,0,0,-2,1,0,0,0,0,0)\,,\\
\tilde{C_4} &\quad : \quad \ell_4 = (0,1,-1,1,-2,0,0,2,-1,0)\,,\\
\tilde{C_5} &\quad : \quad \ell_5 = (0,0,1,0,0,0,0,-2,1,0)\,,\\
\tilde{C_6} &\quad : \quad \ell_6 = (-1,1,0,0,0,-2,1,-1,2,0)\,,\\
\tilde{C_7} &\quad : \quad \ell_7 = (1,0,0,0,0,0,0,1,-2,0)\,.
\end{aligned} \tag{217}$$

The cohomology-valued hypergeometric series in respect to the basis $\{1, \vec{J}, C\}$ of $H^*(X, \mathbb{Q})$ is then

$$\begin{aligned}
w\left(\vec{x}, \frac{\vec{J}}{2\pi i}\right) = \;& 1 + \left(\frac{ix_5}{2\pi} + \frac{3ix_5^2}{4\pi} - \frac{i\log x_1}{2\pi}\right) J_1\\
&+ \left(-\frac{ix_5}{2\pi} + \frac{ix_4x_5}{2\pi} - \frac{3ix_5^2}{4\pi} + \frac{ix_6x_7}{2\pi} - \frac{i\log x_2}{2\pi}\right) J_2\\
&+ \left(-\frac{ix_3}{\pi} - \frac{3ix_3^2}{2\pi} + \frac{ix_4x_5}{2\pi} - \frac{i\log x_3}{2\pi}\right) J_3\\
&+ \left(\frac{ix_3}{2\pi} + \frac{3ix_3^2}{4\pi} + \frac{ix_5}{\pi} - \frac{ix_4x_5}{\pi} + \frac{3ix_5^2}{2\pi} - \frac{ix_7}{2\pi} - \frac{3ix_7^2}{4\pi} - \frac{i\log x_4}{2\pi}\right) J_4\\
&+ \left(-\frac{ix_5}{\pi} - \frac{3ix_5^2}{2\pi} + \frac{ix_7}{2\pi} + \frac{3ix_7^2}{4\pi} - \frac{i\log x_5}{2\pi}\right) J_5\\
&+ \left(-\frac{ix_5}{2\pi} - \frac{3ix_5^2}{4\pi} + \frac{ix_7}{\pi} - \frac{ix_6x_7}{\pi} + \frac{3ix_7^2}{2\pi} - \frac{i\log x_6}{2\pi}\right) J_6\\
&+ \left(\frac{ix_5}{2\pi} + \frac{3ix_5^2}{4\pi} - \frac{ix_7}{\pi} - \frac{3ix_7^2}{2\pi} - \frac{i\log x_7}{2\pi}\right) J_7\\
&+ \left(-\frac{1}{4} + \frac{x_1}{4\pi^2} + \frac{x_1^2}{16\pi^2} + \frac{x_1x_2}{4\pi^2} + \frac{x_3^2}{4\pi^2}\right.\\
&+ \frac{x_5^2}{4\pi^2} - \frac{x_5x_7}{4\pi^2} + \frac{x_7^2}{4\pi^2} - \frac{3\log x_1^2}{8\pi^2} - \frac{3\log x_1\log x_2}{2\pi^2}\\
&- \frac{3\log x_2^2}{4\pi^2} + \frac{x_3\log x_3}{4\pi^2} + \frac{3x_3^2\log x_3}{8\pi^2} - \frac{\log x_1\log x_3}{2\pi^2}\\
&- \frac{\log x_2\log x_3}{2\pi^2} + \frac{x_4x_5\log x_4}{4\pi^2} - \frac{\log x_1\log x_4}{\pi^2}\\
&- \frac{\log x_2\log x_4}{\pi^2} - \frac{\log x_3\log x_4}{4\pi^2} - \frac{\log x_4^2}{4\pi^2} + \frac{x_5\log x_5}{4\pi^2}\\
&+ \frac{x_4x_5\log x_5}{4\pi^2} + \frac{3x_5^2\log x_5}{8\pi^2} - \frac{\log x_1\log x_5}{2\pi^2}\\
&- \frac{\log x_2\log x_5}{\pi^2} - \frac{\log x_3\log x_5}{4\pi^2} - \frac{\log x_4\log x_5}{2\pi^2}\\
&+ \frac{x_6x_7\log x_6}{4\pi^2} - \frac{3\log x_1\log x_6}{4\pi^2} - \frac{3\log x_2\log x_6}{4\pi^2}\\
&- \frac{\log x_3\log x_6}{4\pi^2} - \frac{\log x_4\log x_6}{2\pi^2} - \frac{\log x_5\log x_6}{2\pi^2} - \frac{\log x_6^2}{8\pi^2}\\
&+ \frac{x_7\log x_7}{4\pi^2} + \frac{x_6x_7\log x_7}{4\pi^2} + \frac{3x_7^2\log x_7}{8\pi^2}\\
&- \frac{\log x_1\log x_7}{2\pi^2} - \frac{3\log x_2\log x_7}{4\pi^2} - \frac{\log x_3\log x_7}{4\pi^2}\\
&\left.- \frac{\log x_4\log x_7}{2\pi^2} - \frac{\log x_5\log x_7}{4\pi^2} - \frac{\log x_6\log x_7}{4\pi^2}\right) C\,.
\end{aligned} \tag{218}$$

However we stopped our calculation at the second order in $\vec{x}$. Finally the prepotential is

$$\begin{aligned}
F(\vec{t}) = \;& -\frac{iq_1}{8\pi^3} - \frac{iq_1^2}{64\pi^3} - \frac{iq_1q_2}{8\pi^3} - \frac{iq_1q_5}{8\pi^3}\\
&+ \frac{3t_1}{4} + \frac{3t_1^2}{4} + \frac{t_1^3}{2} + 3t_1t_2 + 3t_1^2t_2 + 3t_1t_2^2\\
&+ t_1t_3 + t_1^2t_3 + 2t_1t_2t_3 + 2t_1t_4 + 2t_1^2t_4\\
&+ 4t_1t_2t_4 + t_1t_3t_4 + t_1t_4^2 + t_1t_5 + t_1^2t_5
\end{aligned}$$

$$
\begin{aligned}
&+4t_1t_2t_5 + t_1t_3t_5 + 2t_1t_4t_5 + \frac{3t_1t_6}{2} + \frac{3t_1{}^2t_6}{2} \\
&+3t_1t_2t_6 + t_1t_3t_6 + 2t_1t_4t_6 + 2t_1t_5t_6 \\
&+\frac{t_1t_6{}^2}{2} + t_1t_7 + t_1{}^2t_7 + 3t_1t_2t_7 + t_1t_3t_7 + 2t_1t_4t_7 \\
&+t_1t_5t_7 + t_1t_6t_7 \\
&+\text{Gclass}[t_2, t_3, t_4, t_5, t_6, t_7] + \text{Gistant}[q_2, q_3, q_4, q_5, q_6, q_7] \,. \qquad (219)
\end{aligned}
$$

5. Conclusion

We have discussed a possible extension of the minimal supersymmetric Standard Model (MSSM). This is only one of the recently proliferated alternative possible models, which embed the Standard Model or more generally $SU(5)$ or $SO(10)$ GUT models in a string theory context. In a bottom-up approach it is not a priori important to start from a GUT model, as remarked in [VW]. We have indeed adopted the same spirits thus reviewing firs the Verlinde-Wijnholt model in section 2.. This has been our starting point. This is a model based on a D3 brane worldvolume theory, where the brane is located at the orbifold singularity of the orbifold $X = \mathbb{C}^3/\Delta_{27}$. We have seen that at a firs sight such orbifold appears to capture the main features of the geometry of a cone over a del Pezzo surface of degree 8. Indeed, at low energy an open string theory with a D3-brane placed near the orbifold singularity can be identifie with the one the one with a D3-brane at the apex of the cone over del Pezzo surface. The advantage of such viewpoint is that the worldsheet CFT of strings on fla space orbifolds can be solved exactly, and the D-brane boundary conditions are exactly known [DoMo, DiGo].

We have indeed shown that the equivalence proposed in [VW] can be made very precise from a mathematical point of view. In particular, we have proven that the resolution of the orbifold singularity is given by a quasi del Pezzo dP_8 surface, thus providing a proof of the isomorphism conjectured in [VW]. This is only an initial step in the concrete realization of the geometric dual of the minimal quiver extension of the minimal supersymmetric Standard Model. Indeed, in order to reproduce the MSSM one needs to implement mutations and symmetry breaking procedures which require the should correspond to a partial resolution of the del Pezzo singularity with non isolated A_2 singularities. Moreover, the complete string theory model would require to embed the noncompact model in compact Calabi-Yau manifold. We expect that the more natural way to construct the desired partial resolution of del Pezzo singularity should pass through the orbifold geometry and our explicit map between them.

However, we have not tried to face such a formidable challenge here, instead opting for illustrating some advanced topics in D-brane dynamics. To this aim, we have considered a simpler toric version of the orbifold singularity, the one corresponding to the abelianization $\mathbb{Z}_3 \times \mathbb{Z}_3$ of Δ_{27}. After repeating a parallel analysis as for the non abelian model, we have then considered a series of four possible resolutions of the singularity, all related by flops and we have shown how can one can use mirror symmetry to determine the explicit sheaf description of the brane configuration and the corresponding prepotential, including all instanton corrections.

Acknowledgments

We are indebted to Bert van Geemen for many explanations and discussions. We are also grateful to Lidia Stoppino, Enrico Schlesinger and Stefano Guerra for very useful discussions. Finally we gratefully acknowledge Giuseppe Berrino who helped us to improve our exposition.

References

[ACT] D. Allcock, J. A. Carlson and D. Toledo, "The complex hyperbolic geometry of the moduli space of cubic surfaces," *J. Alg. Geom.* **11** (2002), 659–724.

[AD] M. Artebani, I. Dolgachev, "The Hesse pencil of plane cubic curves", *L'Ens. Math.* **55** (2009), 235-273.

[Asp] P. S. Aspinwall, "D-branes on Calabi-Yau manifolds", in Progress in String Theory, *TASI 2003 Lecture Notes (2005)*, World Scientific [arXiv:hep-th/0403166].

[AKO] D. Auroux, L. Katzarkov, D. Orlov, "Mirror symmetry for del Pezzo surfaces: vanishing cycles and coherent sheaves", *Invent. Math.* **166** (2006), 537–582.

[BHV1] C. Beasley, J. J. Heckman and C. Vafa, "GUTs and Exceptional Branes in F-theory - I," *JHEP* **0901** (2009) 058.

[BHV2] C. Beasley, J. J. Heckman and C. Vafa, "GUTs and Exceptional Branes in F-theory - II: Experimental Predictions," *JHEP* **0901** (2009) 059.

[BJL] D. Berenstein, V. Jejjala and R. G. Leigh, "The standard model on a D-brane," *Phys. Rev. Lett.* **88**, 071602 (2002).

[BBGW] R. Blumenhagen, V. Braun, T. W. Grimm and T. Weigand, "GUTs in Type IIB Orientifold Compactifications" *Nucl. Phys. B* **815** (2009) 1.

[BC] V. Bouchard and R. Cavalieri, "On the mathematics and physics of high genus invariants of $[C^3/Z_3]$," *Adv. Theor. Math. Phys.* Vol. 13, No.3:695-719, 2009.

[BMMVW] M. Buican, D. Malyshev, D. R. Morrison, H. Verlinde and M. Wijnholt, "D-branes at singularities, compactification and hypercharge," *JHEP* **0701**, 107 (2007).

[BLZ] B. A. Burrington, J. T. Liu and L. A. P. Zayas, "Finite Heisenberg groups from nonabelian orbifold quiver gauge theories," *Nucl. Phys. B* **794**, 324 (2008)

[CaCo-1] S. L. Cacciatori and M. Compagnoni, "On the geometry of $\mathbb{C}^3/D_27$ and del Pezzo surfaces," *JHEP* **1005** (2010) 078.

[CaCo-2] S. L. Cacciatori and M. Compagnoni, "D-Branes on C_6^3 part I: prepotential and GW-invariants," *Adv. Theor. Math. Phys.* Volume 13, Number 5 (2009), 1371-1443.

[Ca] A. Canonaco, "Exceptional sequences and derived autoequivalences", arXiv:0801.0173.

[CoDo] F.R. Cossec, I.V. Dolgachev, "Enriques surfaces. I.", *Progress in Mathematics,* **76**. Birkhäuser Boston, Inc., Boston, MA, 1989.

[CMS] G. Cliff, D. McNeillya, F. Szechtman, "Character field and Schur indices of irreducible Weil characters", *Journal Group Theory* **7** (2004) 39-64.

[CR] A. Craw, M. Reid, "How to calculate A-Hilb $\mathbb{C}^3$", *Geometry of toric varieties,* 129–154, Semin. Congr., 6, Soc. Math. France, Paris, 2002.

[DG] O. DeWolfe and S. B. Giddings, "Scales and hierarchies in warped compactifica tions and brane worlds", *Phys. Rev. D* **67**, 066008 (2003).

[DiGo] D. E. Diaconescu and J. Gomis, "Fractional branes and boundary states in orbifold theories", *JHEP* **0010** (2000) 001.

[DGK] I. Dolgachev, B. van Geemen and S. Kondō, "A complex ball uniformization on the moduli space of cubic surfaces via periods of $K3$ surfaces," *J. reine angew. Math.* **588** (2005), 99–148.

[DW1] R. Donagi and M. Wijnholt, "Model Building with F-Theory," arXiv:0802.2969 [hep-th].

[DW2] R. Donagi and M. Wijnholt, "Breaking GUT Groups in F-Theory," arXiv:0808.2223 [hep-th].

[Dou] M. R. Douglas, "D-branes, categories and N = 1 supersymmetry", *J. Math. Phys.* **42** (2001) 2818.

[DoMo] M. R. Douglas and G. W. Moore, "D-branes, Quivers, and ALE Instantons," arXiv:hep-th/9603167.

[Ful] W. Fulton, "Introduction to toric varieties", Annals of Mathematics Studies, 131, Princeton University Press, Princeton, NJ, (1993).

[Ge1] P. Gérardin, "Three Weil representations associate to finit fields" *Bull. Amer. Math. Soc.* **82** (1976) 268-270.

[Ge2] P. Gérardin, "Weil representations associated to finit fields" *J. Algebra* **46** (1977) 54-101.

[GKP] S. B. Giddings, S. Kachru and J. Polchinski, "Hierarchies from flu es in string compactifications" *Phys. Rev. D* **66**, 106006 (2002).

[Har] R. Hartshorne, "Algebraic geometry", *Graduate Texts in Mathematics*, No. **52**. Springer-Verlag, New York-Heidelberg, 1977.

[Hes1] O. Hesse, "Uber die Elimination der Variabeln aus drei algebraischen Gleichungen vom zweiten Grade mit zwei Variabeln", *Journal fr die reine und angewandte Mathematik* **10** (1844), 68-96.

[Hes2] O. Hesse, “Uber die Wendepunkte der Curven dritter Ordnun”, *Journal fr die reine und angewandte Mathematik* **28** (1844), 97-106.

[H& al] K. Hori *et al.*, “Mirror symmetry”, *Providence, USA: AMS (2003) 929 p*

[Hos1] S. Hosono, “Central charges, symplectic forms, and hypergeometric series in local mirror symmetry”, ”Mirror Symmetry V”, Proceedings of BIRS workshop on Calabi-Yau Varieties and Mirror Symmetry, December 6-11, 2003, [arXiv:hep-th/0404043].

[Jo] C. Jordan, “Mémoire sur les equations différentielles linéairesa intégrale algébrique”, *Journal für die reine und angewandte Mathematik* **84** (1878), 89-215.

[KKLT] S. Kachru, R. Kallosh, A. Linde and S. P. Trivedi, “De Sitter vacua in string theory,” *Phys. Rev. D* **68**, 046005 (2003).

[Ka] R. L. Karp, “On the C^m/Z_m fractional branes”, arXiv:hep-th/0602165.

[KaNo] B.V. Karpov, D. Yu. Nogin, “Three-block exceptional sets on del Pezzo surfaces”, (Russian. Russian summary) *Izv. Ross. Akad. Nauk Ser. Mat.* **62** (1998), no. 3, 3–38; translation in *Izv. Math.* **62** (1998), no. 3, 429–463.

[Lig] G. Ligozat, “Courbes modulaires de genre 1”, *Bull. Soc. Math. France, Mém*. **43**. *Supplément au Bull. Soc. Math. France Tome* **103**, Paris, (1975).

[Mald] J. M. Maldacena, “The large N limit of superconformal fiel theories and supergravity”, *Adv. Theor. Math. Phys.* **2**, 231 (1998) [*Int. J. Theor. Phys.* **38**, 1113 (1999)].

[Maly] D. Malyshev, “Del Pezzo singularities and SUSY breaking”, arXiv:0705.3281 [hep-th].

[MaVe] D. Malyshev and H. Verlinde, “D-branes at Singularities and String Phenomenology”, *Nucl. Phys. Proc. Suppl.* **171** (2007) 139.

[Mas] H. Maschke, “Aufstellung des vollen formensystems einer quaternaren Gruppe von 51840 linearen substitutionen”, *Math. Ann.*, **33**, 317-344 (1889).

[MoPl] D. R. Morrison and M. R. Plesser, “Non-spherical horizons. I”, *Adv. Theor. Math. Phys.* **3** (1999) 1.

[Oda] T. Oda, “Convex bodies and algebraic geometry. An introduction to the theory of toric varieties”, Springer-Verlag, Berlin, (1988).

[Po] J. Polchinski, “String theory”, *Cambridge, UK: Univ. Pr. (1998) 531 p*

[S] N. Seiberg, “Electric - magnetic duality in supersymmetric nonAbelian gauge theories”, *Nucl. Phys. B* **435** (1995) 129.

[SS] H.C. Shaub, H.E. Schoonmaker, “The Hessian Configuratio and Its Relation to the Group of Order 216”, *Amer. Math. Monthly* **38** (1931), 388–393.

[VW] H. Verlinde and M. Wijnholt, “Building the Standard Model on a D3-brane,” *JHEP* **0701**, 106 (2007).

[We] A. Weil, “Sur certains groupes d’opérateurs unitaires”, *Acta Math.* **111** (1964) 143-211.

[Yo] H. Yoshida, “Weil’s representations of the symplectic groups over finit fields” *J. Math. Soc. Japan* **31** (1979) 399-426.

In: New Developments in the Standard Model
Editor: Ryan J. Larsen
ISBN: 978-1-61209-989-7

Chapter 5

FERMION CONDENSATE AS HIGGS SUBSTITUTE

G. Cynolter* and E. Lendvai†
Theoretical Physics Research Group of Hungarian Academy of Sciences, Eötvös University, Budapest, Pázmány Péter stny 1/A, Hungary

Abstract

We propose and analyze an alternative model of dynamical electroweak symmetry breaking. In the Standard Model of electroweak interactions the elementary Higgs fiel and the Higgs sector are replaced by vector-like fermions and their interactions. The new fermions are a weak doublet and a singlet. They have kinetic terms with covariant derivatives and gauge invariant four-fermion interactions. The model is a low energy effective one with a natural cutoff in the TeV regime. Due to the quartic fermion couplings the new fermions form condensates. The new fermions mix in one condensate and the mixing breaks the electroweak symmetry. The condensates contribute to the masses of the new femions, which may or may not have mass terms in the original Lagrangian. Gap equations are derived for the masses of the new fermions and the conditions are presented for mass generations and electroweak symmetry breaking. In the spectrum there are two neutral fermions and a charged one with mass between the neutral ones. The new sector can be described by three parameters, these are the two neutral masses and the mixing angle. These parameters are further constrained by the unitarity of two particle scattering amplitudes, providing an upper bound for the lighter neutral mass depending on the cutoff of the model. The standard chiral fermions get their masses via interactions with the condensing new fermions, but there is no mixing between the standard and the new fermions. There is an effective composite scalar in the model at low energies, producing the weak gauge boson masses in effective interactions. The ρ parameter is one at leading order. The model can be constrained by one-loop oblique corrections. The Peskin-Takeuchi S and T parameters are calculated in the model. The parameters of the model are only slightly constrained, the T parameter requires the new neutral fermion masses not to be very far from each other, allowing higher mass difference for higher masses and smaller mixing. The S

*E-mail address: cyn@general.elte.hu
†E-mail address: lendvai@general.elte.hu

parameter gives practically no constraints on the masses. The new fermions can give positive contributions to T allowing for a heavy Higgs in the precision electroweak tests. It is shown that the new fermions will be copiously produced at the next generation of linear colliders and cross sections are presented for the Large Hadron Collider. An additional nice feature of the model is that the lightest new neutral fermion is an ideal and natural dark matter candidate.

1. Introduction

The Standard Model of particle physics successfully describes known collider experiments reaching the permille level in case of some observables. The only missing particle of the Standard Model is the elementary Higgs boson. In the minimal Standard Model a weak doublet (hypercharge Y=1) scalar fiel is postulated with an *ad hoc* scalar potential to trigger electroweak symmetry breaking. This provides a very economical and simple description. Three Goldstone bosons are eaten up by the $W^{\pm}$, Z gauge bosons providing their correct masses, but the remaining single CP-even neutral Higgs scalar has evaded the experimental discovery so far. There exist experimental constraints on the mass of the Higgs boson. The LEP2 experiment has put a lower bound $M_H > 114.4$ GeV [1] and there is an exclusion window from the combined D0 and CDF measurements at the Tevatron [2] between 158 and 175 GeV. The precision data favour a light Higgs with a central value below the direct LEP2 bound. Including the results of the direct searches both at LEP2 and the Tevatron the upper limit is driven to $M_H \leq 147$ at 95 % C.L. from electroweak precision tests [1]. The Gfitte group has arrived at similar upper bounds $M_H \leq 159$ GeV (155 GeV) with or without the information of the direct Higgs searches [3].

Beside the missing experimental discovery, theories with elementary scalars are burdened with theoretical problems, such as triviality and the most severe one, the gauge hierarchy problem. Elementary scalars are unstable to radiative corrections and without fin tuning the Standard Model must be cut off at few TeV.

There are mainly two ways to solve these problems in particle physics, either impose new symmetries to protect the scalars or eliminate elementary scalars from the theory.

Supersymmetry is the number one candidate for beyond the Standard Model physics, it protects the quadratically unstable Higgs mass, the contribution of the superpartners cancel each other. The Minimal Supersymmetric Standard Model is very attractive considering that electroweak symmetry breaking is triggered radiatively, there are ideal dark matter candidates and gauge couplings unify better in supersymmetric Grand Unifie Theories than in standard GUTs. However supersymmetric theories involve a huge parameter space, all known particles are doubled and no satisfactory mechanism has been worked out for supersymmetry breaking. None of the predicted new superpartners have been found in any of the experiments and supersymmetry may start to lose its appeal. Another shortcoming is that with no discovery the superpartner masses and the scale of supersymmetry breaking are pushed higher and higher reformulating the fin tuning problem at a percent level.

There are strong indications, expectations and a “no lose theorem” that the LHC will reveal the physics of electroweak symmetry breaking. Either the LHC will fin one or more Higgs bosons, it could be the Standard Model one or a scalar coming from an extended Higgs sector, such as the MSSM or the LHC will discover some sign of new, possibly strong dynamics that unitarizes the scattering of longitudinal gauge bosons in the TeV regime. These observations motivate to study alternative models of electroweak symmetry breaking without elementary scalars.

The other main solution to the hierarchy problem employs the mechanism of dynamical symmetry breaking. The original technicolor idea [4, 5, 6] of fermion condensation is already more than thirty years old, it is based on real phenomena of QCD. Technicolor still gives motivation for new research, see a recent review [7], Chivukula et al. in [1] and references therein. New chiral fermions are postulated which are charged under the new technicolor gauge group, the new interaction becomes strong condensing the techni-fermions charged under the weak $SU_L(2)$. To provide fermion masses extended technicolor gauge interactions (ETC) [8, 9] must be included. The tension between sizeable quark masses and avoiding fl vor changing neutral currents led to introduce walking, near conformal dynamics [10, 11]. These ideas and the phase diagram of strongly interacting models triggered activity in lattice studies [12], and further new technicolor models were constructed based on adjoint or two index symmetric representations of the new fermions [13]. The heavy top quark is natural in top condensate models [14, 15], and there are extra dimensional realizations, too [16, 17].

Inspired by the discretized variants of higher dimensional theories "little Higgs" [18] models provide a new class of composite Higgs models, and they attracted considerable interest solving the “little hierarchy problem” [19] allowing to raise the cutoff of the theory up to 10 TeV without excessive fin tuning [20, 21]. Little Higgs models realize the old idea that the Higgs is a pseudo Goldstone boson of some spontaneously broken global symmetry [22]. Contrary to supersymmetric models divergent fermion (boson) loops cancel fermion (boson) loops. Little Higgs models still require large fin tuning unless they posses custodial symmetry at the price of highly extended gauge groups. There are various models where the Higgs is composite [23], the idea was recently realized in extra dimensions [24]. Higgsless models [25] do not utilize a scalar Higgs boson, but using the AdS/CFT correspondence these are extra dimensional "duals" of walking technicolor theories.

In this chapter we present a recently proposed alternative symmetry breaking model of electroweak interactions [26]. The complete symmetry breaking sector is built from a new doublet and a singlet vector-like fermions, the Higgs is a composite state of the new fermions. Using vector-like fermions is advantageous compared to chiral ones as the constraints from precision electroweak measurements are much weaker. Vector-like fermions appear in several extensions of the Standard Model. They are present in extra dimensional models with bulk fermions e.g [27], in little Higgs theories [18, 20, 21], in models of so called improved naturalness consistent with a heavy Higgs scalar [28], in simple fermionic models of dark matter [29, 30], in dynamical models of supersymmetry breaking using gauge medation, topcolor models [31]. Vector-like fermions were essential ingredients of our proposal, in which a nontrivial condensate of new vector-like fermions breaks the electroweak symmetry and provides masses for the standard particles [26].

In the Fermion Condensate Model the Higgs sector is replaced by the interactions of a new doublet $\Psi_D = \begin{pmatrix} \Psi_D^+ \\ \Psi_D^0 \end{pmatrix}$ and a singlet Ψ_S hypercharge 1 vector-like (non-chiral) fermion field After electroweak symmetry breaking Ψ_D^+ fiel corresponds to a positively charged particle and Ψ_D^0 to a neutral one. The new fermions are postulated to have effective non-renormalizable four-fermion interactions and the model is a low energy effective one, valid up to some intrinsic, physical cutoff, that is not be taken to infinit . Therefore we are not forced to add additional terms to calculate at lowest orders following [32], including extra terms will defin a different model. The ultraviolet completion of the model is not yet specified but as usual the four-fermion terms are expected to originated from some spontanously broken gauge interactions. The key point is that the four-fermion interactions become strong at low energies and generate condensates of the new fermions including a mixed condensate of Ψ_D and $\bar{\Psi}_S$, $\left\langle\overline{\Psi}_S\Psi_D\right\rangle_0 \neq 0$. Gap equations are derived for the condensates and the condition of symmetry breaking is determined. The new fermions get contributions to their masses from the condensates. The vacuum solution of the model has a nontrivial weak $SU_L(2)$ quantum number and it spontaneously breakes the electroweak symmetry in a dynamical way. This symmetry breaking scheme was already utilized in our earlier works [33, 34]. The nontrivial condensate further generates mixing between the neutral component of the doublet and the singlet. The Ψ_D doublet has a standard kinetic terms with the usual covariant derivative and after the mixing the weak gauge bosons ($W^{\pm}$, Z) get their masses from the symmetry breaking condensate. The proposed model contains three new particles, two neutral and one charged fermions. The solution of the gap equations shows that the mass of the charged fermion is between the two neutral ones. The lighter neutral particle is an ideal dark matter candidate. The most important constraints on the parameters of the model are coming from the solution of the gap equation and the requirement of perturbative unitarity in two particle elastic scattering processes. Generally the new charged fermion tends to be nearly degenerate with the heavier neutral one. Perturbative unitarity sets an upper bound on the lighter neutral fermion depending on the range of validity of the model (the cutoff), it is $M_1 \leq 230$ GeV for $\Lambda = 3$ TeV.

Any beyond the Standard Model physics must face the tremendous success of the Standard Model in high energy experiments, it must have evaded direct detection and fulfil the electroweak precision tests. LEP1 and LEP2 mesurements have set a direct lower bound [1] for a heavy charged strongly not interacting fermion (lepton) $M_+ > 100$ GeV and without assumptions $M_0 > 45$ GeV for neutral one. Oblique radiative correction which proved to be fatal in case of the original technicolor models are nearly harmless. The starting vector-like doublet and singlet gives no contribution to the Peskin-Takeuchi S and T oblique parameters [35] and the deviations are always proportional to the mixing among the new neutral fermions. Small enough but nonvanishing mixing will break the electroweak symmetry but gives small S and T . Finally the symmetry breaking solutions of the gap equations are so specially constrained that lead to a miniscule S and T parameters.

The rest of the chapter is organized as follows. In section 2 we present the proposed dynamical symmetry breaking model, then the gap equations are derived and solved, the solutions are further constrained by perturbative unitarity in section 4. In section 5 the interactions relevant in phenomenology and direct constraints from the LEP experiment are calculated. In section 6 we calculate the oblique electroweak parameters and section 7 contains the numerical results and figures The cross sections for the LHC and the next generation of linear colliders are presented before the conclusion, and one appendix flashe a new regularization method developed and used by us during this work.

2. The Fermion Condensate Model

Recently self-interacting vector-like fermions were introduced [26] in the Standard Model instead of an elementary standard scalar Higgs. The new colourless Dirac fermions are an extra neutral weak $SU(2)$ singlet $(T = Y = 0)$ and a doublet

$$\Psi_S, \qquad \Psi_D = \begin{pmatrix} \Psi_D^+ \\ \Psi_D^0 \end{pmatrix}, \tag{1}$$

with hypercharge 1. Similar fermions are often dubbed leptons, because they do not participate in strong interactions, and widely studied in the literature as we discussed in the introduction. A model with similar fermion content were studied by Maekawa [36, 37]. There is a new Z_2 symmetry acting only on the new fermions, which protects them from mixings with the standard model quarks and leptons, the new fermions may interact only in pairs. The lightest new fermion is stable providing an ideal weakly interacting dark matter candidate.

The new Lagrangian with gauge invariant kinetic terms and invariant 4-fermion interactions of the new fermions is L_Ψ,

$$\begin{aligned} L_\Psi \quad = \quad & i\overline{\Psi}_D D_\mu \gamma^\mu \Psi_D + i\overline{\Psi}_S \partial_\mu \gamma^\mu \Psi_S - m_{0D}\overline{\Psi}_D\Psi_D - m_{0S}\overline{\Psi}_S\Psi_S + \\ & +\lambda_1 \left(\overline{\Psi}_D\Psi_D\right)^2 + \lambda_2 \left(\overline{\Psi}_S\Psi_S\right)^2 + 2\lambda_3 \left(\overline{\Psi}_D\Psi_D\right)\left(\overline{\Psi}_S\Psi_S\right), \end{aligned} \tag{2}$$

m_{0D}, m_{0S} are bare masses and D_μ is the covariant derivative

$$D_\mu = \partial_\mu - i\frac{g}{2}\underline{\tau}\,\underline{A}_\mu - i\frac{g'}{2}B_\mu, \tag{3}$$

where $\underline{A}_\mu, B_\mu$ and $g,\ g'$ are the usual weak gauge boson field and couplings, respectively. The left handed and the right handed fermions are assumed to be gauged under the same gauge $SU_L(2)$ group. Additional four-fermion couplings are possible but the extra term will not fundamentally change the symmetry breaking and mass generation. We will show in what follows that for couplings λ_i exceeding the critical value the four-fermion interactions of (2) generate condensates

$$\left\langle \overline{\Psi}^0_{D\alpha}\Psi^0_{D\beta}\right\rangle_0 = a_1\delta_{\alpha\beta}, \tag{4}$$

$$\left\langle \overline{\Psi}^+_{D\alpha}\Psi^+_{D\beta}\right\rangle_0 = a_+\delta_{\alpha\beta}, \tag{5}$$

$$\left\langle \overline{\Psi}_{S\alpha}\Psi_{S\beta}\right\rangle_0 = a_2\delta_{\alpha\beta}, \tag{6}$$

$$\left\langle \overline{\Psi}_S\Psi_D\right\rangle_0 = \left\langle \begin{pmatrix} \overline{\Psi}_S\Psi_D^+ \\ \overline{\Psi}_S\Psi_D^0 \end{pmatrix} \right\rangle_0 \neq 0. \tag{7}$$

The formation of the charged condensate (5) firs appeared in [38] and is more general then the condensates in [26]. The non-diagonal condensate in (7) spontaneously breaks the group $SU_L(2) \times U_Y(1)$ to $U_{em}(1)$ of electromagnetism. With the gauge transformations of Ψ_D the condensate (7) can always be transformed into a real lower component

$$\left\langle \overline{\Psi}_{S\alpha}\Psi^0_{D\beta}\right\rangle_0 = a_3\delta_{\alpha\beta}, \quad \left\langle \overline{\Psi}_{S\alpha}\Psi^+_{D\beta}\right\rangle_0 = 0, \tag{8}$$

where a_3 is real. The composite operator $\overline{\Psi}_S\Psi_D$ resembles the standard scalar doublet.

Assuming invariant four-fermion interactions for the new and known fermions,

$$L_f = g_f\left(\overline{\Psi}^f_L\Psi^f_R\right)\left(\overline{\Psi}_S\Psi_D\right) + g_f\left(\overline{\Psi}^f_R\Psi^f_L\right)\left(\overline{\Psi}_D\Psi_S\right), \tag{9}$$

the condensate (8) generates masses to the standard femions. In the linearized, mean fiel approximation the electron mass, for example, is

$$m_e = -4g_e a_3. \tag{10}$$

Up type quark masses can be generated via the charge conjugate fiel $\widetilde{\Psi}_D = i\tau_2\left(\Psi_D\right)^\dagger$. Introducing nondiagonal quark bilinears, the Kobayashi-Maskawa mechanism emerges. As in the Standard Model, from (10) we see that for two particles $m_i/m_j = g_i/g_j$, the masses are proportional to the unconstrained generalized Yukawa coefficients

The masses of the weak gauge bosons arise from the effective interactions of the auxiliary composite $Y = 1$ scalar doublet,

$$\Phi = \begin{pmatrix} \Phi^+ \\ \Phi^0 \end{pmatrix} = \overline{\Psi}_S\Psi_D. \tag{11}$$

Φ develops a gauge invariant kinetic term in the low energy effective description

$$L_H = h\left(D_\mu\Phi\right)^\dagger\left(D^\mu\Phi\right), \tag{12}$$

where D_μ is the usual covariant derivative (3).

The coupling constant h sets the dimension of L_H, $[h] = -4$ in mass dimension, we assume $h > 0$. (12) is a non-renormalizable Lagrangian and it provides the weak gauge boson masses and some of the interactions of the new fermions with the standard gauge bosons.

The terms with Φ^0 in L_H can be written as

$$\begin{aligned} h^{-1}L_H \;=\;& \frac{g^2}{2}W^-_\mu W^{+\mu}\Phi^{0\dagger}\Phi^0 + \frac{g^2}{4\cdot\cos^2\theta_W}Z_\mu Z^\mu\Phi^{0\dagger}\Phi^0 + \\ &+\left[\partial^\mu\Phi^{0\dagger}\partial_\mu\Phi^0 - \frac{i}{2}\frac{g}{\cos\theta_W}\left(\partial^\mu\Phi^{0\dagger}\right)\Phi^0 Z_\mu + \frac{i}{2}\frac{g}{\cos\theta_W}\Phi^{0\dagger}Z_\mu\left(\partial^\mu\Phi^0\right)\right] \end{aligned} \tag{13}$$

in terms of the standard vector boson fields

In the linearized approximation in (13) we put

$$h\,\Phi^{0\dagger}\Phi^0 \to h\left\langle\Phi^{0\dagger}\Phi^0\right\rangle_0 = h\left(16a_3^2 - 4a_1a_2\right) = \frac{v^2}{2}, \tag{14}$$

leading to the standard masses

$$m_W = \frac{gv}{2}, \qquad m_Z = \frac{gv}{2\cos\theta_W}. \tag{15}$$

v^2 is, as usual, $\left(\sqrt{2}G_F\right)^{-1}$, $v = 254$ GeV . The tree masses naturally fulfil the important relation $\rho_{\text{tree}} = 1$. This relation is the direct consequence of the extra global (custodial)

$SU(2)$ symmetry [39] of the Lagrangian (12) and of the vacuum expectation value of the composite scalar field The complete symmetry breaking sector, the Lagrangian (2) does not show this extra global symmetry, because there are mass-like terms breaking the symmetry of global chiral rotations. However, this symmetry breaking does not influenc the $W^{\pm}$, Z mass ratio. The idea is that there is a compositness scale at the order of the cutoff Λ, where the vacuum expectation values of the new fermions and composite fiel Φ is formed, which decouples from the original fermions at lower energies. This way the composite scalar fiel Φ can have separate global custodial symmetry and the new fermions can only influenc the ϱ parameter via suppressed loop corrections.

3. Gap Equations

Once the condensates (4-7) are formed, dynamical mass terms are generated in the Lagrangian (2) beside the bare mass terms.

$$L_{\psi} \to L_{\Psi}^{\text{lin}} = -m_{+}\overline{\Psi_D^{+}}\Psi_D^{+} - m_1\overline{\Psi_D^{0}}\Psi_D^{0} - m_2\overline{\Psi}_S\Psi_S - m_3\left(\overline{\Psi^0}_D\Psi_S + \overline{\Psi}_S\Psi_D^0\right), \tag{16}$$

with

$$m_{+} = m_{0D} - 6\lambda_1 a_{+} - 8\left(\lambda_1 a_1 + \lambda_3 a_2\right) = m_1 + 2\lambda_1\left(a_{+} - a_1\right), \tag{17}$$

$$m_1 = m_{0D} - 6\lambda_1 a_1 - 8\left(\lambda_1 a_{+} + \lambda_3 a_2\right), \tag{18}$$

$$m_2 = m_{0S} - 6\lambda_2 a_2 - 8\lambda_3\left(a_1 + a_{+}\right), \tag{19}$$

$$m_3 = 2\lambda_3 a_3. \tag{20}$$

If $m_3 = 0$ ($\lambda_3 = 0$ or $a_3 = 0$) then (16) is diagonal, the original gauge eigenstates

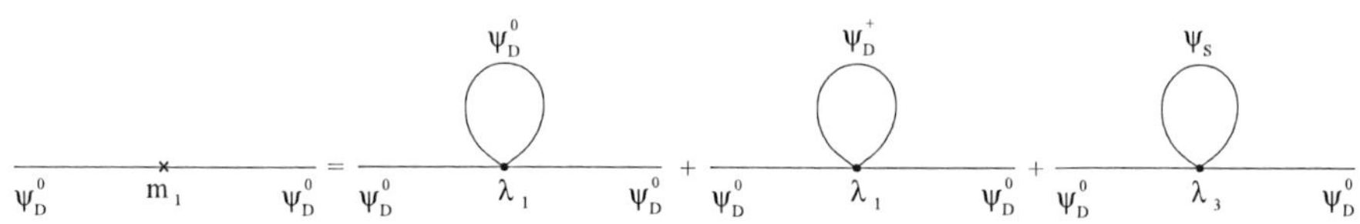

Figure 1. Feynman graphs for the gap equation (18). Similar graphs corresponding to (17,19) with exchanged legs and lines.

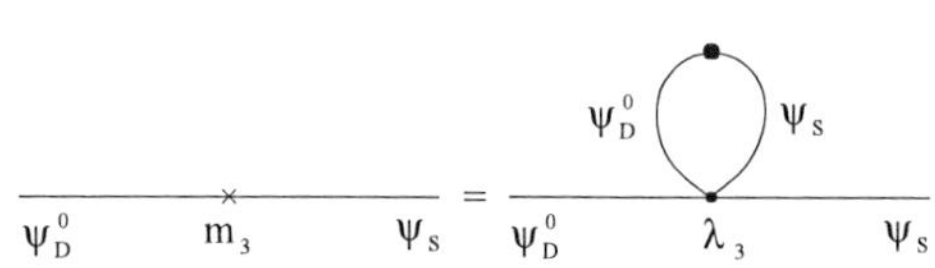

Figure 2. Feynman graphs for the gap equation (20).

are the physical fields the electroweak symmetry is not broken, $\lambda_3 a_3$, the non-diagonal condensate triggers the mixing and symmetry breaking. If $m_3 \neq 0$ (16) is diagonalized via unitary transformation to get physical mass eigenstates

$$\begin{aligned}\Psi_1 &= c\,\Psi_D^0 + s\,\Psi_S, \\ \Psi_2 &= -s\,\Psi_D^0 + c\,\Psi_S,\end{aligned} \tag{21}$$

where $c = \cos\phi$ and $s = \sin\phi$, ϕ is the mixing angle. As Ψ_S is real, only the real components of Ψ_D^0 take part in the mixing. The masses of the physical fermions Ψ_1, Ψ_2 are

$$2M_{1,2} = m_1 + m_2 \pm \frac{m_1 - m_2}{\cos 2\phi}. \tag{22}$$

The mixing angle is define by

$$2m_3 = (m_1 - m_2)\tan 2\phi. \tag{23}$$

Again we see, once $m_3 = 0$ the mixing angle vanishes (for $m_1 \neq m_2$), $M_1 = m_1$ and $M_2 = m_2$. The physical masses wil be equal ($M_1 = M_2$) only if $m_1 = m_2$, the original neutral fermions are degenerate in mass and then the mixing angle is meaningless from the point of view of mass matrix diagonalization.

It follows that the physical eigenstates themselves form condensates since

$$\begin{aligned}c^2\left\langle\overline{\Psi}_{1\alpha}\Psi_{1\beta}\right\rangle_0 + s^2\left\langle\overline{\Psi}_{2\alpha}\Psi_{2\beta}\right\rangle_0 &= a_1\delta_{\alpha\beta}, \\ s^2\left\langle\overline{\Psi}_{1\alpha}\Psi_{1\beta}\right\rangle_0 + c^2\left\langle\overline{\Psi}_{2\alpha}\Psi_{2\beta}\right\rangle_0 &= a_2\delta_{\alpha\beta}, \\ cs\left\langle\overline{\Psi}_{1\alpha}\Psi_{1\beta}\right\rangle_0 - cs\left\langle\overline{\Psi}_{2\alpha}\Psi_{2\beta}\right\rangle_0 &= a_3\delta_{\alpha\beta}.\end{aligned} \tag{24}$$

There is no non-diagonal condesate as Ψ_1, Ψ_2 are independent. Combining the equations of (24) one find

$$a_3 = \frac{1}{2}\tan 2\phi\,(a_1 - a_2)\,. \tag{25}$$

For $a_1 = a_2$, $a_3 \neq 0$ is not possible for $\cos 2\phi \neq 0$. As is seen, (25) is equivalent to $\left\langle\overline{\Psi}_{1\alpha}\Psi_{2\beta}\right\rangle_0 = 0$. Comparing (25) to (23) yields

$$m_1 - m_2 = 2\lambda_3\,(a_1 - a_2)\,. \tag{26}$$

Using the equations (17-20) we are lead to a consistency condition

$$(\lambda_3 - \lambda_1)\left(a_1 + \frac{4}{3}a_+\right) = (\lambda_3 - \lambda_2)\,a_2, \tag{27}$$

$\lambda_1 \neq \lambda_2$ goes with $a_1 + \frac{4}{3}a_+ \neq a_2$.

The equations (17-20) can be formulated as gap equations [32] in terms of the physical field expressing both the masses and the condensates with Ψ_1, Ψ_2 and $\Psi_+ \equiv \Psi_D^+$. Assuming vanishing original masses, $m_{0S} = 0$, $m_{0D} = 0$, the complete set of gap equations are

$$c\cdot s\,(M_1 - M_2) = 2\lambda_3\,c\cdot s\,(I_1 - I_2)\,, \tag{28}$$

$$c^2M_1 + s^2M_2 = -\lambda_1\left(6\left(c^2I_1 + s^2I_2\right) + 8I_+\right) - 8\lambda_3\left(s^2I_1 + c^2I_2\right), \tag{29}$$

$$s^2M_1 + c^2M_2 = -6\lambda_2\left(s^2I_1 + c^2I_2\right) - 8\lambda_3\left(c^2I_1 + s^2I_2 + I_+\right), \tag{30}$$

$$M_+ = -\lambda_1\left(8\left(c^2I_1 + s^2I_2\right) + 6I_+\right) - 8\lambda_3\left(s^2I_1 + c^2I_2\right). \tag{31}$$

The main task of the present work is to explore the structure of the gap equations. There are four algebraic equations for four variables $M_1,\ M_2,\ M_+,\ c^2 = \cos^2\phi$. As in almost all approximation $I_i \sim M_i$, (28-31) show gap equation characteristics, $M_i = 0$ is always a symmetric solution, which is stable for small $|\lambda_i|$. Increasing $|\lambda_i|$ also an energetically favoured [40] massive solution emerges as in the original Nambu Jona-Lasinio model. Now we explore the parameter space λ_i to fin acceptable phyical masses.

Let the condensates be approximated by free fiel propagators

$$\langle\overline{\Psi}_{i\alpha}\Psi_{i\beta}\rangle = \frac{\delta_{\alpha\beta}}{4}I_i = -\frac{\delta_{\alpha\beta}}{8\pi^2}M_i\left(\Lambda^2 - M_i^2\ln\left(1+\frac{\Lambda^2}{M_i^2}\right)\right), \quad i = 1,2,+, \tag{32}$$

where $M_+ = m_+$. Here Λ is a four-dimensional physical cutoff, it sets the scale of the new physics responsible for the non-renormalizable operators. From the point of view of symmetry breaking, the Λ cutoff can be chosen arbitrary large (below the GUT or Planck scale), but higher Λ implies stronger fin tuning of λ_3, see (32), to keep the new fermion masses in the electroweak range. To avoid fin tuning and allow reasonable fermion masses Λ is expected to be a few TeV, typically around 3 TeV [26].

For the electroweak symmetry breaking the most important equation is (28), it triggers mixing between the different representations of the weak gauge group. Applying (32) it reads

$$0 = (M_1 - M_2)\,c\cdot s\left(\frac{1}{\lambda_3} + \frac{\Lambda^2}{\pi^2} - \frac{M_1^3\ln\left(1+\frac{\Lambda^2}{M_1^2}\right) - M_2^3\ln\left(1+\frac{\Lambda^2}{M_2^2}\right)}{M_1 - M_2}\right). \tag{33}$$

(33) always has a symmetric solution $(M_1 - M_2)\,c\cdot s = 0$, implying $\sin 2\phi = 0$ for $M_1 \neq M_2$, there is essentialy no mixing. $M_1 = M_2$ was discussed after (23). If $|\lambda_3|$ is greater than a critical value $|\lambda_3^c| = \frac{\pi^2}{\Lambda^2}$ there also exists a symmetry breaking solution ($M_1 \neq M_2$), which always has lower energy if it exists [40]. Equation (33) has a solution with moderate masses ($M_{1,2} < 0.7\Lambda$) if λ_3 is negative. In the small mass limit the parantheses in (33) simplifie to $\frac{1}{\lambda_3} + \frac{\Lambda^2}{\pi^2} - (M_1^2 + M_1M_2 + M_2^2)\left(\ln(\Lambda^2) - \ln\left(\tilde{M}^2\right)\right)$ where $\tilde{M} \simeq max(M_1, M_2)$. If $|\lambda_3|$ is slightly larger than its critical value, then we generally get small masses compared to Λ, $M_1^2 + M_1M_2 + M_2^2 \ll \Lambda^2$. The critical coupling agrees with the original Nambu-Jona Lasinio value, only a factor of two coming from the definitio in the Lagrangian (2). If $|\lambda_3| < |\lambda_3^c|$ then the parantheses does not vanish in (33), the condensate a_3 is not formed and $(M_1 - M_2)\,c\cdot s = 0$. The physical solution is $c\cdot s = 0$, there is no meaningful mixing, $\Psi_S,\ \Psi_D$ are the physical mass eigenstates, and the electroweak symmetry is not broken.

Despite the complicated structure of the non-linear equations (28-31) we get a relatively simple gap equation for λ_1, similar to (33), from (17) $2\lambda_1(a_1 - a_+) = m_1 - m_+$. In the physical field we have

$$M_+ - c^2M_1 - s^2M_2 = 2\lambda_1\left(I_+ - c^2I_1 - s^2I_2\right). \tag{34}$$

It includes four unknowns, therefore it cannot be analyzed directly. We get a useful restriction solving (28) and (29) for λ_1 and substituting it to (34), relating M_1, M_2, M_+ and c^2 independently of the λ_i's. Requiring that $0 \leq c^2 \leq 1$ we get

$$M_1 \leq M_+ \leq M_2. \tag{35}$$

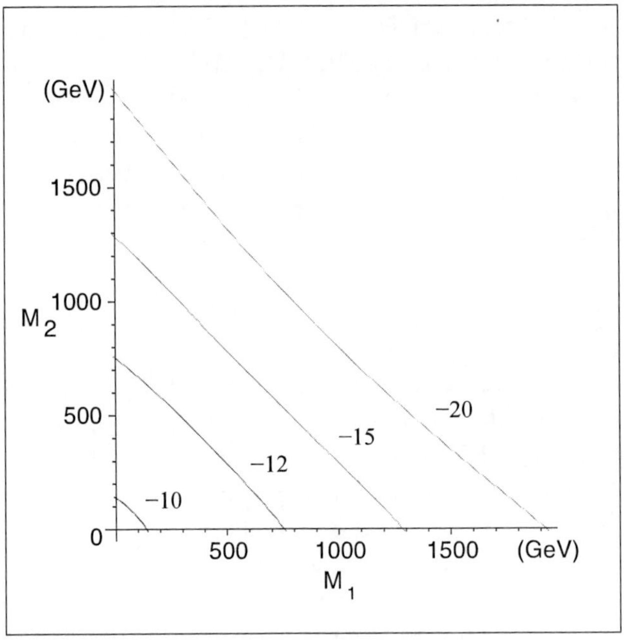

Figure 3. Constant λ_3 contours in the M_1-M_2 plane for $\lambda_3 = \{-10, -12, -15, -20\} \cdot 1/\Lambda^2$, $\Lambda = 3$ TeV.

As a result of the logaritmic terms in I_i, M_+ is nonlinear in c^2, while $m_1 = c^2 M_1 + s^2 M_2$. We remark that though (28) and (34) are very similar, for moderate masses λ_3 is always negative, while λ_1 is positive (also $\lambda_2 > 0$). In the $c^2 = 0$ (1) limit $M_+ = M_2$ (M_1) and there are cancellations in (28-31). Turning back to the symmetric solution of (33) the relation (35) gives $M_+ = M_1 = M_2$ and the rest of the gap equations set the common mass equal to zero unless the special relation $6(\lambda_3 - \lambda_2) = 8(\lambda_3 - \lambda_1)$ holds to provide cancellations.

To fin the critical value for λ_1 and λ_2 we considered the limit $M_+ \to M_2 = M$ and $M_1 \to 0$ then

$$\lambda_1 = \frac{1}{7}\frac{\pi^2}{\Lambda^2 - M^2 \ln\left(1 + \frac{\Lambda^2}{M^2}\right)}, \qquad \lambda_2 = \frac{4}{3}\frac{\pi^2}{\Lambda^2 - M^2 \ln\left(1 + \frac{\Lambda^2}{M^2}\right)}. \tag{36}$$

We get the same NJL type expression if we take the limit $M_+ \to M_2 = M$ and $M_1 \to 0$. (36) provides massive solutions if $\lambda_1 \geq \frac{1}{7}\frac{\pi^2}{\Lambda^2}$ and $\lambda_2 \geq \frac{4}{3}\frac{\pi^2}{\Lambda^2}$. Numerical scans show that these are the minimal, critical values for the couplings and can be approximated in special limits. Numerical solutions are shown in Table 1. for cutoff $\Lambda = 3$ TeV. The role of M_1 and M_2 can be exchanged together with $c^2 \leftrightarrow s^2$, therefore we have chosen $M_1 < M_2$ without the loss of generality. As the cutoff is not too high, 3 TeV, there is no serious fin tuning in the λ_i's to fin relatively small masses.

To understand the signs and roughly the factors in $\lambda_{1,2}^c$ consider the limit $M_1 \simeq M_2 \simeq M_+ \simeq M$. If $M \ll \Lambda$ then $\lambda_3 \simeq \lambda_3^c = -\frac{\pi^2}{\Lambda^2}$, though in the exact limit (28) becomes singular. We get from (28-31) the relation $14\lambda_1 = 6\lambda_2 + 8\lambda_3$ and a single gap equation (

$I = I_M$ in (32))

$$M = -\left(14\lambda_1 + 8\lambda_3\right) I. \tag{37}$$

Small mass solution requires $\tilde{\lambda} = 14\lambda_1 + 8\lambda_3$ to be close to it's critical value $2\pi^2/\Lambda^2$ and provides rough estimates $\lambda_1 \sim \frac{5}{7}\frac{\pi^2}{\Lambda^2}$ and also $\lambda_2 \sim 3\frac{\pi^2}{\Lambda^2}$ to generate small masses. Numerical solutions also provide general (M_+ not close to M_1 or M_2) small masses for couplings close to these values, see Table 1.

Table 1. Solutions of the gap equations for the cutoff Λ =3 TeV, λ_i are given in units of $\frac{\pi^2}{\Lambda^2}$. In the second column λ_2 violates perturbative unitarity.

$\lambda_1\left(\frac{\pi^2}{\Lambda^2}\right)$	0.546	0.740	0.496	0.380	0.502	0.468	0.419
$\lambda_2\left(\frac{\pi^2}{\Lambda^2}\right)$	2.540	3.11 (!)	2.403	2.120	2.457	2.455	2.451
$\lambda_3\left(\frac{\pi^2}{\Lambda^2}\right)$	-1.031	-1.041	-1.042	-1.070	-1.083	-1.178	-1.330
M_1 (GeV)	100	148	100	100	150	200	200
M_2 (GeV)	150	150	200	300	300	500	800
M_+ (GeV)	149	149	190	290	290	490	790

In the strongest small mass limit one neglects the logarithmic terms in the condensates (32), and equations (28-31) reduce to a linear homogeneous system of equations [38]. Finally we get two relations for the masses, $M_+ = m_1 = c^2 M_1 + s^2 M_2$ and $\frac{m_1}{m_2} = \frac{1-6\lambda_2\Lambda^2/\pi^2}{16\lambda_3\Lambda^2/\pi^2} = \frac{8\lambda_3\Lambda^2/\pi^2}{1-14\lambda_1\Lambda^2/\pi^2}$.

The solutions of the gap equations are further constrained by perturbative unitarity.

4. Perturbative Unitarity

In this section we apply tree-level partial wave unitarity to two-body scatterings of the new fermions following the arguments of the pioneering work by Lee et al. [41], where perturbative unitarity has been employed to constrain the Standard Model Higgs mass. Perturbative unitarity is a powerful tool, it can be used to build up the bosonic sector of the Standard Model, moreover it was essential to build higgsless models of electroweak symmetry breaking in extra dimensional fiel theories [25]. The method was used to constrain the parameters in the dynamical symmetry breaking vector condensate model in [42].

Consider the amplitudes of two particle $\left(\Psi_D^{(+)}, \Psi_D^{(0)} \text{ or } \Psi_S\right)$ elastic scattering processes and impose $|\Re a_0| \leq 1/2$ for the $J = 0$ partial wave amplitudes. The contact graph gives the dominant contribution, neglecting the fermion masses for the $\Psi_D^{(+)}\Psi_D^{(-)}$ scattering gives an upper bound on λ_1 coupling, $|\lambda_1|\, s \leq 8\pi$, where s is the maximal center of mass energy $\left(M_+^2 \ll s \leq \Lambda^2\right)$.

We cannot always use the small mass limit, as the solution of the gap equations provide higher λ_i's for significantl higher masses. Therefore we have calculated different helicity amplitudes [43] for non-vanishing masses. For $\Psi_a(1)\bar{\Psi}_a(2) \to \Psi_a(3)\bar{\Psi}_a(4)$, $(a = 0, s, +)$, $M = \lambda_i\left[\left(\bar{v}_2 u_1\right)\left(\bar{u}_3 v_4\right) - \left(\bar{u}_3 u_1\right)\left(\bar{v}_2 v_4\right)\right]$, where $\lambda_{i=1,2,3}$ are the only relevant four-fermion

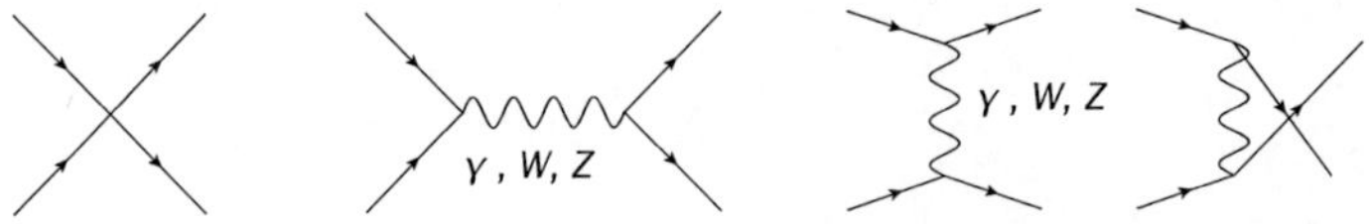

Figure 4. Feynman graphs of 2-particle elastic scattering.

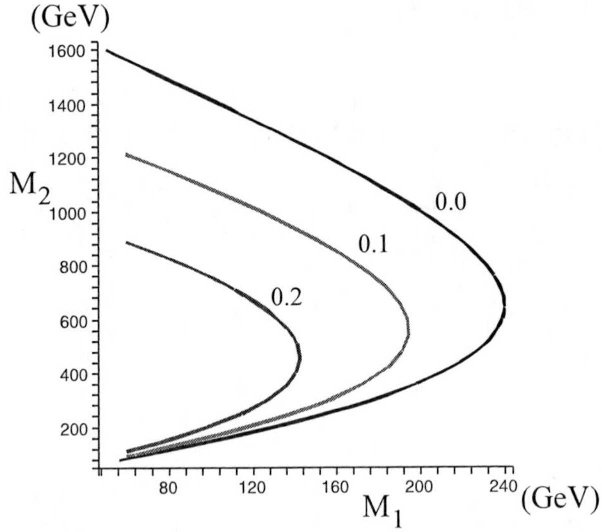

Figure 5. The maximum value of $c^2 = \cos^2\Phi$ on the M_1, M_2 plane from the gap equation and unitarity. c^2 can be higher inside the curves.

couplings. We consider $\Psi_S = s\Psi_1 + c\Psi_2$ scattering as a linear combination in the coupled Ψ_1, Ψ_2 channels to employ only λ_2 (and simiarly $\Psi_D^{(0)}$ to constrain λ_1). The contributions of the γ, Z exchange graphs are negligible $\left(\mathcal{O}\left(g^2\right) \ll 8\pi\right)$ because of the extra propagator. There are three different helicity channels, we give the representative helicity amplitudes, these are maximal for the back to back scattering ($\theta^{\{13\}}_{\text{scattering}} = \pi$)

$$M\left((+-) \to (+-)\right) = \lambda_i\left(s - 4M_i^2\right), \tag{38}$$

$$M\left((++) \to (--)\right) = \lambda_i s, \tag{39}$$

$$M\left((+-) \to (-+)\right) = \lambda_i 4M_i^2. \tag{40}$$

For other scattering angles $|M|$ is smaller than in (39), for example the maximum for $\theta = 0$ is $\lambda_i 4M_i^2$. The mass dependent unitarity bound agrees with the firs estimate

$$\lambda_i s \leq 8\pi, \tag{41}$$

where i=1,2,3 and $s \leq \Lambda^2$ is the center of mass energy. The unitarity constraints are most stringent for λ_2, even the equal small mass limit (37) would set $\lambda_2 \simeq 3\pi^2/\Lambda^2$, which is above the maximum value allowed by unitarity $8\pi/\Lambda^2 \simeq 2.55 \cdot \frac{\pi^2}{\Lambda^2}$. As an example we show a non-physical nearly equal mass solution in the second column of Table 1., which is not allowed by perturbative unitarity. (41) implies an absolute upper bound on the smaller neutral mass, $M_1 < 240$ GeV for $\Lambda = 3$ TeV. Perturbative unitarity for λ_2 and the solution of the gap equations generally push up the charged mass close to M_2 and sets the mixing angle $\sin\phi$ close to 1 in (21) meaning that there is only a small mixing, Ψ_2 is mostly composed of Ψ_D^0 and there is only a small mass splitting in the doublet Ψ_D after symmetry breaking. This observation will be important to estimate electroweak oblique corrections. The allowed M_1, M_2 masses and the maximum value of c^2 is shown in Figure 5. The maximum value of the cosine of mixing angle is determined from the condition that λ_2 should stay below the unitarity bound (41). The charged fermion mass must be relatively close to the mass of the heavier neutral one. The mixing angle ϕ is relatively close to $\cos\phi \sim 0$, the mixing is weak, see the curve on the right in Figure 6. Ψ_2 is mostly composed of Ψ_D^0 and there is only a small mass splitting in the doublet Ψ_D after symmetry breaking.

5. Interactions with $W^{\pm}$,Z and Constraints from the Z Decay

The collider phenomenology and radiative corrections (see section 6) in the model are coming from the doublet kinetic term in (2) taking into account the mixing (21)

$$\begin{aligned} L_I &= \overline{\Psi_D^+}\gamma^\mu\Psi_D^+\left(\frac{g'}{2}B_\mu + \frac{g}{2}W_{3\mu}\right) + \\ &+ \left(c^2\overline{\Psi}_1\gamma^\mu\Psi_1 + s^2\overline{\Psi}_2\gamma^\mu\Psi_2 - sc\left(\overline{\Psi}_1\gamma^\mu\Psi_2 + \overline{\Psi}_2\gamma^\mu\Psi_1\right)\right)\left(\frac{g'}{2}B_\mu - \frac{g}{2}W_{3\mu}\right) + \\ &+ \left[\frac{g}{\sqrt{2}}W_\mu^+\left(c\overline{\Psi_D^+}\gamma^\mu\Psi_1 - s\overline{\Psi_D^+}\gamma^\mu\Psi_2\right) + h.c.\right]. \end{aligned} \quad (42)$$

The interactions between the new and the standard fermions in L_f (9) turns out to be very weak. Indeed, from (10) and (14) we have an upper bound for g_e, $g_e \leq \sqrt{2}h\frac{m_e}{v} = \sqrt{2}hg_e^{SM}$, which is suppressed by two factors of the scale of new physics compared to the standard model value g_e^{SM}.

We will explore the consequences of these interactions in the decay of the Z boson and in the precision electroweak tests of the standard model in the next section.

The proposed new fermions could not be seen in the high energy experiments so far, because of their large masses and/or small couplings to ordinary particles. The mixing in the doublet reduces the coupling to the gauge bosons, but the new charged fermion is not affected. From the LEP1 and LEP2 measurements there is lower bound for the mass of a heavy charged lepton, valid here $M_+ > 100$ GeV [1]. For the neutral component of the doublet (without mixing) there are smaller lower bounds; without further assumptions $M_2 > 45$ GeV. Using the relation (35) M_2 is at least 100 GeV with or without mixing. The mixing generates small, but non-vanishing coupling between the Z boson and the new lighter neutral fermion (e.g. the remnant of the singlet, it has c^2 part of a doublet). Therefore if it is light enough it contributes to the invisible width of the Z boson

$$\Gamma(Z \to \bar{\Psi}_1\Psi_1) = \frac{\sqrt{2}G_F M_Z^3}{6\pi}\left(\frac{c^4}{4}\right)\sqrt{1 - \frac{4M_1^2}{M_Z^2}}. \quad (43)$$

The Z width is experimentally known at high precision and the pull factor is rather small

$$\Gamma(Z) = (2.4952 \pm 0.0023)\text{GeV}. \quad (44)$$

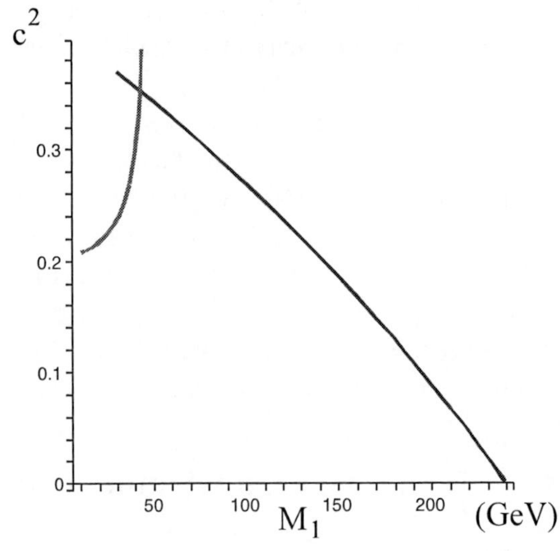

Figure 6. The maximum value of the $c^2 = \cos^2\Phi$ vs. the lighter neutral mass M_1. The right curve is derived from the gap equation and unitarity. The upper left curve is from the width of the Z boson.

We estimate the maximum possible room for new physics as 3σ in the experimental Z width, $\Gamma_Z^{\mathrm{new}} < 7$ MeV. In [44] the minimum value of $\Gamma_Z^{\mathrm{theory}}$ (at maximum $\sin^2\theta_W$ and minimum M_Z^2 and α_S) was compared to the maximal experimental value, and gave a similar 3σ window for new physics. We see that M_1 masses well below $M_Z/2$ are still allowed for rather small mixing, see the (red) curve on the left on Figure 6.

6. Electroweak Precision Parameters

The new fermions have direct interactions with the standard fermions (9) and gauge bosons (42). The four-fermion couplings of the new particles to the light fermions are weak; weaker than the corresponding ones in the Standard Model [26]. The new couplings to the gauge bosons are the gauge couplings suppressed only by the $\mathcal{O}(1)$ mixing factors. Therefore the couplings to the light fermions which participate in the precision experiments, are suppressed compared to the couplings to the gauge bosons. The new fermions thus mainly contribute to the gauge boson self energies in the precision experiments. In most of the solutions of the gap equation [38] $M_+, M_2 \gg M_Z$. Expecting further $M_1 > M_Z$ we can give a good estimate of the effects of new physics in terms of the general S, T and U parameters introduced by Peskin and Takeuchi [35]. We get a rough estimate of the loop effects if the mass of the lighter neutral fermion is not far above the Z mass.

The two relevant parameters, S and T define via the gauge boson self energies

$$\alpha(M_Z)T = \frac{\Pi_{WW}^{\mathrm{new}}(0)}{M_W^2} - \frac{\Pi_{ZZ}^{\mathrm{new}}(0)}{M_Z^2}, \tag{45}$$

$$\frac{\alpha(M_Z)}{4s_W^2 c_W^2}S = \frac{\Pi_{ZZ}^{\mathrm{new}}(M_Z^2) - \Pi_{ZZ}^{\mathrm{new}}(0)}{M_Z^2} - \frac{c_W^2 - s_W^2}{c_W s_W}\frac{\Pi_{Z\gamma}^{\mathrm{new}}(M_Z^2)}{M_Z^2} - \frac{\Pi_{\gamma\gamma}^{\mathrm{new}}(M_Z^2)}{M_Z^2}, \tag{46}$$

where $s_W^2 = \sin^2\theta_W(M_Z)$ and $c_W^2 = \cos^2\theta_W(M_Z)$ are $\sin^2$ ($\cos^2$) of the weak mixing angle. Barbieri et al. [45] revised the definitio of the oblique parameters. The Π functions are define from the transverse gauge boson vacuum polarization amplitudes expanded around zero $\Pi_{ab}(q^2) \simeq \Pi_{ab}(0) + q^2\Pi'_{ab}(0) + 1/2 \cdot q^2\Pi''_{ab}(0) + ...$, (a,b = 1,3,Y) up to second order. The 12 coefficient defin 7 parameter at the end. The definition of the old parameters are

$$\frac{\alpha(M_Z)}{4s_W^2c_W^2}S = \Pi'^{\rm new}_{3Y}(0), \quad (47)$$

$$\alpha(M_Z)T = \frac{1}{M_W^2}\left(\Pi^{\rm new}_{33}(0) - \Pi^{\rm new}_{11}(0)\right), \quad (48)$$

$$\frac{\alpha(M_Z)}{4s_W^2}U = \Pi'^{\rm new}_{33}(0) - \Pi'^{\rm new}_{11}(0). \quad (49)$$

These parameters (with the extra 4 -V, X, Y and W) fall into three groups according to their symmetry properties [45]. The Peskin-Takeuchi S parameter is custodially symmetric but weak isospin breaking. The T and U parameters break both the custodial and the weak isospin symmetry. It is reasonable to expect (and the actual calculation justifie the assumption) that the parameters with the same symmetry properties are related to each other. Since U mainly differs from T by an extra derivation of the Π functions, $U \sim \frac{M_W^2}{M_{\rm new}^2}T$ is expected, where $M_{\rm new}$ is the mass scale of new physics. When there is a gap between $M_{\rm new}$ and M_W it is reasonable to keep only the lowest derivative terms with a given symmetry property, S and T. If there is no special fin tuning U is expected to be less important than T and S is kept as the leading effect in its symmetry class.

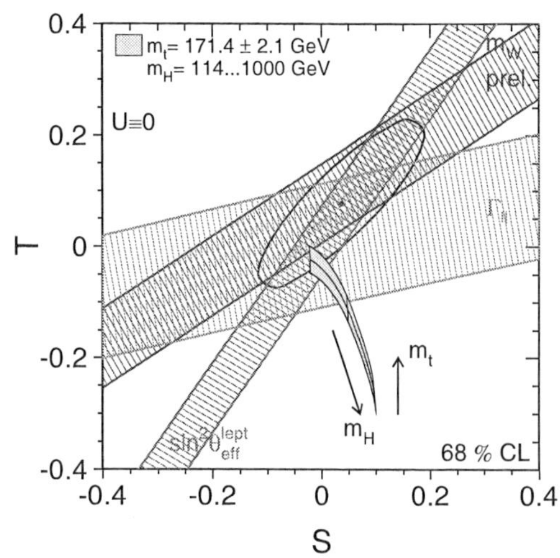

Figure 7. Experimental constraints and Standard Model predictions for S and T [46].

The experimental data determines S, T and U [1]

$$S = -0.10 \pm 0.10\ (-0.08), \quad (50)$$

$$T = -0.08 \pm 0.11\ (+0.09), \quad (51)$$

$$U = +0.15 \pm 0.11\ (+0.01), \quad (52)$$

where the central value assumes $M_H = 117$ GeV. The difference is shown for $M_H = 300$ GeV in the parentheses. The various experimental constraints and the dependence on the top and Higgs mass can be seen in Figure 7. In our model the Higgs mass of the fi is understood as the contribution of a composite Higgs particle with the given mass.

The contributions of the new sector to the gauge boson vacuum polarizations are fermion loops with generally two non-degenerate masses m_a and m_b [47]. In the low energy effective model we have preformed the calculation with a 4-dimensional Euclidean momentum cutoff Λ. The coupling constants are define in the usual manner $L_I \sim V_\mu \bar{\Psi}\left(g_V\gamma^\mu + g_A\gamma_5\gamma^\mu\right)\Psi$

$$\Pi(q^2) = \frac{1}{4\pi^2}\left(g_V^2\,\tilde{\Pi}_V + g_A^2\,\tilde{\Pi}_A\right). \tag{53}$$

The electroweak parameters depend on the values and derivatives of the Π functions at $q^2 = 0$

$$\begin{aligned}\tilde{\Pi}_V(0) &= \frac{1}{4}(m_a^2+m_b^2) - \frac{1}{2}(m_a-m_b)^2\ln\left(\frac{\Lambda^2}{m_a m_b}\right) - \\ &-\frac{m_a^4+m_b^4-2m_a m_b\left(m_a^2+m_b^2\right)}{4\left(m_a^2-m_b^2\right)}\ln\left(\frac{m_b^2}{m_a^2}\right).\end{aligned} \tag{54}$$

The firs derivative is

$$\begin{aligned}\tilde{\Pi}'_V(0) &= -\frac{2}{9} - \frac{4m_a^2m_b^2 - 3m_am_b\left(m_a^2+m_b^2\right)}{6\left(m_a^2-m_b^2\right)^2} + \frac{1}{3}\ln\left(\frac{\Lambda^2}{m_am_b}\right) + \\ &+\frac{\left(m_a^2+m_b^2\right)\left(m_a^4-4m_a^2m_b^2+m_b^4\right)+6m_a^3m_b^3}{6\left(m_a^2-m_b^2\right)^3}\ln\left(\frac{m_b^2}{m_a^2}\right).\end{aligned} \tag{55}$$

For completeness we give the second derivative, too. It can be used to calculate further precision parameters [45, 48], two extra parameters were introduced by Barbieri et al., and it is presented for extra vector-like fermions in [49],

$$\tilde{\Pi}''_V(0) = \frac{\left(m_a^2+m_b^2\right)\left(m_a^4-8m_a^2m_b^2+m_b^4\right)}{8\left(m_a^2-m_b^2\right)^4} + \frac{m_am_b\left(m_a^4+10m_a^2m_b^2+m_b^4\right)}{6\left(m_a^2-m_b^2\right)^4} - \tag{56}$$

$$-\frac{m_a^3m_b^3\left(3m_am_b-2m_a^2-2m_b^2\right)}{2\left(m_a^2-m_b^2\right)^5}\ln\left(\frac{m_b^2}{m_a^2}\right). \tag{57}$$

We get the functions for axial vector coupling by flippin exactly one of the masses in the previous results ($m_a \to m_a$ and $m_b \to -m_b$). The method of our calculation has nice properties: it has no quadratic divergence as expected; it fulfill gauge invariance in two aspects, $\Pi_V(m_a, m_a, 0) = 0$ and the complete Π function is transverse, the coefficient of the $g_{\mu\nu}$ and $-p_\mu p_\nu/p^2$ parts are equal.

The values of the vacuum polarizations for identical masses ($m_b = m_a$) are smooth limits and agree with direct calculation.

$$\tilde{\Pi}_V(0) = 0, \quad \tilde{\Pi}'_V(0) = -\frac{1}{3} + \frac{1}{3}\ln\left(\frac{\Lambda^2}{m_a^2}\right), \quad \tilde{\Pi}''_V(0) = \frac{2}{15}\frac{1}{m_a^2}. \tag{58}$$

The S parameter is then given by (for the sake of simplicity the index V is omitted)

$$S = \frac{1}{\pi}\left(+\tilde{\Pi}'(M_+, M_+, 0) - c^4\tilde{\Pi}'(M_1, M_1, 0) - s^4\tilde{\Pi}'(M_2, M_2, 0) - 2s^2c^2\tilde{\Pi}'(M_2, M_1, 0)\right). \tag{59}$$

The firs three terms cancel the divergent contribution of the last one.

The T parameter related to $\Delta\rho$ is

$$\begin{aligned} T &= \frac{1}{4\pi s_W^2 M_W^2}\Big[+\tilde{\Pi}(M_+,M_+,0)+c^4\tilde{\Pi}(M_1,M_1,0)+s^4\tilde{\Pi}(M_2,M_2,0)+ \\ &\quad +2s^2c^2\tilde{\Pi}(M_2,M_1,0)-2c^2\tilde{\Pi}(M_+,M_1,0)-2s^2\tilde{\Pi}(M_+,M_2,0)\Big]. \end{aligned} \tag{60}$$

For completeness we give the U parameter in the model

$$\begin{aligned} U &= -\frac{1}{\pi}\Big[+\tilde{\Pi}'(M_+,M_+,0)+c^4\tilde{\Pi}'(M_1,M_1,0)+s^4\tilde{\Pi}'(M_2,M_2,0)+ \\ &\quad +2s^2c^2\tilde{\Pi}'(M_2,M_1,0)-2c^2\tilde{\Pi}'(M_+,M_1,0)-2s^2\tilde{\Pi}'(M_+,M_2,0)\Big]. \end{aligned} \tag{61}$$

The gauge boson self-energies are calculated from a renormalizable part of a non-renormalizable theory, hence dimensional regularization can be used to calculate the general vacuum polarization function with two fermions of different masses circulating in the loop [49, 50].

7. Numerical Constraints from Precision Tests

There are 3 free parameter in the model to confront with the experiments. These can be chosen the three dimensionful four-fermion couplings $\lambda_{1,2,3}$, or more practically the two physical neutral masses M_1, M_2 and the mixing angle, $c^2 = \cos^2\phi$. For the cutoff $\Lambda \simeq 3$ TeV there is a maximum value for the masses, $M_1 \leq 240$ GeV. c^2 has an upper bound depending on the mass M_1, see Figure 5. The mass of the charged fermion is given by the solution of the gap equations, the value of M_+ is close to, but not equal to $c^2M_1 + s^2M_2$.

If there is no real mixing $c^2 = 0$; or if $M_1 = M_2 = M_+$, then there is one degenerate vector-like fermion doublet and a decoupled singlet, and S and T vanish explicitely. In this case the new sector does not violate $SU_L(2)$ and there is an exact custodial symmetry. Increasing the mass difference in the remnants of the original doublet by increasing the $|M_1 - M_2|$ mass difference and/or moving away from the non-mixing case $c^2 = 0$, we get a higher S and T. For small violation of the symmetries S and T are expected to be small. Numerical evaluation shows that for the new masses in the range allowed by the LEP bound, gap equations and unitarity the U parameter is indeed an order of magnitude smaller than the T parameter and generally smaller than S. U is always in the experimental window. In case of relatively small masses the oblique parameters are understood as rough estimates, but still in agreement with experiment.

Generally the S parameter depends only on the masses of the new particles and the mixing angle. For the solutions of the gap equations fulfillin perturbative unitarity the S parameter is always positive and far below the 95 % C.L. For a given M_1, M_2 S increases with increasing c^2 and maximal for the highest allowed c^2. This maximum value of the S parameter is plotted against M_2 for three given M_1 in Figure 8. The small value of S does not constrain the parameters of the model.

The value of the T parameter is always positive. The T parameter (60) sensitive to the differences and ratios of the masses $M_{1,2,+}$. T still varies for a given (M_1, M_2) pair depending on M_+ or equally on c^2; T is maximal for largest mass difference, for the largest

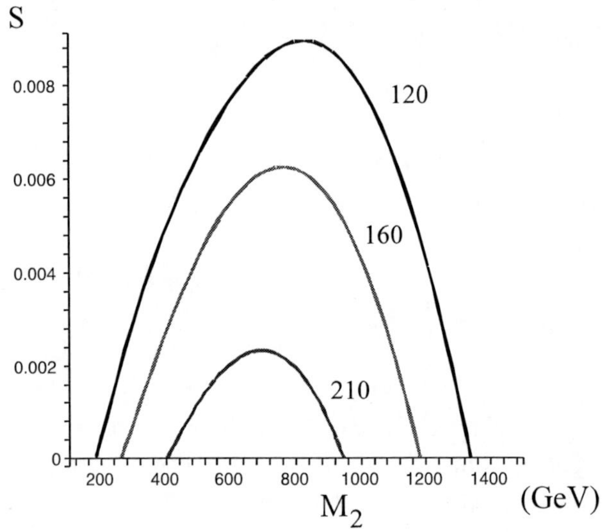

Figure 8. The maximum value of the S parameter vs. M_2 for $M_1 = 120,\ 160,\ 210$ GeV. The 95 % C.L. bounds [-0.296, 0.096] are outside the figure

c^2 allowed by the gap equations and perturbative unitarity. The T parameter can always be in agreement with experiment for any (M_1, M_2) pair for small mixing, for $c^2 = 0$ the T parameter vanishes identically. We plotted the worst case in the (M_1, M_2) plane, the possible maximum value of the T parameter; it is given by the maximum $M_2 - M_+$ mass difference, which is determined at the maximally allowed value of c^2.

If the Higgs is heavy, e.g. $M_H = 300$ GeV (50, 51) the central value of S decreases and T increases compared to the light Higgs case. The S parameter is still in agreement with the predictions of the model. Incrasing the Higss mass the Standard Model moves away in the (S,T) plane from the experimentally allowed ellipse, see [46]. The negative contribution $(-.09)$ of the heavy Higgs to the T parameter can be compensated by the positive T contribution of the new fermions with considerable mass difference. For example $(160,\ 800)$ GeV and the largest mixing $c^2 \sim 0.115$ allowed by the gap equations and unitarity gives $\Delta T \simeq 0.1$. Even heavier Higgs boson can be compensated as can be read off from Figure 9. Non-degenerate vector-like fermions with reasonable mixing allow a space for heavy Higgs in the precision tests of the Standard Model.

8. Collider Signatures

In this section we study the production of the new fermions at LHC and the planned linear collider. We focus on the preoduction of the new charged femions with mass M_+, we denote it by D^+ and its antiparticle by D^-.

Since the light standard fermions are coupled very weakly to the new fermions producing pairs of new fermions is expected to be more considerable from virtual γ and Z exchanges, that is we consider the Drell-Yan mechanism, $p\overset{(-)}{p} \to D^+D^- + X$ via quark-antiquark annihilation . The new fermion can only be produced in pairs because of the Z_2

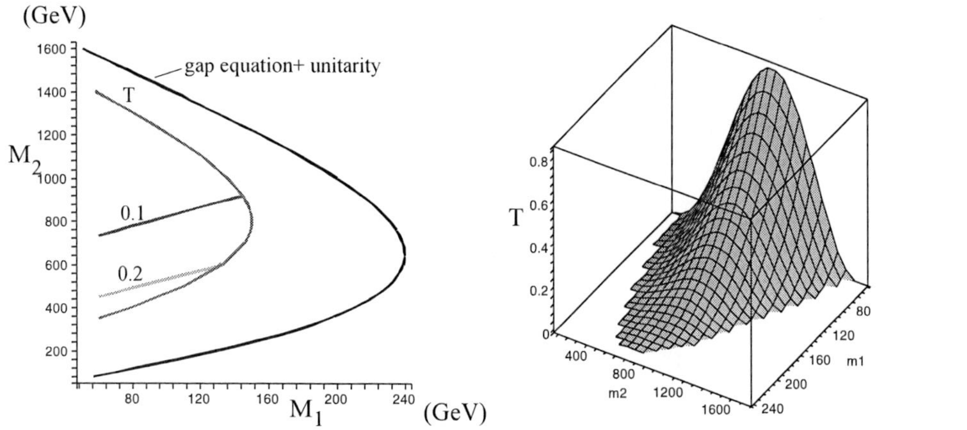

Figure 9. Constraints on the (M_1, M_2) plane. The solution of the gap equations respecting perturbative unitarity are inside the outer curve. The inner curve shows the region, where the T parameter gives the maximum value of c^2 at 95 % C.L.. Below the 0.1 and 0.2 line c^2 can exceed 0.1 and 0.2. The right panel shows the maximum value of T vs. (M_1, M_2).

symmetry of the original Lagrangian (2).

The Drell-Yan cross section for the above hadronic collisions can be written as

$$\sigma(p\overset{(-)}{p} \to D^+D^- + X) = \int_{\tau_0}^{1} d\tau \int_{\tau}^{1} \frac{dx}{2x} \sum_i \sigma(q_i\bar{q}_i \to D^+D^-) \cdot$$
$$\left(f_i^1(x,\hat{s}) f_{\bar{i}}^2(\tau/x,\hat{s}) + f_{\bar{i}}^1(x,\hat{s}) f_i^2(\tau/x,\hat{s})\right), \quad (62)$$

where x and τ/x are the parton momentum fractions, $\hat{s} = \tau s$ is the square of the centre of mass energy of $q_i\bar{q}_i$, s is the same for the hadronic initial state, $f_i^1(x,\hat{s})$ means the number distribution of i quarks in hadron 1 at the scale $\hat{s}$ and the sum runs over the quark fl vours u,d,s,c. In the computation the MSTW parton distribution functions [51] were used.

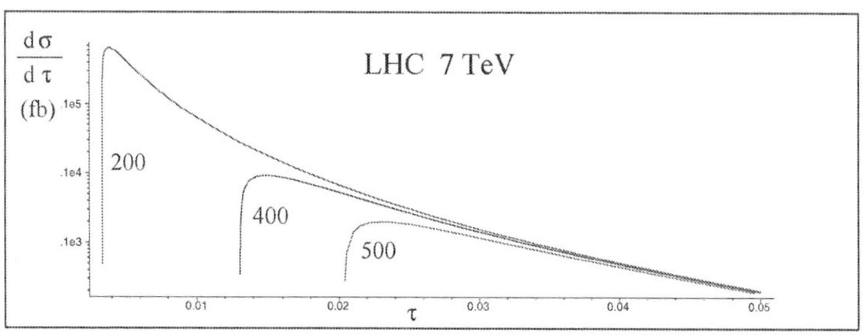

Figure 10. The differential Drell-Yan production cross section of $\Psi_D^-\Psi_D^+$ at the 7 TeV LHC.

The angle integrated, colour averaged annihilation cross section $\sigma(q_i\bar{q}_i \to D^+D^-)$ is calculated at the lowest order in the gauge couplings, and QCD corrections are neglected. We hope this approximation shows the order of magnitude of the cross section. We give the result of the charged fina state as there is no unknown mixing angle in the estimates. The D^+D^- pairs appear via $\gamma + Z$ exchange, the relevant interactions are in (2). The cross section at the parton level is similar to the $\sigma(q_i\overline{q}_i \to \mu^+\mu^-)$ cross section with increased masses (case of a fourth family heavy lepton), see Figure 10.

The total cross sections for different masses are shown in Table 2, and the expected number of events are very low at the delivered integrated luminosity 35 pb $^{-1}$.

Table 2. Total production Drell-Yan cross section of D^+D^- at the 7 TeV LHC

M_+(GeV)	200	400	500
σ (fb)	215	9.3	2.6

The new charged fermion $D^+(\Psi_D^+)$ may leave a charged track or a misplaced vertex if it decays in very short time to the lighter neutral new fermion described by Ψ_1. Finally the lighter neutral fermion expected to disappear leaving back missing energy and momentum, making it difficul to select this model from other sources of dark matter candidates. If new vector-like fermions can mix with the standard femions and decay to standard particles one can search the new particles in jetmass distributions [52] and can cope with the huge background. We expect a higher yield at the 10-14 TeV LHC with the high design luminosity.

A cleaner signal is expected at the next generation of linear collider.To test the model at the forthcoming accelerators we consider the productions of new fermion pairs in electron-positron annihilation. It is most useful to investigate the case of a charged new fermion pair, we denote this D^+D^-.

The contact graph from (10) yields the cross section

$$\sigma\left(e^+e^- \to D^+D^-\right) = \frac{g_e^2}{16\pi} s\sqrt{1 - 4\frac{m_+^2}{s}}\left(1 - \frac{5}{2}\frac{m_+^2}{s}\right), \tag{63}$$

where s is the centre of mass energy squared. The cross section is negligible at moderate s. For example at $h \sim (2TeV)^{-4}$, $\sqrt{s} = 1$TeV it is still at the order of 10^{-13} fb.

We expect a higher number of events from the photon and Z exchange processes $e^+e^- \to \gamma, Z \to D^+D^-$. The usual Standard Model coupling at the e^+e^-Z vertex is

$$i\frac{g}{2\cos\theta_W}\gamma_\mu\left(g_V + \gamma_5 g_A\right), \text{ where } \quad g_V = -\frac{1}{2} + 2\sin^2\theta_W, \; g_A = -\frac{1}{2}. \tag{64}$$

By making use of (42) one obtains the cross section

$$\sigma\left(e^+e^- \to D^+D^-\right) = \frac{1}{16\pi}\sqrt{1 - 4\frac{m_+^2}{s}}\frac{1}{s}\left|M\right|^2 \quad ,$$

$$\begin{aligned} |M|^2 \quad = \quad & \frac{4}{3}e^4\frac{s + 2m_+^2}{s} + \frac{2}{3}\frac{e^4}{\sin^2\theta_W\cos^2\theta_W}g_V\frac{s + 2m_+^2}{s - m_Z^2} + \\ & + \frac{1}{12}\frac{e^4}{\sin^4\theta_W\cos^4\theta_W}\left(g_V^2 + g_A^2\right)s\frac{s + 2m_+^2}{\left(s - m_Z^2\right)^2}, \end{aligned} \tag{65}$$

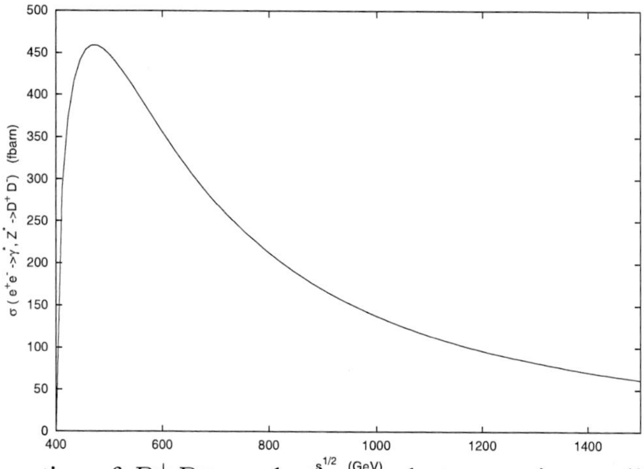

Figure 11. Cross section of D^+D^- production at electron-positron collider vs. $\sqrt{s}$ for $m_+ = 200$GeV.

where the three terms in $|M|^2$ are coming from photon exchange , photon-Z interference and pure Z exchange. Similar cross section belongs to the neutral pair productions, too. The cross section rises fast after the threshold, at high energies it falls off as $1/s$ reflectin that all the interactions are renormalizable in the process. The cross section is given in

Table 3. Cross section of D^+D^- production at $\sqrt{s}$ =500 GeV

m_+(GeV)	100	150	200
$\sigma\,(e^+e^- \to D^+D^-)$ (fb)	560	535	450

Table 3. for a few masses and plotted versus $\sqrt{s}$ in Fig. 11. for $M_+ = 200$ GeV. At a linear collider of $\sqrt{s} = 500$GeV (TESLA) and integrated luminosity 50 fb^{-1}/year a large number of events is expected.

Table 4. Cross section of D^+D^- production at $\sqrt{s}$ =1500 GeV

m_+(GeV)	100	200	400	700
$\sigma\,(e^+e^- \to D^+D^-)$ (fb)	62	61	60	32

The cross section at $\sqrt{s} = 1500$GeV is an order of magnitude smaller (see Table 4.) but with an integrated luminosity of 100 fb^{-1} per annum a large number of events appears and higher mass range can be searched for.

9. Conclusion

In this chapter we have investigated a new dynamical symmetry breaking of the electroweak symmetry based on four-fermion interactions of new hypothetical doublet and singlet vector-like fermions. Four-fermion interactions are postulated involving the standard and the new fermions. Gap equations were derived and we have found the conditions for dynamical symmetry breaking, in the vacuum diagonal and neutral non-diagonal condensates are formed. The lightest new particle is neutral and perturbative unitarity sets an upper bound for its mass depending on the cutoff. This particle is an ideal dark matter candidate. In the low energy effective theory limit the Higgs is a composite particle. The S and T oblique parameters were calculated and presented. The solutions of the gap equations provide masses that are always in the experimental window of the S parameter. The T parameter measures the deviation from custodial symmetry and constrains the mixing. The experimental data gives an upper bound for the mixing angle, but there is always a room for this type of new physics. This alternative of the Standard Model nicely accommodates a composite heavy Higgs in the precision electroweak test of the Standard Model. The vector-like quarks can easily compensate the negative contribution of a heavy Higgs invalidating the light Higgs preference of the present precision tests. We have presented the Drell-Yan cross section for the production of the new charged fermion at the 7 TeV LHC, the expected number of events is rather small with the 35 pb^{-1} luminosity delivered in 2010. The cross sections for linear electron-positron colliders are higher and are more promising for a potential discovery. Vector-like fermions appear in several beyond the Standard Model researches and can elegantly accomodate a heavy Standard Model like Higgs and provide a competitive dark matter candidate.

Acknowledgment

The authors dedicate this chapter to the late George Pócsik for collaboration on the early phases of this work.

Appendix A. Regularization with Momentum Cutoff

There are low energy theories, like the fermion condensate model, which have an intrinsic cutoff, i.e. the upper bound of the model. The naive calculation of divergent Feynman graphs with a momentum cutoff is thought to break continuous symmetries of the model. In this case the gauge invariance of the two point function with two different fermion masses in the loops can be reconstructed by subtractions leading to finit ambiguity. To avoid these problems we used dimensional regularization in $d = 4 - 2\epsilon$ and identifie the poles at $d = 2$ with quadratic divergencies while the poles at $d = 4$ with logarithmic divergencies [53]. Carefully calculating the one and two point Passarino-Veltman functions in the two schemes the divergencies are the following in the momentum cutoff regularization

$$4\pi\mu^2\left(\frac{1}{\epsilon - 1} + 1\right) = \Lambda^2, \tag{66}$$

$$\frac{1}{\epsilon} - \gamma_E + \ln\left(4\pi\mu^2\right) + 1 = \ln\Lambda^2, \tag{67}$$

where μ is the mass-scale of dimensional regularization. The finit part of a divergent quantity is define by

$$f_{\text{finite}} = \lim_{\epsilon\to 0}\left[f(\epsilon) - R(1)\left(\frac{1}{\epsilon - 1} + 1\right) - R(0)\left(\frac{1}{\epsilon} - \gamma_E + \ln 4\pi + 1\right)\right], \tag{68}$$

where $R(1)$, are the residues of the poles at $\epsilon = 1,\ 0$ respectively.

We have found that contrary to the expectations the ambiguity of the cutoff regularization scheme is coming from the replacement of

$$l_\mu l_\nu \rightarrow g_{\mu\nu} l^2/4 \tag{69}$$

and not from shifting the loop-momentum (l) .

In [54] we have worked out a symmetry preserving regularization in four dimensions. The key point is that tracing and divergent integration are not commutative. Under divergent integrals regulated by momentum cutoff in the new method the following identificatio will respect gauge and Lorentz symmetry during the calculation

$$\int_{\Lambda\, reg} d^4 l_E \frac{l_{E\mu} l_{E\nu}}{(l_E^2 + m^2)^{n+1}} := \frac{1}{2n} g_{\mu\nu}^{(E)} \int_{\Lambda\, reg} d^4 l_E \frac{1}{(l_E^2 + m^2)^n}, \qquad n = 1, 2, ... \tag{70}$$

This identificatio is Lorentz invariant, in gauge theories (70) guarantees the validity of the Slavnov-Taylor identities. It is shown in [55] that the ABJ triangle anomaly involving one γ_5 can be correctly calculated with this regularization.

References

[1] K. Nakamura *et al* (Particle Data Group) 2010 *J. Phys. G.: Nucl. Part. Phys.* **37** 0075021 .

[2] CDF and D0 Collaboration, *Phys. Rev. Lett.* **104** (2010) 061802, updated in arXiv:1007.4587 [hep-ex].

[3] M. Goebel, PoS *ICHEP2010* (2010) 570, arXiv:1012.1331 [hep-ph].

[4] S.Weinberg, *Phys. Rev. D* **13**, 974 (1976).

[5] S.Weinberg, *Phys. Rev. D* **19**, 1277 (1979).

[6] L.Susskind, *Phys. Rev. D* **20**, 2619 (1979).

[7] C. T. Hill and E. H. Simmons, *Phys. Rept.* **381** (2003) 235, [Erratum-ibid. **390** (2004) 553].

[8] E. Eichten and K. D. Lane, *Phys. Lett. B* **90** (1980) 125.

[9] S. Dimopoulos and L. Susskind, *Nucl. Phys. B* **155** (1979) 237.

[10] B. Holdom, *Phys. Rev. D* **24** (1981) 1441.

[11] K. Yamawaki, M. Bando and K. Matumoto, *Phys. Rev. Lett.* **56** (1986) 1335

[12] T. Appelquist, G. T. Fleming and E. T. Neil, *Phys. Rev. Lett.* **100** (2008) 171607.

[13] F. Sannino and K. Tuominen, *Phys. Rev. D* **71** (2005) 051901.

[14] W. A. Bardeen,C.T. Hill and M. Lindner, *Phys.Rev. D* **41** 1647 (1990).

[15] C.T. Hill, *Phys.Lett. B* **266**, 419 (1991).

[16] H. C. Cheng, B. A. Dobrescu and C. T. Hill, *Nucl. Phys. B* **589** (2000) 249.

[17] Y. Bai, M. Carena and E. Ponton, *Phys. Rev. D* **81** (2010) 065004.

[18] N. Arkani-Hamed, A.G. Cohen and H. Georgi, *Phys.Lett.* **B513**, 232 (2001).

[19] L. Giusti, A. Romanino and A. Strumia, *Nucl. Phys. B* **550** (1999) 3.

[20] N. Arkani-Hamed, A.G. Cohen, T. Gregoire and J.G. Wacker, *JHEP* **0208**, 020 (2002).

[21] N. Arkani-Hamed, A.G. Cohen, E. Katz, A.E. Nelson,T. Gregoire, Jay G. Wacker, *JHEP* **0208**, 021 (2002).

[22] H. Georgi and D. Pais, *Phys. Rev.* **D10** 539 (1974), ibid **D12** 508 (1975).

[23] D. B. Kaplan, H. Georgi and S. Dimopoulos, *Phys. Lett. B* **136**, 187 (1984).

[24] K. Agashe, R. Contino and A. Pomarol, *Nucl. Phys. B* **719** (2005) 165.

[25] C. Csaki, C. Grojean, H. Murayama, L. Pilo and J. Terning, *Phys. Rev. D*46 (1992) 381.

[26] G. Cynolter, E. Lendvai and G. Pócsik, *Eur. Phys. J.* **C46**: 545 (2006).

[27] T. Appelquist, H-C. Cheng, B. A. Dobrescu, *Phys. Rev. D*64 (2001) 035002.

[28] Riccardo Barbieri, Lawrence J. Hall, Vyacheslav S. Rychkov, *Phys. Rev.* **D74**: 015007 (2006).

[29] R. Enberg, P.J. Fox, L.J. Hall, A.Y. Papaioannou, M. Papucci, *JHEP* **0711**: 014 (2007); Rakhi Mahbubani, Leonardo Senatore, *Phys. Rev.* **D73**: 043510 (2006).

[30] F. D'Eramo, *Phys.Rev.***D76**:083522 (2007).

[31] W.A. Bardeen, C.T. Hill and M. Lindner, *Phys.Rev. D*41 1647 (1990); C.T. Hill, *Phys.Lett. B266*, 419 (1991); M. Lindner and D. Ross, *Nucl.Phys. B*370, 30 (1992); Bogdan A. Dobrescu and Christopher T. Hill, *Phys. Rev. Lett.* **81**, 2634 (1998).

[32] Y. Nambu and G. Jona-Lasinio, *Phys. Rev.* **122**, 345 (1961); Y. Nambu and G. Jona-Lasinio, *Phys. Rev.* **124**, 246 (1961).

[33] G. Cynolter, E. Lendvai and G. Pocsik, Eur. *Phys. J. C*38, 247 (2004).

[34] C. P. Hays, L. Bruchers, R. Santos, A. Gutierrez-Rodriguez, M. S. Berger, G. Cynolter and H. N. Long, "Search for the Higgs Boson," ISBN 1-59454-861-7, 2006.

[35] M. E. Peskin and T. Takeuchi, *Phys. Rev. D* **46,** 381 (1992).

[36] N. Maekawa, *Phys. Rev. D* **52** (1995) 1684.

[37] N. Maekawa, *Prog. Theor. Phys.* **93** (1995) 919.

[38] G. Cynolter, E. Lendvai, *J. Phys G* **34**, 1711 (2007).

[39] P. Sikivie *et al.*, *Nucl. Phys. B* **173** 189, (1980).

[40] S. P. Klevansky, *Rev. Mod. Phys.* **64**, No. 3 (1992).

[41] B.Lee, C.Quigg and H.Thacker, *Phys. Rev. D* **16**, 1519 (1977); D.Dicus and V.Mathur, *Phys. Rev. D* **7**, 3111 (1973).

[42] G. Cynolter, A. Bodor and G. Pocsik, *Heavy Ion Phys.* **7**, 245 (1998).

[43] T. Appelquist, Michael S. Chanowitz, *Phys. Rev. Lett.* **59**, 2405 (1987), Erratum-ibid. **60**, 1589 (1988).

[44] G. Pocsik, E. Lendvai and G. Cynolter, *Acta Phys. Polon. B* **24** (1993) 1495.

[45] R. Barbieri, A. Pomarol, R. Rattazzi and A. Strumia, *Nucl. Phys. B* **703**, 127 (2004).

[46] LEP Electroweak Working Group homepage, http://lepewwg.web.cern.ch/LEPEWWG.

[47] G. Cynolter, E. Lendvai and G. Pocsik, *Mod. Phys. Lett. A* **24** (2009) 2331.

[48] I. Maksymyk, C.P. Burgess and David London, *Phys.Rev. D***50**, 529 (1994); G. Altarelli, R. Barbieri and S. Jadach, *Nucl. Phys. B***369** 3 (1992).

[49] G. Cynolter and E. Lendvai, *Eur. Phys. J. C* **58**, 463 (2008).

[50] L. Lavoura and J. P. Silva, *Phys. Rev. D***47**, 2046 (1993).

[51] A. D. Martin, W. J. Stirling, R. S. Thorne and G. Watt, *Eur. Phys. J. C* **63** (2009) 189.

[52] Witold Skiba, David Tucker-Smith, *Phys.Rev. D***75**:115010 (2007).

[53] K. Hagiwara, S. Ishihara, R. Szalapski and D. Zeppenfeld, *Phys. Rev. D* **48** (1993) 2182.

[54] G. Cynolter and E. Lendvai, *CEJP* DOI: 10.2478/s11534-011-0039-y, [arXiv:1002.4490 [hep-ph]].

[55] G. Cynolter and E. Lendvai, accepted for pulication in *Mod. Phys. Lett. A arXiv:1012.4648 [hep-ph].*

In: New Developments in the Standard Model
Editor: Ryan J. Larsen

ISBN: 978-1-61209-989-7

Chapter 6

LEPTON FLAVOR VIOLATION SHEDDING LIGHT ON CP-VIOLATION

Yasaman Farzan
School of Physics, Institute for Research in Fundamental Sciences (IPM), Tehran, Iran

Abstract

Search for Lepton Flavor Violating (LFV) rare decay $\mu \to e\gamma$ has played a key role in forming the standard model. Null result for searches was a hint for the fact that more than one type of neutrinos exist and ν_e and ν_μ are two distinct particles. Search for $\mu \to e\gamma$ is still a powerful tool to look for unknown physics. In fact, the present bound on Br($\mu \to e\gamma$) already probes energy scales that is beyond the reach of the LHC. The MEG experiment at PSI of Switzerland is currently collecting data to probe values of Br($\mu \to e\gamma$) two orders of magnitude below the present bound. If Br ($\mu \to e\gamma$) is close to the present bound, the MEG collaboration will enjoy collecting large amount of data. Considering that decaying muons are almost 100 polarized, it would be possible to study the angular distribution of the fina particles and derive information on the parity structure of the underlying theory. Recently, it is shown that by measuring the polarization of the fina particles in $\mu \to e\gamma$ as well as in other lepton fl vor violating processes such as $\mu \to eee$ or $\mu - e$ conversion on nuclei, one can derive information on the CP-violating parameters of the underlying theory. Remembering that CP-violation is one of the key ingredients in explaining the fundamental question of matter anti-matter asymmetry of the universe, the importance of such a measurement becomes more evident. We review this novel method in this chapter.

1. Introduction

As is well-known, leptons in the Standard Model (SM) come in three generations or fl vors: (a) firs generation: $(\nu_{eL}\ e_L)$, e_R; (b) second generation: $(\nu_{\mu L}\ \mu_L)$, μ_R; (c) third generation: $(\nu_{\tau L}\ \tau_L)$, τ_R. Electron, being the lightest charged particle, is strictly stable within the SM and its mainstream extensions that preserve the electric charge. In fact, experimental searches set a very stringent lower bound on the lifetime of the electron: $\tau_e \gtrsim 10^{26}$ years [1]. However, μ and τ, being heavier, decay with lifetime respectively equal to 2.2×10^{-6} sec and 3×10^{-13} sec [2].

After a series of experiments, in 1896 J. J. Thomson and his collaborators identifie the particle that we know today as the electron and established the ratio of its mass to its electric charge [3]. Muon was firs detected in 1936 while studying the cosmic ray and subsequently confirme in 1937 [4]. During 1974 to 1977, the tau lepton was discovered by Martin Perl at SLAC.

Within the SM, only left-handed neutrinos exist. In the following, we shall collectively show the left-handed doublets as $L_\alpha = (\nu_{L\alpha}\; l_{L\alpha})$ where $\alpha = e, \mu, \tau$. Similarly, we show the right-handed charged leptons collectively as $l_{R\alpha}$. The leptons do not participate in the strong interactions. The gauge interaction terms of the leptons are listed below.
Electromagnetic interaction:

$$\sum_{\alpha\in\{e,\mu,\tau\}} eA_\mu \bar{l}_\alpha \gamma^\mu l_\alpha = \sum_{\alpha\in\{e,\mu,\tau\}} eA_\mu \left[\bar{l}_{L\alpha}\gamma^\mu l_{L\alpha} + \bar{l}_{R\alpha}\gamma^\mu l_{R\alpha}\right] , \tag{1}$$

W-boson interaction:

$$\sum_{\alpha\in\{e,\mu,\tau\}} \frac{e}{\sqrt{2}\sin\theta_W} \left[W_\mu^+ \bar{\nu}_{L\alpha}\gamma^\mu l_{L\alpha} + W_\mu^- \bar{l}_{L\alpha}\gamma^\mu \nu_{L\alpha}\right] \tag{2}$$

Z-boson interaction:

$$\frac{e}{\cos\theta_W \sin\theta_W} \sum_{\alpha\in\{e,\mu,\tau\}} Z_\mu \left[a_L^e \bar{l}_{L\alpha}\gamma^\mu l_{L\alpha} + a_R^e \bar{l}_{R\alpha}\gamma^\mu l_{R\alpha} + a_L^\nu \bar{\nu}_{L\alpha}\gamma^\mu \nu_{L\alpha}\right] \tag{3}$$

in which $a_L^e = -1/2 + \sin^2\theta_W$, $a_R^e = \sin^2\theta_W$ and $a_L^\nu = 1/2$. The Yukawa interaction has to be added to these interactions:

$$\frac{m_\alpha}{v} \bar{l}_{R\alpha} \Phi \cdot L_\alpha \, . \tag{4}$$

After the electroweak symmetry breaking, this term gives mass to the charged leptons; however, within the "old" SM, neutrinos remain massless. The remarkable point is that none of the above terms mix the generations. That is all the terms are fl vor diagonal. As a result, within the old SM, Lepton Flavor Violating (LFV) processes such as $\mu \to e\gamma$, $\mu \to eee$ or $\tau \to eee$ are strictly forbidden. In fact, observation of neutrino oscillation ($\nu_\alpha \to \nu_\beta$ where $\alpha \neq \beta$) shows that neutrinos have to be massive and their mass terms have to violate lepton fl vor to explain the data. The charged leptons, electron (muon and tau-lepton) being mass eigenstates, do not oscillate into each other [5]. If the only source of LFV is the neutrino mass, the rate of LFV processes such as $\mu \to e\gamma$ will be extremely small and below the reach of the experiments in the foreseeable future [6]. Any signal for the LFV processes such as $\mu \to e\gamma$ will be the herald of new physics. The main focus of this chapter is to study what kind of information can be derived from such observation.

Lepton fl vor conserving observable quantities are also sensitive probes of new physics. Examples include the electric and magnetic dipole moments of the charged leptons, the muon lifetime as well as the detailed distributions of the fina particles in the main decay mode of the muon (*i.e.,* the Michel mode). We will briefl review these possibilities.

2. Muon Lifetime and Fermi Constant

Heavy particles with electroweak interactions can affect low energy observable quantities through loop effects. As a result, by measuring these quantities with precision better than a few percent, new physics can be probed. In fact, even before producing the top quark at Tevatron, through precision measurement methods, the top mass was determined with 30 % accuracy (*see e.g.,* [7]). Through a similar method an upper bound is set on the SM Higgs mass which is $m_h < 165$ GeV at 95 % [7]. The method employs parameters known as oblique parameters which were firs introduced to test the technicolor model [8] which is now ruled out by the electroweak precision data. One of the major inputs in the precision analysis and in calculating the oblique parameters is the Fermi constant which can be extracted from the muon lifetime. Taking into account the quantum electrodynamic correction, the muon decay rate is given by [9, 10]

$$\Gamma_\mu = \frac{G_F^2 m_\mu^5}{192\pi^3} F(\frac{m_e^2}{m_\mu^2}) \left(1 + \frac{3}{5}\frac{m_\mu^2}{m_W^2}\right) \left(1 + \frac{\alpha(m_\mu)}{2\pi}(\frac{25}{4} - \pi^2)\right) , \tag{5}$$

where $F(x) = 1 - 8x + 8x^3 - x^4 - 12x^2 \log x$. Thus, inserting the measured value of the decay rate [2],

$$\Gamma_\mu^{-1} = \tau_\mu = (2.197034 \pm 0.000021) \times 10^{-6} \text{ sec}$$

and the values of m_e and m_μ which are known with very high accuracy, G_F can be extracted

$$G_F = 1.16637(1) \times 10^{-5} \text{ GeV}^{-2} . \tag{6}$$

Notice that unlike the hadron sector no mixing parameter appears in the decay rates of the charged leptons so measuring the muon decay rate readily gives G_F. G_F measured by this method can be also used as an input to derive mixing parameter V_{ud} in the CKM matrix from the neutron decay rate.

3. Dipole Moments of the Charged Leptons

Using non-relativistic approximation, it is straightforward to show that an elementary charged Dirac fiel with electric charge of q, spin of $\vec{s}$ and mass of M has a magnetic dipole moment equal to $q\vec{s}/M$ at the tree level. The magnetic dipole moment is usually expressed as

$$\vec{\mu} = g\frac{q}{2M}\vec{s}$$

where at tree level $g = 2$. Loop correction slightly shift g from 2. The magnetic dipole moments of both the electron and the muon are measured with breath-takingly high accuracy [2]. The precise value of the electron magnetic dipole moment is used to extract $\alpha = e^2/4\pi$ [11] which is in turn an input for extracting the oblique parameters discussed in the previous section.

Although the precision of the electron magnetic dipole moment is higher, the muon magnetic dipole moment is usually more sensitive to the correction from new physics. The reason is as follows. The dipole moments is a chirality flippin quantity so if the chirality breaking parameters are all given by the fermion mass (as in the case of the SM and most of its extensions), the loop correction of new physics to dipole moment will be proportional to M. Since the muon is heavier than the electron, its magnetic dipole moment is more sensitive to new physics.

According to the most recent measurement of the muon magnetic moment in the Brookhaven national lab [12, 2]

$$\frac{g-2}{2} = (11659209 \pm 6) \times 10^{-10} .$$

The latest prediction of the SM, using the state-of-the-art techniques especially in calculating the hadronic vacuum polarization, shows that there is a small but non-negligible discrepancy between the SM prediction and the observed quantity [12, 13, 14]

$$\frac{g_\mu^{\rm obs.} - g_\mu^{\rm SM}}{2} = (246 \pm 80) \times 10^{-11} .$$

A wide range of models have been advocated as the explanation for this discrepancy. In particular, the supersymmetric models are extensively studied in this regard. For a recent review, see [14].

It is worth remembering that the electric dipole moments of the charged leptons (d_e, d_μ and d_τ) are also of interest in the quest for new physics. In fact, so far only upper bounds have been derived on the Electric Dipole Moments (EDMs) of the elementary particles [2]. Notice that while magnetic dipole moments preserve CP, nonzero EDMs for elementary particle with spin $\vec{s}$ violate CP. In other words, under time reversal, the potential term for magnetic dipole moment ($\vec{s} \cdot \vec{B}$) remains invariant but the term for EDM ($\vec{s} \cdot \vec{E}$) changes sign. Searches for the electric dipole moments are still in progress. The prediction of the SM for EDMs is far below the sensitivity of experiments in any foreseeable future [15]. New physics can introduce new sources of CP-violation leading to sizeable d_e or d_μ. If the future searches report a nonzero result, it will be a strong evidence for new physics. In this regard, correlation between LFV and CP-violating parameters will be a valuable source of information on the underlying physics [16].

4. New Physics in Charged Lepton Decay

The main decay mode of the muon is $\mu \to e\bar{\nu}_e\nu_\mu$. This decay mode is called the normal or Michel decay mode. The branching ratio of this decay mode is about 99 %. So far two additional decay modes are observed [2]: Br($\mu^+ \to e^+\nu_e\bar{\nu}_\mu\gamma$) $= (1.4 \pm 0.4)\%$ and Br($\mu^+ \to e^+\nu_e\bar{\nu}_\mu e^- e^+$) $= (3.4 \pm 0.4) \times 10^{-5}$. The $\mu \to e\nu\nu\gamma$ decay mode is called the radiative decay mode because a photon is "radiated" from the muon decay. Notice that the "radiative decay mode" can take place at the tree level. All these three decay modes are allowed within the SM.

It is possible to measure the energy spectrum of the emitted e^+ in the Michel decay mode of μ^+. If the initial μ^+ is polarized, it will be possible to study the angular distribution of e^+ relative to the polarization of μ^+. It is also feasible to measure the polarization of the fina positron. These distributions are sensitive to the form of the effective Lagrangian leading to the muon decay. All these distributions have been experimentally extracted.

These measurements have played a pivotal role in determining the $V - A$ form of the interaction we know today [10].

Let us take the most general effective Lagrangian leading to $\mu \to e\nu\nu$ which results from integrating out the heavy states [10, 17]

$$\begin{aligned}
\mathcal{L}_{\mu\to e\nu\bar{\nu}} &= -\frac{4G_F}{\sqrt{2}}\Big[g^S_{RR}(\bar{e}_R\nu_{eL})(\bar{\nu}_{\mu L}\mu_R) + g^S_{RL}(\bar{e}_R\nu_{eL})(\bar{\nu}_{\mu R}\mu_L) \\
&+ g^S_{LR}(\bar{e}_L\nu_{eR})(\bar{\nu}_{\mu L}\mu_R) + g^S_{LL}(\bar{e}_L\nu_{eR})(\bar{\nu}_{\mu R}\mu_L) \\
&+ g^V_{RR}(\bar{e}_R\gamma^\mu\nu_{eR})(\bar{\nu}_{\mu R}\gamma_\mu\mu_R) + g^V_{RL}(\bar{e}_R\gamma^\mu\nu_{eR})(\bar{\nu}_{\mu L}\gamma_\mu\mu_L) \\
&+ g^V_{LR}(\bar{e}_L\gamma^\mu\nu_{eL})(\bar{\nu}_{\mu R}\gamma_\mu\mu_R) + g^V_{LL}(\bar{e}_L\gamma^\mu\nu_{eL})(\bar{\nu}_{\mu L}\gamma_\mu\mu_L) \\
&+ \frac{g^T_{RL}}{2}(\bar{e}_R\sigma^{\mu\nu}\nu_{eL})(\bar{\nu}_{\mu R}\sigma_{\mu\nu}\mu_L) + \frac{g^T_{LR}}{2}(\bar{e}_L\sigma^{\mu\nu}\nu_{eR})(\bar{\nu}_{\mu L}\sigma_{\mu\nu}\mu_R) + \text{H.c.}\Big]\,. \qquad (7)
\end{aligned}$$

In the above formula, we have used the same notation as that in [10]. Although in the SM only left-handed neutrinos exist, the general Lagrangian in Eq. (7) also includes possible interactions of right-handed neutrinos. No right-handed neutrino has been so far detected. However, if right-handed neutrinos lighter than m_μ exist and couple to μ and e as in Eq. (7), they can be emitted in the muon decay. Remember that in the experiments devoted to studying $\mu \to e\nu\nu$, neutrinos are not directly detected and regardless of being right-handed or left-handed, they appear as missing energy. Integrating out the SM W^+ boson yields only the g^V_{LL} term but going beyond the SM, other terms in Eq. (7) can also appear. Taking the general effective Lagrangian, it has been shown that after summing over the polarization of the fina charged lepton, the partial decay rate of the muon in its rest frame to the leading order in m_e/m_μ is given by the following formula [10]

$$\frac{d^2\Gamma(\mu^\pm \to e^\pm\nu\nu)}{dE\, d\cos\theta} = \frac{G_F^2 m_\mu^3}{32\pi^3}\sqrt{E^2 - m_e^2}\left(F_{IS}(\frac{2E}{m_\mu}) \pm \mathbb{P}_\mu \cos\theta F_{AS}(\frac{2E}{m_\mu})\right) \qquad (8)$$

where E is the energy of the fina charged lepton and θ is the angle between its momentum and the polarization of the initial muon and

$$F_{IS}(x) = x(1-x) + \frac{2}{9}\rho(4x^2 - 3x - x_0^2) + \eta x_0(1-x) \qquad (9)$$

and

$$F_{AS}(x) = \frac{1}{3}\xi\sqrt{x^2 - x_0^2}\left(1 - x + \frac{2}{3}(4x-3)\delta\right) \qquad (10)$$

where $x_0 = 2m_e/m_\mu$. Parameters ρ, η, ξ and δ appearing in F_{IS} and F_{AS} are the famous Michel parameters [18] whose numerical values depend on the parameters of the effective Lagrangian leading to $\mu \to e\nu\nu$ [1]. It is remarkable that regardless of the form of the effective Lagrangian, the shape of the spectrum is given by Eq. (8). The details are only encoded in the numerical values of the Michel parameters. The predictions of the SM are $\rho = \delta = 3/4$, $\eta = 0$ and $\xi = 1$ which agree with the observation. Any deviation would indicate physics beyond the SM.

If, in addition to the spectrum and the angular distribution of the fina charged lepton, its polarization is also measured, more information can be extracted about the parameters of Eq. (7). To describe the polarization of the fina $e^\pm$, f ve "new" Michel parameters are

[1]Notice that although the Michel parameter ρ has the same symbol as the famous ρ parameter (*i.e.*, $m_W^2/m_Z^2\cos^2\theta_W$), their definition as well as numerical values are quite different.

introduced [19]. Let us take the momentum of the emitted charged lepton to be $\vec{P}_e$ and the polarization of the muon to be $\vec{\mathbb{P}}_\mu$. The polarizations of e in all three directions $\hat{z} = \vec{P}_e/|\vec{P}_e|$, $\hat{y} = \hat{z} \times \vec{\mathbb{P}}_\mu/|\hat{z} \times \vec{\mathbb{P}}_\mu|$ and $\hat{x} \equiv \hat{y} \times \hat{z}$ contain independent and new information. Notice that under the time reversal $\hat{z}$ and $\hat{x}$ as well as the polarization change direction but $\hat{y}$ does not. As a result the transverse polarization of e in the $\hat{y}$ direction is a measure of CP or T-violation. The nine Michel parameters (four old plus f ve new parameters) combined with the measured value of the muon lifetime provide information on the couplings in the effective Lagrangian. If we drop the terms involving the right-handed neutrinos (keeping only the g^S_{RR} and g^V_{LL} couplings), these 10 observable quantities over-constrain the effective Lagrangian. However, if we involve the right-handed neutrinos, there will be more parameters in the effective Lagrangian than there are observable quantities in the muon decay. The radiative decay mode can in principle provide more information [10].

In the above discussion, it is implicitly assumed that all the "missing energy"in the muon decay is carried away by a pair of neutrinos. Suppose that a hypothetical light neutral particle such as Majoron or SLIM (introduced in [20]) is emitted along the neutrinos. The above formalism is not suitable to account for such a possibility. Nevertheless, emission of an extra neutral particle will leave its imprint in the spectrum of the fina charged particle. In particle the endpoint of the spectrum of the fina lepton will be shifted from $(m_\mu^2 + m_e^2)/(2m_\mu)$ to

$$\mathrm{Max}\left[\frac{m_\mu^2 + m_e^2 - m_S^2}{2m_\mu}, \frac{(m_\mu - m_S)^2 + m_e^2}{(m_\mu - m_S)}\right]$$

where m_S is the mass of the hypothetical new particle that escapes detection. For $m_\mu - m_S \gg m_e$, the end-point is given by $(m_\mu^2 - m_S^2)/(2m_\mu)$. Such an emission will also affect the total rate of $\mu \to e +$ missing energy [21].

5. Lepton Flavor Violating Processes

Search for lepton fl vor violating rare decay has a history as long as the muon itself [22]. Null results for search for this process has played a crucial role in shaping the SM as we know it today. From the spectrum of the fina charged lepton in $\mu \to e\nu\nu$, it was known that more than one neutral particles have to be emitted along e. As it was discussed in [23], if both of these particles were the same, the process $\mu \to e\gamma$ would be possible at one loop level with a branching ratio of order of 10^{-4}. However, searches did not fin such a decay mode [24]. The null result was an indication for the fact that more than one fl vor of neutrino exists. In 1962 direct detection of ν_μ confirme this hypothesis [25]. This discovery led to the Nobel prize in 1988. Null result for searches for $\mu \to e\gamma$ also indicates that the muon is not "an excited electron;" otherwise like atomic states or heavier baryons made up of u and d quark (such as $N(1440)$ or $N(1535)$), we would expect radiative decay.

Search for $\mu \to e\gamma$ has been continuing since the late 40s [22]. The present bound is [2]

$$\mathrm{Br}(\mu \to e\gamma) < 1.2 \times 10^{-11}\,.$$

Currently the MEG experiment at PSI is running. This experiment aims at probing $\mathrm{Br}(\mu \to e\gamma)$ as low as $O(10^{-13})$. As discussed in sect. 1., if the only source of LFV is the neutrino mass matrix, Br($\mu \to e\gamma$) will be so small that cannot be probed in the foreseeable future. Within beyond SM theories, new sources for LFV might exist. Let us take the scale of new physics, m_{NEW} and denote the LFV source by a dimensionless parameter as $\tilde{\kappa}_{\alpha\beta}$. Within the MSSM $\tilde{\kappa}_{\alpha\beta}$ might be $\tilde{m}^2_{\alpha\beta}/m^2_{NEW}$ where $\tilde{m}^2_{\alpha\beta}$ is the LFV mass term for sleptons. The order of magnitude of the one loop contribution to $\mu \to e\gamma$ can then be estimated as

$$\mathrm{Br}(\mu \to e\gamma) \sim \frac{\alpha \left(g^e_{\mathrm{NEW}} g^\mu_{\mathrm{NEW}}\right)^2 \tilde{\kappa}^2_{\mu e}}{G_F^2 m^4_{NEW}}, \tag{11}$$

where g^e_{NEW} and g^μ_{NEW} are respectively the couplings of the new particles to the electron and the muon. The present bound Br($\mu \to e\gamma$) $< 1.2 \times 10^{-11}$ leads to the following lower bound on m_{NEW}:

$$m_{NEW} > 50 \left(\tilde{\kappa}_{\mu e} g^e_{\mathrm{NEW}} g^\mu_{\mathrm{NEW}}\right)^{1/2} \text{ TeV}. \tag{12}$$

In other words, for $m_{NEW} \sim 100$ GeV, this bound implies $\tilde{\kappa}_{\mu e} g^e_{\mathrm{NEW}} g^\mu_{\mathrm{NEW}} \lesssim 5 \times 10^{-6}$. If the MEG experiment does not fin any signal and sets the bound Br($\mu \to e\gamma$) $< 10^{-13}$, the lower bound would be $m_{NEW} > 145 \left(\tilde{\kappa}_{\mu e} g^e_{\mathrm{NEW}} g^\mu_{\mathrm{NEW}}\right)^{1/2}$ TeV. For comparison just remember that the center of mass energy at the LHC at its second stage will be only 14 TeV. Probing such high scales will be beyond the reach of the accelerators in the foreseeable future. There is also a strong bound on muon decaying into an electron and a pair of photons: Br($\mu \to e\gamma\gamma$) $< 7.2 \times 10^{-11}$ [2]. Remember that Br($\mu \to e\gamma\gamma$) is suppressed by an extra factor of $\alpha/4\pi$ so in general Br($\mu \to e\gamma$) yields a stronger bound on new physics. The present bound on the LFV mode $\mu^- \to e^- e^+ e^-$ is even stronger Br($\mu^- \to e^- e^+ e^-$) $< 1.0 \times 10^{-12}$. Models can be divided into two classes:

- Models such as R-parity violating MSSM within which both even and odd numbers of new particles can appear in each vertex. In this case, $\mu \to eee$ can take place at tree level so the bound on Br($\mu \to eee$) strongly constrains the LFV parameters of the model.

- Models with a Z_2 symmetry that forbids having odd numbers of new particles at each vertex. Examples are MSSM with R-parity or models in [20]. For these models, $\mu \to eee$ can take place only through a one loop diagram. Being a three-body process, it is typically suppressed by an extra factor of $(\alpha/4\pi)\log(m_\mu^2/m_e^2)$ relative to Br($\mu \to e\gamma$) [27]. As a result, Br($\mu \to e\gamma$) yields a stronger bound.

The set-up of the MEG experiment is such that the initial muon is at rest and the detectors will search for back to back e and γ [26]. As a result, MEG will not unfortunately be able to improve the bound on Br($\mu \to eee$). Probing smaller values of $\mu \to eee$ would be a powerful tool to shed light on new physics. If Br($\mu \to eee$) turns out to be close to the present bound but no signal for $\mu \to e\gamma$ is found, models within which new particles appear only in even numbers would be disfavored.

Another process that has been investigated is $\mu - e$ conversion on nuclei: $\mu^- N \to e^- N$. When a μ^- beam is shot at a target, the muons eventually lose energy and form "muonic" atom with the material in the target. Muons then decay in the background coulomb fiel of the nucleon or go through the "muon capture", $\mu N \to \nu_\mu N'$. If the lepton fl vor is violated, the LFV conversion $\mu N \to eN$ can also take place. Since 1950 [28], experiments have searched for such a process and all so far have reported only null results. Various nuclei have been used: lead, gold, Titanium and etc [29]. Stringent upper bounds have been derived on $\Gamma(\mu^- N \to e^- N)$; *see* bounds and references in [10]. There are also proposals to improve the present bound by several orders of magnitude [30, 31].

New physics can also induce LFV rare decays of the τ lepton. Strong bounds have also been derived on such decays. In particular, the recent data from the B-factories set the following bounds on the radiative LFV decay modes [2]

$$\text{Br}(\tau \to e\gamma) < 3.3 \times 10^{-8} \quad \text{and} \quad \text{Br}(\tau \to \mu\gamma) < 4.4 \times 10^{-8} .$$

There are also strong bounds on the three-body decay modes: Br($\tau \to \mu\mu\mu$), Br($\tau \to e\mu\mu$), Br($\tau \to ee\mu$) and Br($\tau \to eee$) with different combinations of the charges of the fina states [2]. Notice that in the case of τ^-, processes of type $\tau^- N \to \mu^- N$ or $\tau^- N \to e^- N$ are not important because τ^- has a very short lifetime and decays before forming "tauonic" atoms. Instead, since here the mass of the tau lepton is higher, hadronic decays such $\tau \to e\pi$ or $\tau \to eK^0_S$ are kinematically accessible. There are also strong bounds on the branching ratios of such LFV decay modes [2]. In the rest of this chapter, we shall focus only on muon decays; however, one should bear in mind that similar discussions also hold for the tau lepton. If the future superfl vor factory find a signal for LFV decay modes of τ, revisiting these arguments will be imperative.

6. Effective Lepton Flavor Violating Lagrangian

Consider a general beyond SM theory with heavy ($m \gtrsim 100$ GeV) states that lead to low energy LFV processes such as $\mu \to e\gamma$, $\mu \to eee$ and $\mu N \to eN$. After integrating out the heavy states, the low energy effects of the new physics providing the LFV sources can be described by an effective Lagrangian. The general form of the effective Lagrangian responsible for $\mu \to e\gamma$ is

$$\mathcal{L} = \frac{A_R}{m_\mu}\bar{\mu}_R\sigma^{\mu\nu}e_L F_{\mu\nu} + \frac{A_L}{m_\mu}\bar{\mu}_L\sigma^{\mu\nu}e_R F_{\mu\nu} + \frac{A_R^*}{m_\mu}\bar{e}_L\sigma^{\mu\nu}\mu_R F_{\mu\nu} + \frac{A_L^*}{m_\mu}\bar{e}_R\sigma^{\mu\nu}\mu_L F_{\mu\nu} , \qquad (13)$$

where $\sigma^{\mu\nu} = \frac{i}{2}[\gamma^\mu, \gamma^\nu]$ and $F_{\mu\nu}$ is the photon fiel strength: $F_{\mu\nu} = \partial_\mu A_\nu - \partial_\nu A_\mu$. Notice that terms of form $F(q^2)A_\alpha\bar{\mu}\gamma^\alpha e$ cannot lead to $\mu \to e\gamma$ because the Ward identity implies at $q^2 \to 0$, $F(q^2) = 0$. The effective Lagrangian leads to $\Gamma(\mu \to e\gamma) = m_\mu(|A_L|^2 + |A_R|^2)/(4\pi)$. Thus, by measuring Br($\mu \to e\gamma$), the value of $(|A_L|^2 + |A_R|^2)$ can be extracted. However, the difference $|A_L|^2 - |A_R|^2$ which is a measure of parity violation in the underlying theory cannot be extracted from measuring Br($\mu \to e\gamma$) alone.

The couplings in Eq. (13) also lead to $\mu \to eee$ and $\mu - e$ conversion but these two processes receive contributions from other effective terms, too. The additional effective couplings leading to $\mu \to eee$ can be written as

$$\begin{aligned} \mathcal{L}_{\text{eff}} &= B_1(\bar{\mu}_L e_R)(\bar{e}_R e_L) + B_2(\bar{\mu}_R e_L)(\bar{e}_L e_R) + \\ & C_1(\bar{\mu}_L e_R)(\bar{e}_L e_R) + C_2(\bar{\mu}_R e_L)(\bar{e}_R e_L) + \\ & G_1(\bar{\mu}_R\gamma^\nu e_R)(\bar{e}_R\gamma_\nu e_R) + G_2(\bar{\mu}_L\gamma^\nu e_L)(\bar{e}_L\gamma_\nu e_L) + \text{H.c.} \end{aligned} \qquad (14)$$

These terms lead to

$$Br(\mu \to eee) = \frac{1}{32G_F^2}\left[|B_1|^2 + |B_2|^2 + 8(|G_1|^2 + |G_2|^2)\right.$$

$$
\begin{aligned}
&+ \quad \frac{|C_1|^2+|C_2|^2}{2} + 32(4\log\frac{m_\mu^2}{m_e^2}-11)\frac{|A_R|^2+|A_L|^2}{m_\mu^4} \\
&- \quad 64\frac{\mathrm{Re}[A_R G_2^* + A_L G_1^*]}{m_\mu^2} + 32\frac{\mathrm{Re}[A_R B_1^* + A_L B_2^*]}{m_\mu^2}\Bigg] \qquad (15)
\end{aligned}
$$

It is convenient to write the effective Lagrangian leading to $\mu N \to eN$ as

$$
\begin{aligned}
\mathcal{L}_{eff} &= \frac{e}{\sin\theta_W \cos\theta_W m_Z^2}\sum_{q\in\{u,d\}}(H_L\bar{e}_L\gamma^\mu\mu_L + H_R\bar{e}_R\gamma^\mu\mu_R)(Z_L^q\bar{q}_L\gamma_\mu q_L + Z_R^q\bar{q}_R\gamma_\mu q_R) \\
&- \sum_{q\in\{u,d\}}\frac{Q_q e}{p^2}\left(B_L^*\bar{e}_L\gamma_\mu\mu_L + B_R^*\bar{e}_R\gamma_\mu\mu_R + i\frac{A_R^*}{m_\mu}\bar{e}_L\sigma^{\mu\nu}p_\nu\mu_R + i\frac{A_L^*}{m_\mu}\bar{e}_R\sigma^{\mu\nu}p_\nu\mu_L\right)(\bar{q}\gamma_\mu q) \\
&+ \sum_{q\in\{u,d\}} C_L(\bar{\mu}_L e_R)(\bar{q}_L q_R) + C_R(\bar{\mu}_R e_L)(\bar{q}_R q_L) + \text{H.c.} \qquad (16)
\end{aligned}
$$

where $p = p_\mu - p_e$. A_L and A_R are the same couplings appearing in Eq. (13). Notice that the couplings depend on p^2. In particular, the Ward identity implies that for $p^2 \to 0$, $B_L(p^2)$ and $B_R(p^2)$ both vanish. If the γ and Z boson exchange penguin diagrams dominate, the firs two lines will be dominant with $Z^q_{L(R)} = T^3_q - Q_q \sin^2\theta_W$. In the following, we will ignore C_L and C_R.

The above effective couplings lead to

$$\Gamma(\mu N \to eN) = \frac{S}{2}(|K_L|^2 + |K_R|^2)$$

where nuclear form factor is accounted by the numerical factor S [33] and

$$K_R \equiv aH_L + b(A_R^* + B_L^*) \qquad (17)$$

and

$$K_L \equiv aH_R + b(A_L^* + B_R^*) \qquad (18)$$

in which a and b are given by the numbers of protons (Z) and neutrons (N) inside the nucleus:

$$a = \frac{e\left[Z(1/2 - 2\sin^2\theta_W) - N/2\right]}{2m_Z^2 \sin\theta_W \cos\theta_W} \quad \text{and} \quad b = \frac{eZ}{m_\mu^2}\,. \qquad (19)$$

By measuring $\Gamma(\mu N \to eN)$, the combination $|K_L|^2 + |K_R|^2$ can be extracted. In principle, by measuring the conversion rate on various nuclei with different N and Z, independent combinations of the effective couplings can be extracted, opening the possibility of independently extracting the values of $|H_L|^2 + |H_R|^2$, $|A_R + B_L|^2 + |A_L + B_R|^2$ and $Re[H_L(A_R + B_L) + H_R(A_L + B_R)]$. However, in practice this derivation will be quite challenging because although the different nuclei in question have quite different numbers of protons and neutrons, the ratio Z/N and consequently a/b is almost the same for all of them. Thus, changing the nucleus does not help much to resolve degeneracies unless precision measurement is done which in turn would be plagued by uncertainties in the nuclear form factors embedded in the numerical factor S.

The sizes of these effective coupling are determined by the underlying new physics. Depending on the relative sizes of these effective couplings, the ratios of the rates of the different LFV processes will be different. Determining the ratio pattern of the rates of these LFV processes helps us to discriminate between different models. In particular, Br($\mu \to e\gamma$)/Br($\mu \to eee$) gives a clue. For example as shown in [32], this pattern can help us to distinguish between Minimal Supersymmetric SM (MSSM) and Littlest Higgs model with T-parity (LHT).

7. Physics with Polarized Muon Source

At the MEG experiment, muons are produced by decay of charged pions which in turn are produced by shooting protons at a target. The pions are stopped at the target before they decay so at the production they are completely polarized [34]. The muon produced by the pion decay will have energy equal to $(m_\pi^2 + m_\mu^2)/(2m_\pi)$. The produced muons come to rest at a second target where they eventually decay. The set-up is such that muons maintain their polarization so we can practically consider them 100 % polarized at decay [34]. The initial pions which are produced by shooting protons at the firs target can be both π^+ or π^-. However, only μ^+ from the π^+ decay is directed towards the second target. As mentioned in the previous section, μ^- forms bound state with nuclei at the target so it is not suitable for the purpose of studying the decay of the muon at rest.

With the polarized muon source, it is possible to study the angular distribution of the decay products relative to the polarization of the initial muon [35]:

$$\sum_{\vec{s}_{e^+},\vec{s}_\gamma} \frac{d\Gamma[\mu^+(P_{\mu^+}) \to e^+(P_{e^+},\vec{s}_{e^+})\gamma(P_\gamma,\vec{s}_\gamma)]}{d\cos\theta} =$$
$$\frac{m_\mu}{8\pi}\left[|A_L|^2(1+\mathbb{P}_\mu\cos\theta) + |A_R|^2(1-\mathbb{P}_\mu\cos\theta)\right] , \qquad (20)$$

where $\mathbb{P}_\mu$ is the polarization of the initial muon and θ is the angle between the muon polarization and the momentum of the emitted electron. The sum is on the spins of the fina particles. Combining the measured value of the forward-backward asymmetry,

$$R_1 \equiv \frac{\int_0^1 (d\Gamma/d\cos\theta)d\cos\theta - \int_{-1}^0 (d\Gamma/d\cos\theta)d\cos\theta}{\int_{-1}^1 (d\Gamma/d\cos\theta)d\cos\theta} = \frac{|A_L|^2 - |A_R|^2}{|A_L|^2 + |A_R|^2} \qquad (21)$$

with the value of Br($\mu \to e\gamma$), the values of $|A_L|$ and $|A_R|$ can be separately derived. However, after these measurements, the relative phase $\arg[A_L A_R^*]$ still remains unknown.

The angular distributions of the fina particles in the three body decay of the polarized muon ($\mu \to eee$) also contain new information on the effective couplings in Eqs. (13) and (14) (*see* Ref. [10]). Combining the information on the angular and energy distributions of the fina particles, it will be in principle possible to extract $|G_1|^2 - |G_2|^2 + (-|C_1|^2 + |C_2|^2)/16$, $|B_1|^2 - |B_2|^2$, $|A_L|^2 - |A_R|^2$, $\mathrm{Re}[A_L G_2^* - A_R G_1^*]$ and $\mathrm{Re}[A_L B_1^* - A_R B_2^*]$ as well as the CP-odd quantities $\mathrm{Im}[A_L G_2^* + A_R G_1^*]$ and $\mathrm{Im}[A_L B_1^* + A_R B_2^*]$ [38]. However, information on quantities such as $\arg[A_R A_L^*]$ or $\arg[B_1 B_2^*]$ will be still missing.

Let us now discuss the case of $\mu - e$ conversion on nuclei. The polarization of μ^- bound around nuclei is about 16 % or lower [36]. Some methods have however proposed to repolarize the muon [37]. The angular distribution of the emitted electron relative to the polarization of the initial muon is given by the following formula

$$\frac{d\Gamma(\mu N \to eN)}{d\cos\theta} = S\left[\frac{1-\mathbb{P}_\mu\cos\theta}{2}|K_R|^2 + \frac{1+\mathbb{P}_\mu\cos\theta}{2}|K_L|^2\right] , \qquad (22)$$

where K_R and K_L are define in Eqs. (17,18) [33]. Thus, by studying the angular distribution of the fina electron, the absolute values of K_L and K_R can be determined but their relative phase which is a physical CP-violating observable cannot be derived by this method. Similarly to Eq. (21), it is convenient to defin asymmetry R_2 as follows [41]

$$R_2 \equiv \frac{|K_L|^2 - |K_R|^2}{|K_L|^2 + |K_R|^2}. \tag{23}$$

8. Polarization of the Final Particles in the LFV Processes

As discussed in the previous section, studying the angular distribution of the emitted fina particles relative to the polarization of the initial muon provides new information but it is not enough to determine all the physical parameters in the effective Lagrangian. As has been recently shown in a series of papers [38, 39, 40, 41, 42], the transverse polarization of the fina particles provides complementary information.

First, let us comment on the feasibility of measuring the polarization. There are established techniques to measure the polarization of the positrons in the energy range of interest which is based on studying the azimuthal distribution of the photon pair created by annihilation of the positrons on the polarized electrons at rest [43]. As discussed in [38], by similar techniques, the transverse polarization of the electrons can also be measured (*i.e.,* by studying the azimuthal distribution of the scattered electrons on polarized electrons at rest.). Measuring the polarization of the photon is going to be more challenging [44]. In the following, we will discuss what combinations can be derived from the polarizations of the fina particles in $\mu \to e\gamma$, $\mu \to eee$ and $\mu N \to eN$.

8.1. $\mu^+ \to e^+\gamma$

Under the CP transformation, the couplings in Eq. (13) transform as follows

$$A_L \to \eta A_L^* \qquad \text{and} \qquad A_R \to \eta A_R^*$$

where η is a pure phase which can be removed by redefinin the μ and/or e fields As discussed in the previous section, by measuring $\Gamma(\mu \to e\gamma)$ and the angular distribution of the fina particles, only the absolute values of A_L and A_R can be extracted. It is shown in [39] that by measuring the polarization of the fina particles in $\mu \to e\gamma$, the relative phase of A_L and A_R can also be derived. To fulfil this task, it is necessary to simultaneously measure the polarizations of both fina particles. The formulation described in the following demonstrates this possibility.

Let us defin the longitudinal and traverse directions as follows

$$\hat{T}_3 \equiv \frac{\vec{p}_{e^+}}{|\vec{p}_{e^+}|}, \quad \hat{T}_2 \equiv \frac{\vec{p}_{e^+} \times \hat{T}_3}{|\vec{p}_{e^+} \times \hat{T}_3|} \quad \text{and} \quad \hat{T}_1 \equiv \hat{T}_2 \times \hat{T}_3\,. \tag{24}$$

The polarization of the emitted positron in arbitrary direction T is given by

$$\langle s_T \rangle \equiv \frac{\sum_{\vec{s}_\gamma} \left\{ d\Gamma \left[\mu^+ \to e^+(\vec{s}_{e^+} = \frac{1}{2}\hat{T})\gamma(\vec{s}_\gamma) \right] - d\Gamma \left[\mu^+ \to e^+(\vec{s}_{e^+} = -\frac{1}{2}\hat{T})\gamma(\vec{s}_\gamma) \right] \right\}}{\sum_{\vec{s}_\gamma, \vec{s}_{e^+}} d\Gamma \left[\mu^+ \to e^+(\vec{s}_{e^+})\gamma(\vec{s}_\gamma) \right]}. \tag{25}$$

Polarization of the photon is given by

$$\langle P_T \rangle \equiv \frac{\sum_{\vec{s}_{e^+}} d\Gamma \left[\mu^+ \to e^+(\vec{s}_{e^+})\gamma(\vec{\epsilon} \parallel \hat{T})\right]}{\sum_{\vec{s}_\gamma \vec{s}_{e^+}} d\Gamma \left[\mu^+ \to e^+(\vec{s}_{e^+})\gamma(\vec{\epsilon})\right]} \tag{26}$$

in which $\vec{\epsilon}$ is the photon polarization vector. As shown in [41]

$$\langle s_{T_1} \rangle = \langle s_{T_2} \rangle = 0 \quad \langle s_{T_3} \rangle = \frac{|A_R|^2(1-\mathbb{P}_\mu \cos\theta) - |A_L|^2(1+\mathbb{P}_\mu \cos\theta)}{|A_R|^2(1-\mathbb{P}_\mu \cos\theta) + |A_L|^2(1+\mathbb{P}_\mu \cos\theta)}$$

and

$$\langle P_{T_1} \rangle = \langle P_{T_2} \rangle = \frac{1}{2}.$$

Thus, measuring only the spin of one of the fina particles, it will not be possible to derive $\arg[A_L A_R^*]$. Notice that $\langle s_{T_3} \rangle$ is sensitive to $|A_R|/|A_L|$. Thus, to derive $|A_R|/|A_L|$, it is enough to measure $\langle s_{T_3} \rangle$ emitted in an arbitrary direction (*i.e.*, at a given value of θ). This method of deriving $|A_R|/|A_L|$ is an alternative to the method based on measuring the angular distribution of the fina particles in $\mu \to e\gamma$ which we discussed earlier.

However, the double correlation is sensitive to the relative phase:

$$\langle P_{T_1} s_{T_1} \rangle = -\langle P_{T_2} s_{T_1} \rangle = \frac{-\mathbb{P}_\mu \mathrm{Re}[A_L^* A_R] \sin\theta}{|A_R|^2(1-\mathbb{P}_\mu \cos\theta) + |A_L|^2(1+\mathbb{P}_\mu \cos\theta)} \tag{27}$$

and

$$\langle P_{T_1} s_{T_2} \rangle = -\langle P_{T_2} s_{T_2} \rangle = \frac{\mathbb{P}_\mu \mathrm{Im}[A_L^* A_R] \sin\theta}{|A_R|^2(1-\mathbb{P}_\mu \cos\theta) + |A_L|^2(1+\mathbb{P}_\mu \cos\theta)}. \tag{28}$$

Thus,

$$\frac{\langle P_{T_1} s_{T_2} \rangle}{\langle P_{T_1} s_{T_1} \rangle} = \frac{\langle P_{T_2} s_{T_2} \rangle}{\langle P_{T_2} s_{T_1} \rangle} = -\frac{\mathrm{Im}[A_L^* A_R]}{\mathrm{Re}[A_L^* A_R]}.$$

8.2. $\mu^+ \to e^+ e^- e^+$

As shown in [38, 39, 40], the polarization of the fina particles in $\mu \to eee$ also contains information on the CP-violating phases. Two special cases are of particular interest:

Case 1 in which the A_L and A_R couplings dominate $\mu \to eee$. If $\mu \to eee$ takes place only at one-loop level, it is likely that A_L and A_R will be dominant. The reason is that for the configuratio that the energy of one of the fina positrons (in the muon rest frame) is around $m_\mu/2$, the photon propagator goes on-shell, leading to a significan enhancement. As a result, in this case a significan percent of the decays leads to a positron with energy in the vicinity of $m_\mu/2 - O(m_e^2/m_\mu)$. That is $d\Gamma(\mu^+ \to e_1^+ e_2^+ e^-)/dE_{e_1^+}$ has a narrow peak around $E_{e_1^+} = m_\mu/2 - O(m_e^2/m_\mu)$. It is shown in [39, 41] that the polarization of this energetic positron (e_1^+) is

$$\langle s_{T_1} \rangle = \frac{\mathbb{P}_\mu \sin\theta\,(\cos 2\phi \mathrm{Re}[A_R A_L^*] + \sin 2\phi \mathrm{Im}[A_R A_L^*])}{|A_L|^2(1+\mathbb{P}_\mu \cos\theta) + |A_R|^2(1-\mathbb{P}_\mu \cos\theta)}$$

and

$$\langle s_{T_2} \rangle = \frac{\mathbb{P}_\mu \sin\theta\,(-\cos 2\phi \mathrm{Im}[A_R A_L^*] + \sin 2\phi \mathrm{Re}[A_R A_L^*])}{|A_L|^2(1+\mathbb{P}_\mu \cos\theta) + |A_R|^2(1-\mathbb{P}_\mu \cos\theta)},$$

where $\hat{T}_2 = -(\vec{s}_\mu \times \vec{p}_{e_1^+})/|\vec{s}_\mu \times \vec{p}_{e_1^+}|$ and $\hat{T}_1 = (\hat{T}_2 \times \vec{p}_{e_1^+})/|\hat{T}_2 \times \vec{p}_{e_1^+}|$. The angle between the momentum of the energetic positron and the muon polarization is denoted by θ. Finally, ϕ is the azimuthal angle determining the direction of the emitted electron measured in the $\hat{T}_1$ and $\hat{T}_2$ plane. Thus, to derive $\arg[A_R A_L^*]$, in addition to measuring the polarization of e^+, studying the azimuthal distribution of the electron is also required. For $\mu \to eee$ dominated by A_L and A_R,

$$\frac{\mathrm{Br}(\mu^+ \to e^+e^+e^-)}{\mathrm{Br}(\mu^+ \to e^+\gamma)} \simeq \frac{\alpha}{3\pi}[\log(m_\mu^2/m_e^2) - 11/4] \simeq 0.0061 \tag{29}$$

which considering $\mathrm{Br}(\mu^+ \to e^+\gamma) < 1.2 \times 10^{-11}$, implies quite low statistics in a MEG like set-up, making such measurement rather futuristic.

case 2: Let us now discuss the case that four-Fermi couplings in Eq. (14) dominate. As discussed earlier, within the models that odd numbers of new particles can appear in vertices, the effective couplings in Eq. (14) can be created at the tree level. If the four-Fermi couplings dominate $\mu \to eee$, the relation in Eq. (29) does not hold and Br($\mu \to eee$) can saturate the present bound. As shown in Ref. [40], the transverse polarizations of the fina e^+ and e^- are sensitive to different combinations of the couplings so they yield independent information:

$$\frac{\langle s_{T_2^-} \rangle}{\langle s_{T_1^-} \rangle} = \arg[2.6 B_1^* B_2 - 4G_1^* C_1 - 4G_2 C_2^*]$$

and

$$\frac{\langle s_{T_2^+} \rangle}{\langle s_{T_1^+} \rangle} = \arg[B_1 C_2^* + B_2^* C_1 + 24 G_1^* B_1 - 24 G_2 B_2^*]$$

where $T_{1,2}^+$ and $T_{1,2}^-$ are the transverse directions of the positron and electron, respectively. They are define similarly to the case of Eq. (24). In these formulas, the effects of A_L and A_R are dropped because they are severely constrained by the upper bound on Br($\mu \to e\gamma$). Notice that all the transverse polarizations ($\langle s_{T_1^-} \rangle$, $\langle s_{T_2^-} \rangle$, $\langle s_{T_1^+} \rangle$ and $\langle s_{T_2^+} \rangle$) are proportional to $\mathbb{P}_\mu$. That is these polarizations vanish for unpolarized initial muon. However, when we take the ratios, $\mathbb{P}_\mu$ cancels out.

Under the CP transformation,

$$B_1 \to \eta B_1^*,\quad B_2 \to \eta B_2^*,\quad C_1 \to \eta C_1^*,\quad C_2 \to \eta C_2^*,\quad G_1 \to \eta G_1^* \text{ and } G_2 \to \eta G_2^*$$

where η is a pure phase that can be removed by redefinin the electron and muon fields The transverse polarization $\langle s_{T_2^+} \rangle$ and $\langle s_{T_2^-} \rangle$ are sensitive to these CP-violating parameters and vanish at the CP conserving limit.

Notice that unlike the case of $\mu \to e\gamma$, here the challenging photon polarimetry is not involved.

8.3. $\mu - e$ Conversion

Like the case of effective Lagrangian in Eqs. (13,14), the relative phases of the couplings in Eq. (16) are physical quantities that violate CP. It is shown in [42, 41] that the transverse polarization of the fina electron in the $\mu - e$ conversion on nuclei is sensitive to the CP-violating parameters of the effective Lagrangian (16). Following [41], we defin

$$\langle s_{T_i}\rangle \equiv \frac{d\Gamma\left[\mu N \to e(\vec{s}_e = \frac{1}{2}\hat{T}_i)N\right] - d\Gamma\left[\mu N \to e(\vec{s}_e = -\frac{1}{2}\hat{T}_i)N\right]}{\sum_{\vec{s}_e} d\Gamma[\mu N \to eN]}.$$

As shown in [41],

$$\langle s_{T_1}\rangle = \frac{2\mathrm{Re}\,[K_R K_L^*]\,\mathbb{P}_\mu \sin\theta}{|K_R|^2(1-\mathbb{P}_\mu\cos\theta)+|K_L|^2(1+\mathbb{P}_\mu\cos\theta)}, \tag{30}$$

$$\langle s_{T_2}\rangle = \frac{2\mathrm{Im}\,[K_R K_L^*]\,\mathbb{P}_\mu \sin\theta}{|K_R|^2(1-\mathbb{P}_\mu\cos\theta)+|K_L|^2(1+\mathbb{P}_\mu\cos\theta)} \tag{31}$$

where $\hat{T}_2 = (\vec{p}_e \times \vec{s}_\mu)/|\vec{p}_e \times \vec{s}_\mu|$ and $\hat{T}_1 = ((\vec{p}_e \times \vec{s}_\mu) \times \vec{p}_e)/|(\vec{p}_e \times \vec{s}_\mu) \times \vec{p}_e|$. K_R and K_L are define in Eqs. (17,18), respectively.

The advantage of $\mu - e$ conversion over $\mu \to e\gamma$ is that here the challenging photon polarimetry is not involved. The disadvantage is the relatively low polarization of the initial muon.

References

[1] H. O. Back *et al.*, *Phys. Lett. B* **525** (2002) 29; P. Belli *et al.*, *Phys. Rev. D* **61**, 117301 (2000).

[2] K. Nakamura *et al.* [Particle Data Group], *J. Phys. G* **37**, 075021 (2010).

[3] J. J. Thomson, “Cathode Rays,” *Philosophical Magazine* **44** (1897) 293.

[4] S. H. Neddermeyer and C. D. Anderson, *Phys. Rev.* **51** (1937) 884; J. C. Street and E. C. Stevenson, *Phys. Rev.* **52** (1937) 1003.

[5] E. K. Akhmedov, *JHEP* **0709**, 116 (2007). [arXiv:0706.1216 [hep-ph]].

[6] S. T. Petcov, *Sov. J. Nucl. Phys.* **25** (1977) 340 [*Yad. Fiz.* **25** (1977 ERRAT,25,698.1977 ERRAT,25,1336.1977) 641]; S. M. Bilenky, S. T. Petcov and B. Pontecorvo, *Phys. Lett. B* **67** (1977) 309; G. Altarelli, L. Baulieu, N. Cabibbo, L. Maiani and R. Petronzio, *Nucl. Phys. B* **125** (1977) 285 [Erratum-ibid. *B* **130** (1977) 516].

[7] *See* the fourth chapter of R. Barbieri, Pisa, Italy: Sc. Norm. Sup. (2007) 84 p. [arXiv:0706.0684 [hep-ph]].

[8] M. E. Peskin, T. Takeuchi, *Phys. Rev.* **D46** (1992) 381-409.

[9] T. Kinoshita and A. Sirlin, *Phys. Rev.* **113** (1959) 1652; W. J. Marciano and A. Sirlin, *Phys. Rev. Lett.* **61** (1988) 1815.

[10] Y. Kuno, Y. Okada, *Rev. Mod. Phys.* **73** (2001) 151-202 [hep-ph/9909265].

[11] P. J. Mohr and B. N. Taylor, *Rev. Mod. Phys.* **72** (2000) 351.

[12] G. W. Bennett *et al.* [Muon G-2 Collaboration], *Phys. Rev.* **D73** (2006) 072003 [hep-ex/0602035].

[13] M. Davier, A. Hoecker, B. Malaescu *et al.*, *Eur. Phys. J.* **C66** (2010) 1-9 [arXiv:0908.4300 [hep-ph]]; J. Prades, E. de Rafael, A. Vainshtein, [arXiv:0901.0306 [hep-ph]]; J. Prades, *Acta Phys. Polon. Supp.* **3** (2010) 75-86 [arXiv:0909.2546 [hep-ph]]; F. Jegerlehner, A. Nyffeler, *Phys. Rept.* **477** (2009) 1-110 [arXiv:0902.3360 [hep-ph]]; M. Passera, W. J. Marciano, A. Sirlin, *AIP Conf. Proc.* **1078** (2009) 378-381. [arXiv:0809.4062 [hep-ph]].

[14] A. M. Thalapillil, [arXiv:1012.4829 [hep-ph]].

[15] M. J. Booth, [hep-ph/9301293]; A. de Gouvea, S. Gopalakrishna, *Phys. Rev.* **D72** (2005) 093008. [hep-ph/0508148].

[16] Y. Ayazi, Y. Farzan, [arXiv:0809.4930 [hep-ph]]; S. Y. Ayazi, Y. Farzan, *JHEP* **0706** (2007) 013; Y. Farzan, *Phys. Rev.* **D69** (2004) 073009. [hep-ph/0310055]; Y. Farzan, M. E. Peskin, *Phys. Rev.* **D70** (2004) 095001; D. A. Demir, Y. Farzan, *JHEP* **0510** (2005) 068.

[17] W. Fetscher, H. J. Gerber, K. F. Johnson, *Phys. Lett.* **B173** (1986) 102.

[18] L. Michel, *Proc. Phys. Soc.* **A63** (1950) 514-531; C. Bouchiat, L. Michel, *Phys. Rev.* **106** (1957) 170-172.

[19] T. Kinoshita, A. Sirlin, *Phys. Rev.* **108** (1957) 844-850.

[20] C. Boehm, Y. Farzan, T. Hambye *et al.*, *Phys. Rev.* **D77** (2008) 043516 [hep-ph/0612228]; Y. Farzan, *Phys. Rev.* **D80** (2009) 073009 [arXiv:0908.3729 [hep-ph]]; Y. Farzan, M. Hashemi, *JHEP* **1011** (2010) 029. [arXiv:1009.0829 [hep-ph]].

[21] A. P. Lessa, O. L. G. Peres, *Phys. Rev.* **D75** (2007) 094001. [hep-ph/0701068].

[22] E. P. Hincks, B. Pontecorvo, *Phys. Rev.* **73** (1948) 257-258.

[23] G. Feinberg, *Phys. Rev.* **110** (1958) 1482; Berley, Lee and Bardon, *Phys. Rev. Lett.* **2** (1958) 357.

[24] J. Schwinger, *Ann. Phys.* **2** (1957) 407; K. Nishijima, *Phys. Rev.* **108** (1957) 907.

[25] *Phys. Rev. Lett.* **10** (1963) 260;

G. Danby, J. M. Gaillard, K. A. Goulianos, L. M. Lederman, N. B. Mistry, M. Schwartz and J. Steinberger, *Phys. Rev. Lett.* **9** (1962) 36.

[26] http://meg.web.psi.ch/index.html; *see also,* M. Grassi [MEG Collaboration], *Nucl. Phys. Proc. Suppl.* **149** (2005) 369.

[27] Y. Okada, K. -i. Okumura, Y. Shimizu, *Phys. Rev.* **D61** (2000) 094001. [hep-ph/9906446].

[28] Feinberg and Weinberg, *Phys. Rev. Lett.* **3111** (1959) 244; Steinberger and Wolfe, *Phys. Rev.* **100** (1955) 1480.

[29] http://sindrum2.web.psi.ch/

[30] http://mu2e-docdb.fnal.gov/ E. J. Prebys, D. Bogert, D. R. Broemmelsiek *et al.*, "Expression of Interest: A Muon to Electron Conversion Experiment at Fermilab."

[31] Y. Kuno, *Nucl. Phys. Proc. Suppl.* **149** (2005) 376-378.

[32] M. Blanke, A. J. Buras, B. Duling *et al.*, *JHEP* **0705**, 013 (2007). [hep-ph/0702136]. M. Blanke, A. J. Buras, B. Duling *et al.*, *Acta Phys. Polon.* **B41** (2010) 657-683. [arXiv:0906.5454 [hep-ph]].

[33] J. Hisano, T. Moroi, K. Tobe and M. Yamaguchi, *Phys. Rev. D* **53** (1996) 2442 [arXiv:hep-ph/9510309].

[34] B. Jamieson *et al., Phys. Rev. D* **74** (2006) 72007.

[35] Y. Kuno and Y. Okada, *Phys. Rev. Lett.* **77**, 434 (1996) [arXiv:hep-ph/9604296].

[36] V. S. Evseev, in *Muon Physics Vol. III Chemistry and Solids*, 1975, edited by V. W. Hughes and C. S. Wu (Academic Press), p. 236.

[37] K. Nagamine and T. Yamazaki, *Nucl. Phys. A* **219** (1974) 104; Y. Kuno, K. Nagamine and T. Yamazaki, *Nucl. Phys. A* **475** (1987) 615.

[38] Y. Farzan, S. Najjari, *Phys. Lett.* **B690** (2010) 48-56. [arXiv:1001.3207 [hep-ph]].

[39] Y. Farzan, *JHEP* **0707** (2007) 054. [hep-ph/0701106].

[40] Y. Farzan, *Phys. Lett.* **B677** (2009) 282-290. [arXiv:0902.2445 [hep-ph]].

[41] S. Y. Ayazi, Y. Farzan, *JHEP* **0901** (2009) 022. [arXiv:0810.4233 [hep-ph]].

[42] S. Davidson, [arXiv:0809.0263 [hep-ph]].

[43] H. Burkard *et al.*, *Phys. Lett. B* **160** (1985) 343.

[44] P. F. Bloser, S. D. Hunter, G. O. Depaola and F. Longo, arXiv:astro-ph/0308331; F. Adamyan *et. al, Nucl. Ins. and Meth. in Phys. Research A* **546** (2005) 376.

INDEX

D

E

F

M

N

O

P

Q

R

S

T

U

V

W

Y